Phased Array Antenna Handbook

For a complete listing of the *Artech House Antenna Library*,
turn to the back of this book

Phased Array Antenna Handbook

Robert J. Mailloux

Artech House
Boston • London

Library of Congress Cataloging-in-Publication Data
Phased Array Antenna Handbook/Robert J. Mailloux
Includes bibliographical references and index.
ISBN 0-89006-502-0
1. Phased Array Antennas. I. Title
TK6590.A6M35   1993   93-38273
621.382'4–dc20   CIP

A catalogue record for this book is available from the British Library

© 1994 ARTECH HOUSE, INC.
685 Canton Street
Norwood, MA 02062

International Standard Book Number: 0-89006-502-0
Library of Congress Catalog Card Number: 93-38273

10 9 8 7 6 5

With love,
to Patrice, Julie, Denise,
and especially Marlene

Contents

Preface

Any pile of tin with a transmission line exciting it may be called an antenna. It is evident on physical grounds that such a pile of tin does not make a good antenna, and it is worthwhile to search for some distinguishing characteristics that can be used to differentiate between an ordinary pile of tin and one that makes a good antenna.

This fascinating quote, discovered by my friend Phil Blacksmith, is taken out of context from Volume 8 of the MIT Radiation Laboratory series *The Principles of Microwave Circuits* [1]. It is a fitting introduction to a text that attempts to address today's advanced state of antenna array engineering. The present and future of antenna technology is concerned with a degree of pattern control that goes well beyond the simple choice of one or another pile of tin. Present antenna arrays are a union of antenna technology and control technology; and they combine the radiation from thousands of antennas to form precise patterns with beam peak directions that can be controlled electronically, with very low sidelobe levels, and pattern nulls that are moved to suppress radiation from unwanted directions.

Antenna technology remains interesting because it is dynamic. The past years have seen the technology progress from frequency-scanned and electronically steered arrays for scanning in one plane to the precise two-dimensional control using digital systems that can include mutual interactions between elements. Adaptive control has been used to move antenna pattern nulls to suppress interfering signals. Even the basic elements and transmission lines have changed, with a variety of microstrip, stripline, and other radiators replacing the traditional dipoles or slots fed by coaxial line or waveguides. Finally, the state of development in two fields—devices and automation—has brought us to an era in which phased arrays will be produced automatically, not assembled piece by piece, as has been the standard to date. This revolution in fabrication and device integration will dictate entirely new array architectures that emphasize monolithic fabrication with basic new elements and the use of a variety of planar monolithic transmission media.

Using digital processing or analog devices, future arrays will finally have the time-delay capability to make wideband performance possible. They will, in many cases, have reconfigurable apertures to resonate at a number of frequencies or allow the whole array surface to be restructured to form several arrays performing separate functions. Finally, they will need to be reliable and to fail gracefully, so they may incorporate sensing devices to measure the state of performance across the aperture and redundant circuitry to reprogram around failed devices, elements, or subarrays.

Although it contains some introductory material, this book is intended to provide a collection of design data for radar and communication system designers and array designers. Often the details of a derivation are omitted, except where they are necessary to fundamental understanding. This is particularly true in the sections on synthesis, where the subject matter is well developed in other texts. In addition, the book only briefly addresses the details of electromagnetic analysis, although that topic is the heart of antenna research. That subject is left as worthy of more detail than can be given in such a broad text as this.

Chapter 1, "Phased Arrays for Radar and Communication Systems," is written from the perspective of one who wishes to use an array in a system. The chapter emphasizes array selection and highlights those parameters that determine the fundamental measurable properties of arrays: gain, beamwidth, bandwidth, size, polarization, and grating lobe radiation. The chapter includes some information to aid in the trade-off between so-called "active" arrays, with amplifiers at each element, and "passive" arrays, with a single power source. There are discussions of the limitations in array performance due to phase versus time-delay control, transmission feed-line losses, and tolerance effects. Finally, there are discussions of special techniques for reducing the number of controls in arrays that scan over a limited spatial sector and methods for introducing time delay to produce broadband performance in an array antenna. The abbreviated structure of this introductory, "system-level" chapter necessitated frequent references to subsequent chapters that contain more detailed treatment of array design.

Chapter 2 and all the other chapters in the book are written to address the needs of antenna designers. Chapter 2, "Pattern Characteristics and Synthesis of Linear and Planar Arrays," includes the fundamental definitions of the radiation integrals and describes many of the important issues of array design. Element pattern effects and mutual coupling are treated in a qualitative way in this chapter but in more detail in Chapter 6. The primary topics of this chapter are the characteristics of antenna patterns and their directivity. The chapter also addresses several special types of arrays, including those scanned to endfire and thinned arrays.

Chapter 3 is a brief treatment of array synthesis, and it lists basic formulas and references on a wide variety of techniques for producing low sidelobe or shaped antenna patterns. The chapter includes a discussion of pattern optimization techniques, such as those used for adaptive array antennas. Chapter 4 treats arrays on nonplanar surfaces, and Chapter 5 describes the variety of array elements, relevant transmission lines, and array architectures.

Chapters 6 and 7 treat several factors that limit the performance of array antennas. Chapter 6 shows some of the effects of mutual coupling between array elements. This interaction modifies the active array element patterns and can cause significant impedance change with scan. This complex subject is treated with the aid of two appendices. Chapter 7 describes pattern distortion due to random phase and amplitude errors at the array elements and to phase and amplitude quantization across the array.

Chapter 8, the final chapter, summarizes techniques for three kinds of special-purpose arrays: multiple beam systems, arrays for limited sector scan, and arrays with wideband time-delay feeds. A vast technology has developed to satisfy these special needs while minimizing cost, and this technology has produced affordable high-gain electronic scanning systems using scanning arrays in conjunction with microwave quasioptical systems or advanced subarray techniques.

I am grateful to so many colleagues who have contributed directly or indirectly to this text. To Professors R. W. P. King and T. T. Wu, for introducing me to antenna theory and for letting me witness their unflagging enthusiasm for the electromagnetics of antennas and scattering. To Carl Sletten, for his mentorship. To Phil Blacksmith, for demonstrating that there is a simple and fundamental way of looking at complicated concepts, and to Hans Zucker, who has continually introduced me to concepts more complicated than I could fathom. To my Air Force colleagues, for many insightful discussions, especially Allan Schell, Jay Schindler, Peter Franchi, Hans Steyskal, John McIlvenna, Jeff Herd, Boris Tomasic, and Ed Cohen of Arcon. The text is strongly influenced by their publications, our discussions, our collaborations. Their collective expertise spans the whole subject matter of the text, and their individual contributions are apparent in nearly every chapter.

I owe special thanks to Hal Schrank, who reviewed the entire text and offered many helpful suggestions that improved the manuscript.

I am most grateful to my family for their love, consideration, and support, and for the time spent away from them while in their midst—especially to Marlene, for tolerating a dining room table cluttered with references and the developing text; to Patrice, for diligently typing lists of references; and to Julie and Denise, for their encouragement and tolerance.

REFERENCES

[1] *Principles of Microwave Circuits*, Eds. C. G. Montgomery, R. H. Dicke, E. M. Purcell, New York: McGraw-Hill, 1948.

Chapter 1
Phased Arrays in Radar and Communication Systems

1.1 INTRODUCTION

Phased array antennas consist of multiple stationary antenna elements, which are fed coherently and use variable phase or time-delay control at each element to scan a beam to given angles in space. Variable amplitude control is sometimes also provided for pattern shaping. Arrays are sometimes used in place of fixed aperture antennas (reflectors, lenses), because the multiplicity of elements allows more precise control of the radiation pattern, thus resulting in lower sidelobes or careful pattern shaping. However, the primary reason for using arrays is to produce a directive beam that can be repositioned (scanned) electronically. Although arrays with fixed (stationary) beams and multiple stationary beams will be discussed in this text, the primary emphasis will be on those arrays that are scanned electronically.

The radar or communication system designer sees the array antenna as a component (with measurable input and output) and a set of specifications. The array designer sees the details of the array and the physical and electrical limitations imposed by the radar or communications system, and within those constraints seeks to optimize the design. This chapter is written from the perspective of, and for, the system designer. The remainder of the text discusses array design issues.

1.1.1 System Requirements for Radar and Communication Antennas

In accordance with the principle of power conservation, the radiated power density in watts/square meter at a distance R from a transmitter with an omnidirectional antenna is given by

$$S = \frac{1}{4\pi} \frac{P_{\text{rad}}}{R^2} \qquad (1.1)$$

where P_{rad} is the total radiated power (watts), and the power density S is shown here as scalar.

Directive Properties of Arrays

Figure 1.1 shows an array of aperture antennas and indicates the coordinate system used throughout the text. If the antenna has a directional pattern with power density $S(\theta, \phi)$, then the antenna pattern *directive gain* $D(\theta, \phi)$ is defined so that the power density at some distant spherical surface a distance R from the origin is

$$S(\theta, \phi) = \frac{P_{rad} D(\theta, \phi)}{4\pi R^2} \tag{1.2}$$

so that

$$D(\theta, \phi) = \frac{4\pi R^2 S(\theta, \phi)}{P_{rad}} \tag{1.3}$$

or

$$D(\theta, \phi) = \frac{4\pi S(\theta, \phi)}{\int_\Omega S(\theta, \phi) \, d\Omega} \tag{1.4}$$

where the last integral is over the solid angle that includes all of the radiation. In the most general case it is

$$\int_\Omega S(\theta, \phi) d\Omega = \int_0^{2\pi} d\phi \int_0^\pi d\theta \, S(\theta, \phi) \sin \theta \tag{1.5}$$

The expression above (1.4) is the definition of directive gain and implies that the power density used is the total in both polarizations (i.e., the desired or *copolarization*, and the orthogonal or *crossed polarization*).

The *directivity* D_0 is defined as the maximum value of the directive gain:

$$D_0 = \max[D(\theta, \phi)] \tag{1.6}$$

which is a meaningful parameter primarily for antennas with narrow beamwidths (pencil beam antennas).

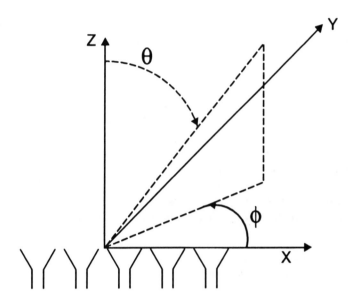

Figure 1.1 Array and coordinate systems.

The radiated power is less than the input power P_{in} by an efficiency factor ϵ_L, which accounts for circuit losses, and by the reflected signal power

$$P_{rad} = \epsilon_L P_{in}(1 - |\Gamma|^2) \qquad (1.7)$$

where Γ is the antenna reflection coefficient measured at the feed transmission line.

The power density in the far field can thus be written in terms of a gain function $G(\theta, \phi)$, with

$$S(\theta, \phi) = \frac{1}{4\pi} \frac{P_{in}}{R^2} G(\theta, \phi) \qquad (1.8)$$

where

$$G(\theta, \phi) = \epsilon_L(1 - |\Gamma|^2)D(\theta, \phi) \qquad (1.9)$$

Again, the peak value of the gain distribution is called the *gain* G_0.

$$G_0 = \max[G(\theta, \phi)] \qquad (1.10)$$

In practice, the maximum directivity of a planar aperture is achieved for uniform amplitude and phase illumination of the aperture (except for the special case of superdirectivity) [1] and is

$$D_{max} = 4\pi \frac{A}{\lambda^2} \tag{1.11}$$

for an aperture with area A at the wavelength λ.

In the case of a planar aperture with a large number of elements, it is also convenient to define a term called *aperture efficiency* ϵ_A,[1] which is not a real efficiency in the sense of measuring power lost or reflected, but relates the directivity to the maximum directivity D_{max}. Thus, the gain G_0 of a planar aperture is often written

$$G_0 = \epsilon_L \epsilon_A (1 - |\Gamma|^2) D_{max} \tag{1.12}$$

The concept of an antenna aperture becomes meaningless for an array with only a few elements or a linear (one-dimensional) array of dipoles or slots, and one must either use the general equation (1.4) or rely on the concept of element pattern gain to evaluate the array directivity and gain. This topic is discussed in more detail in Chapter 2.

Array Noise Characterization

In addition to receiving the desired signal, every antenna system also receives a part of the noise radiated from objects within the angular extent of its radiation pattern. Any physical object at a temperature above zero kelvin has an equivalent *brightness temperature*, or *noise temperature*, T_B, which is less than or approaching the physical temperature. The body radiates a noise signal received by the antenna and contributes to an effective antenna noise temperature. The antenna temperature for a lossless antenna is the integral of the observed brightness temperature $T_B(\theta, \phi)$ weighted by the antenna directive gain, or [2]

---

[1]The term *aperture efficiency* as defined in (1.12) is sometimes called *taper efficiency* and, in early references, as *gain factor*. Expressed in decibels, it is sometimes termed *taper loss* or *illumination loss*. An attempt has been made throughout this text to use *aperture efficiency* in strict accordance with the definition above, and to reserve the term *taper efficiency* to define a less rigorous parameter introduced later in this chapter.

$$T_A = \frac{\int_0^{2\pi} \int_0^{\pi} T_B(\theta, \phi) D(\theta, \phi) \sin\theta \, d\theta \, d\phi}{\int_0^{2\pi} \int_0^{\pi} D(\theta, \phi) \sin\theta \, d\theta \, d\phi} \tag{1.13}$$

The denominator of this expression normalizes the temperature so that a uniform brightness temperature distribution T_B produces an antenna temperature equal to the brightness temperature.

If there were no dissipative or mismatch loss in the antenna, the noise power available at the antenna terminals would be

$$N_A = K T_A \Delta f \tag{1.14}$$

where K is Boltzmann's constant (1.38×10^{-23} J/K) and N_A is in watts. In this expression, Δf is the bandwidth of the receiver detecting the noise signal or the bandwidth of the narrowest band component in the system. Since Δf is constant throughout the system calculations, it is convenient to work with the noise temperature alone.

The antenna temperature measured at the antenna terminals is modified by losses. At the terminals of any real antenna, the noise temperature has two components, as indicated in Figure 1.2. One noise component N_A is due to the pattern itself, which is a function of the brightness temperature distribution that the antenna "sees" within its receiving pattern. A second component is due to dissipative losses within the antenna, couplers, or transmission medium preceding the antenna terminals. Defining a loss factor ϵ as the ratio of power at the output terminals of the transmission line to the total received power (note that $\epsilon \leq 1$, and $10 \log_{10} \epsilon$ is the loss in decibels of the transmission line), then if the lossy material is at the temperature T_L, the effective antenna temperature at the antenna terminal (point A) is [3]

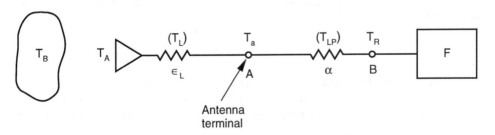

Figure 1.2 Antenna noise temperature modified by line losses.

$$T_a = \epsilon T_A + T_L(1 - \epsilon) \tag{1.15}$$

The effective system temperature relative to the preamplifier input (which is also point A, the antenna terminal) is obtained by adding the input noise temperature of the preamplifier and those of following stages as

$$T_S = T_a + T_1 + \frac{T_2}{g_1} + \frac{T_3}{g_1 g_2} + \cdots \tag{1.16}$$

where T_1 is the preamplifier effective noise temperature, T_n is the effective input temperature of the nth cascaded element, and g_{n-1} is the gain ($g > 1$) or loss ($g < 1$) of any element in the cascade chain (preceding the nth element).

This expression emphasizes the well-known result that for an amplifier system the system temperature is mainly established by the overall antenna temperature T_a plus the preamplifier input noise temperature T_1, since all successive noise contributions are divided by large gain numbers.

The effective noise power received at the preamplifier input over the bandwidth Δf is thus

$$N = K\, T_S \Delta f \tag{1.17}$$

The use of effective input temperature is convenient for very-low-noise components. However, for many radio frequency (RF) devices used in phased array radars and communication systems, the concept of noise figure is more commonly used.

The noise figure of a two-port device with gain g and internally generating noise N_N is defined as

$$F = \frac{(S/N)_{in}}{(S/N)_{out}} = \frac{N_{out}}{g N_{in}} = \frac{g N_{in} + N_N}{g N_{in}} \tag{1.18}$$

The input noise N_{in} is assumed from an ideal generator and so has only the thermal noise

$$N_{in} = K T_0 \Delta f \tag{1.19}$$

The noise contribution N_N at the output of the two-port is due to noise sources in the two-port itself. Its equivalent temperature T is defined as if it were the temperature of a resistor generating noise that is amplified by the gain g of the two-port.

$$N_N = g K T \Delta f \tag{1.20}$$

Thus, the noise factor of the two-port is given from the above as

$$F = 1 + T/T_0 \qquad (1.21)$$

or

$$T = (F - 1)T_0 \qquad (1.22)$$

where T_0 is the ambient temperature (290K) and T is the equivalent two-port noise temperature.

In the case of an attenuator or transmission line at temperature T_{LP}, with transmission line efficiency α and producing a noise power

$$N_N = (1 - \alpha)K\ T_{LP}\Delta f \qquad (1.23)$$

one can use equations (1.18) and (1.23) with $g = \alpha$ to show [2, p. 221] that the noise figure F (referred to the input side of the attenuator) is

$$F = \frac{gN_{\text{in}} + N_N}{gN_{\text{in}}} = \frac{\alpha T_0 + (1 - \alpha)T_{LP}}{\alpha T_0} \qquad (1.24)$$

and the associated noise temperature at the same point is

$$T = (F - 1)T_0 = \{1/\alpha - 1\}T_{LP} \qquad (1.25)$$

If the physical temperature of the attenuator is T_0, then the noise figure F is equal to the inverse of the transmission factor, and the equation below replaces (1.24).

$$F = 1/\alpha \qquad (1.26)$$

Written in terms of the noise figure for a cascade of networks, the system temperature is given (at antenna terminal A) as

$$
\begin{aligned}
T_S &= T_a + T_1 + \frac{(F_2 - 1)}{g_1}\,T_0 + \frac{(F_3 - 1)}{g_1 g_2}\,T_0 + \cdots \\
&= T_a + (F_1 - 1)T_0 + \frac{(F_2 - 1)T_0}{g_1} + \frac{(F_3 - 1)T_0}{g_1 g_2} + \cdots \qquad (1.27) \\
T_S &= T_a + (F_S - 1)T_0
\end{aligned}
$$

As shown, this system temperature is given at antenna terminal A. The effective temperature at other locations in the chain is obtained by multiplying by any amplifier gains (or transmission factor for losses) between this location and the new location. If the transmission line between points A and B is lossy and the expression above is written at the terminal (point B), then the $T_a + (F_S - 1)T$ is multiplied by the transmission factor $\alpha \leq 1$. At that point, the gain is also multiplied (reduced) by the same factor. This convention maintains the system signal-to-noise ratio as a unique system constant.

As an example, notice that the temperature contribution of the loss $[1/\alpha - 1]T_{LP}$ (given above (1.25) at point A) does not have the same form as that of the temperature contribution in the antenna temperature $(1 - \epsilon)T_L$ in (1.15). The former is referred away from the receiver through the attenuation to the antenna terminals, while the latter is referred toward the receiver through the attenuation of the antenna itself and to the antenna terminals. To refer this temperature to the output of the attenuator (point B), one should multiply by the transmission factor of the attenuator, which now makes the two expressions similar.

Sometimes, as a convenience, a term called the *system noise factor* (or *system noise figure*) is defined as (e.g., see Skolnik [4]) $NF = T_S/T_0$. The relationship between this and the usual noise figure definition is given below.

$$NF = T_S/T_0 = T_a/T_0 + (F_S - 1) \tag{1.28}$$

The Receiving Antenna in a Polarized Plane Wave Field

A receiving antenna immersed in an incident wave field receives power roughly proportional to the amount of energy it intercepts. This leads to the concept of an effective area A_E for the antenna, so that if the polarization of the receiving antenna is the same as that of the incident wave, then the received power is given by

$$P_r = A_E \, S(\theta, \phi) \tag{1.29}$$

The maximum value of the effective area is related to the antenna directivity D_0 by [5]

$$A_{E_{max}} = \frac{\lambda^2}{4\pi} D_0 \tag{1.30}$$

and the practical value of the effective aperture accounts for reflection and dissipative loss and is (for the polarization matched case)

$$A_E = \frac{\lambda^2}{4\pi} D_0 \, \epsilon_{ER}(1 - |\Gamma_r|^2) = (\lambda^2/4\pi)G_R \qquad (1.31)$$

where ϵ_{ER} is the loss efficiency for the receiving antenna.

The polarization match between the receiving antenna and the incident wavefront is described in terms of a unit polarization vector of the incident wave $\hat{\rho}_w$ and the receiving antenna $\hat{\rho}_a$. Figure 1.3 illustrates an example of matched and mismatched polarizations.

The dipole, or a thin wire with its axis in the **z**-direction as indicated in Figure 1.3, produces an electric field far from the antenna with only a **θ** component [6]. If an orthogonal set of dipoles were to receive that energy, the dipole oriented in the **φ** direction receives no signal, while the **θ**-oriented dipole receives maximum energy. Most antennas have less ideal polarization characteristics, and so experi-

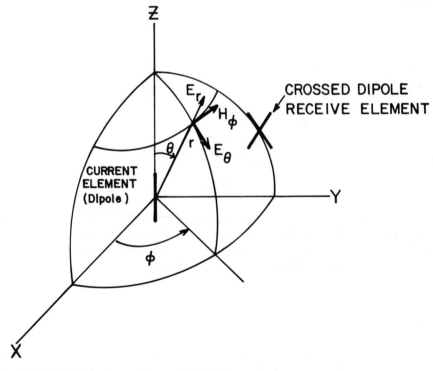

Figure 1.3 Polarization characteristics of ideal dipole antenna.

menters routinely take measurements of both polarizations. A formalism or notation for the description of a polarized wave is summarized here. For a wave traveling in the negative z-direction with electric field components,

$$\mathbf{E} = \hat{\mathbf{x}}\, E_x e^{j(kz + \phi_x)} + \hat{\mathbf{y}}\, E_y e^{j(kz + \phi_y)} \tag{1.32}$$

The polarization unit vector of this wave is defined as [7]

$$\hat{\boldsymbol{\rho}}_w = \frac{\hat{\mathbf{x}} E_x + \hat{\mathbf{y}} E_y}{[E_x^2 + E_y^2]^{1/2}}\, e^{j(\phi_y + \phi_x)} \tag{1.33}$$

The polarization unit vector of the antenna is defined according to the wave it excites or optimally receives. If a transmitting antenna excites a wave with the wave unit vector given above, then its polarization vector is the same as that of the wave.

An antenna that receives a wave has its effective aperture modified by the polarization loss factor ϵ_P, with

$$\epsilon_P = \left| \hat{\boldsymbol{\rho}}_a \cdot \hat{\boldsymbol{\rho}}_w^* \right|^2 \tag{1.34}$$

where the superscript asterisk denotes the complex conjugate.

The total power received is given by

$$
\begin{aligned}
P_r &= SA_E \epsilon_P \\
&= S \frac{\lambda^2}{4\pi} D_0 \epsilon_{ER} \epsilon_P (1 - |\Gamma_r|^2) \\
&= S(\lambda^2/4\pi)\epsilon_P G_R
\end{aligned}
\tag{1.35}
$$

In addition to linearly polarized antennas, circularly polarized antennas are often used for space communication or other applications in which the relative orientations of transmit and receive antennas are unknown. In (1.32), the polarization unit vector is circularly polarized if $E_x = E_y$ and $\theta_y = \theta_x + (\frac{1}{2} + 2n)\pi$ for any integer n.

System Considerations

The concept of an effective aperture for a receiving antenna, coupled with the formulas for power density (1.2) and polarization efficiency, leads to the following expression for the power received.

$$P_r = P_T G_T [\lambda/(4\pi R)]^2 G_R \epsilon_P \tag{1.36}$$

which is known as the *Friis transmission equation*. The term $[\lambda/(4\pi R)]^2$ is the free-space loss factor and accounts for losses due to the spherical spreading of the energy radiated by the antenna.

A similar form defining the received power for a monostatic radar system is given by the following reduced form of the radar range equation:

$$P = \frac{(P_T G_T)}{4\pi} \sigma [\lambda/(4\pi R^2)]^2 G_R \epsilon_P \tag{1.37}$$

where, in this particular case, it is not assumed that $G_T = G_R$. The constant σ is the scattering cross section of the target, which is defined as if the target collects power equal to its cross section multiplied by the incident power and then reradiates it isotropically.

At the receiver input, the sensitivity is determined by the signal-to-noise ratio, that is,

$$\frac{P}{N} = (P_T G_T) \frac{G_R}{T_S} \frac{\epsilon_P}{K\Delta f} \frac{\lambda^2}{[4\pi R]^2} \qquad \text{Communications} \tag{1.38}$$

$$\frac{P}{N} = \frac{(P_T G_T)}{4\pi} \frac{G_R}{T_S} \frac{\epsilon_P}{K\Delta f} \frac{\lambda^2 \sigma}{[4\pi R^2]^2} \qquad \text{Radar} \tag{1.39}$$

Subject to some minimum P/N ratio at the receiver, the range of a radar system varies as the fourth root of $G_T P_T$ (called the *effective isotropic radiated power* (EIRP)) and as the fourth root of the receiver parameters G_R/T_S. These system aspects are of great significance when choosing between passive, active, or mechanically scanned antennas for a particular requirement, as will be illustrated later.

Monopulse Beam Splitting

For radar applications, one of the most important properties of an array is the ability to form a precisely located deep monopulse pattern null for angle tracking. Figure 1.4 shows a 40-dB Bayliss pattern [8] (see Chapter 3), which is a frequently used distribution for monopulse radars. The pattern characteristics of importance to angle tracking are the antenna sum pattern gain and the difference pattern slope. Kirkpatrick [9] is attributed with introducing the measure of difference pattern slope k_m by which various antenna systems are compared. He also showed that the maximum angular sensitivity (difference mode gain slope at boresight) is obtained

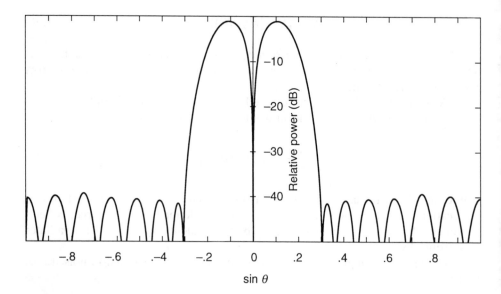

Figure 1.4 Low-sidelobe Bayliss radiation pattern.

for an aperture illumination with a linear amplitude distribution and odd symmetry about the antenna center.

The standard deviation of the error in the estimate of the difference between the antenna boresight axis and the target direction caused by receiver thermal noise is given [10] by the equation below for off-axis targets. This equation, including an error term due to the off-axis angle θ, is due to Sharenson [11]. This formula is accurate when receiver noise dominates (small targets or long range):

$$\Delta_t = \frac{\theta_B[1 + (k_m\theta/\theta_B)^2]^{1/2}}{k_m[B\tau(S/N)(f_r/\beta_n)]^{1/2}} \tag{1.40}$$

where

Δ_t = angle error (same units as θ_B)

θ_B = sum pattern half-power beamwidth

k_m = normalized monopulse difference slope or the error detection slope (in volts/volt/radian)

S /N = signal-to-noise power ratio of the sum beam (single pulse)
β_n = servo bandwidth
B = receiver bandwidth
τ = pulse width
f_r = repetition rate in pulses per second

Skolnik [12] points out that in this normalized form, reasonable values for k_m lie between 1.2 and 1.9, and that the efficiency decreases for the larger number. Antenna characteristics enter into the above expression in two ways: they determine the received sum beam S/N and the error detection slope.

1.2 ARRAY CHARACTERIZATION FOR RADAR AND COMMUNICATION SYSTEMS

The behavior of an array in a radar or communication system is far more complex than that of a passive, mechanically positioned antenna, because the performance characteristics vary with scan angle. This section describes the important array phenomena that determine scanning performance, bandwidth, and sidelobe levels of phased array systems.

1.2.1 Fundamental Results From Array Theory

A thorough mathematical treatment of phased array radiation, including mutual interaction between elements, is formidable. Even the mathematics for a single element can involve a detailed evaluation of vector field parameters, and the array analysis must also include the interactions between each of the elements of the array.

Fortunately, array theory provides the tool to do most array synthesis and design without the need to derive exact electromagnetic models for each element. This section consists primarily of the practical results of array theory; it is intended to introduce the reader to the properties of arrays and, in conjunction with Section 1.2.2, can be used by system designers to determine the approximate array configuration for a given application.

The sketch in Figure 1.5 portrays a generalized distribution of array elements, here shown as small radiating surfaces. Each element radiates a vector directional pattern that has both angle and radial dependence near the element. However, for distances very far from the element, the radiation has the $[\exp(-jkR)]/R$ dependence of a spherical wave multiplied by a vector function of angle $\mathbf{f}_i(\theta, \phi)$, called

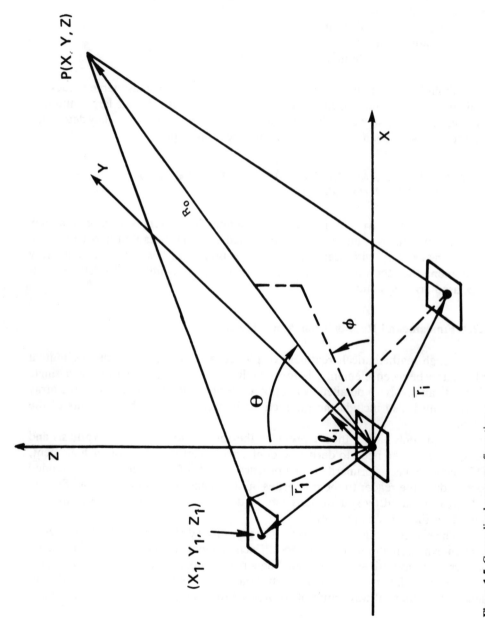

Figure 1.5 Generalized array configuration.

the *element pattern*. Although this vector function $\mathbf{f}_i(\theta, \phi)$ depends on the kind of element used, the *far field* of any *i*th element can be written

$$\mathbf{E}_i(r, \theta, \phi) = \mathbf{f}_i(\theta, \phi) \exp(-jkR_i)/R_i \qquad (1.41)$$

for

$$R_i = [(x - x_i)^2 + (y - y_i)^2 + (z - z_i)^2]^{1/2} \qquad (1.42)$$

and where $k = 2\pi/\lambda$ is the free-space wave number at frequency f.

If the pattern is measured at a distance very far from the array, then the exponential above can be approximated by reference to a distance R measured from an arbitrary center of the coordinate system.

Since

$$R_i \approx R - \hat{\mathbf{r}} \cdot \mathbf{r}_i \qquad (1.43)$$

then

$$\exp(-jkR_i)/R_i = \exp(-jkR)/R \, \exp(+jk\mathbf{r}_i \cdot \hat{\mathbf{r}})$$

for $\mathbf{r}_i$, the position vector of the *i*th element relative to the center of the chosen coordinate system, and $\hat{\mathbf{r}}$, a unit vector in the direction of any point in space (R, θ, ϕ). These vectors are written

$$\mathbf{r}_i = \hat{\mathbf{x}} x_i + \hat{\mathbf{y}} y_i + \hat{\mathbf{z}} z_i \qquad (1.44)$$

$$\hat{\mathbf{r}} = \hat{\mathbf{x}} u + \hat{\mathbf{y}} v + \hat{\mathbf{z}} \cos \theta \qquad (1.45)$$

where $u = \sin \theta \cos \phi$ and $v = \sin \theta \sin \phi$ are the direction cosines. The required distance R for which one can safely use the far-field approximation depends on the degree of fine structure desired in the pattern. Using the distance

$$R = 2L^2/\lambda \qquad (1.46)$$

for L the largest array dimension, is adequate for many pattern measurements, but for measuring extremely low sidelobe patterns or patterns with deep nulled regions, it may be necessary to use $10L^2/\lambda$ or a greater distance [13,14]. Far-field expressions will be used throughout this book unless otherwise stated.

For an arbitrary array, one can generally write the pattern by superposition:

$$\mathbf{E}(\mathbf{r}) = \frac{\exp(-jkR)}{R} \sum_i a_i \mathbf{f}_i(\theta, \phi) \exp(jk\mathbf{r_i} \cdot \hat{\mathbf{r}}) \tag{1.47}$$

The expression above is very general in form because it is written in terms of the unknown element patterns for each element in the presence of the whole array. The coefficients a_i are the applied element weights (voltages or currents) of the incident signals. One could obtain equally valid representations derived directly from actual (unknown) element currents or electric fields instead of the applied weights, but in this case these are subsumed into the element pattern description above. In general, the vector element patterns are different for each element in the array, even in an array of like elements; the difference is usually due to the interaction between elements near the array edge. However, throughout the rest of Chapter 1, it will be assumed that all patterns in a given array are the same. In this case, (1.47) becomes

$$\mathbf{E} = \mathbf{f}(\theta, \phi) \frac{\exp(-jkR)}{R} \sum a_i \exp(+jk\ \mathbf{r_i} \cdot \hat{\mathbf{r}}) \tag{1.48}$$

It is customary to remove the factor $\{\exp(-jkR)/R\}$ because the pattern is usually described or measured on a sphere of constant radius and this factor is just a normalizing constant. Thus, one can think of the pattern as being the product of a vector element pattern $\mathbf{f}(\theta, \phi)$ and a scalar array factor $F(\theta, \phi)$, where

$$F(\theta, \phi) = \sum a_i \exp(jk\mathbf{r_i} \cdot \hat{\mathbf{r}}) \tag{1.49}$$

Scanning and Collimation of Linear and Planar Arrays

Array scanning can be accomplished by applying the complex weights a_i in the form

$$a_i = |a_i| \exp(-jk\mathbf{r_i} \cdot \hat{\mathbf{r}_0}) \tag{1.50}$$

$$\hat{\mathbf{r}_0} = \hat{\mathbf{x}}u_0 + \hat{\mathbf{y}}v_0 + \hat{\mathbf{z}} \cos \theta_0 \tag{1.51}$$

with

$$k = 2\pi/\lambda$$

These weights steer the beam peak to an angular position (θ_0, ϕ_0), because at that location the exponential terms in (1.50) cancel those in (1.49), and the array factor is the sum of the weight amplitudes $|a_i|$. With this choice of weights, the pattern peak is stationary for all frequencies. This required exponential dependence has a linear phase relationship with frequency that corresponds to inserting time delays or lengths of transmission line. These are chosen so that the path length differences for the generalized array locations of Figure 1.5 are compensated in order to make the signals from all elements arrive together at some desired distant point.

More commonly, the steering signal is controlled by phase shifters instead of by switching in actual time delays. In this case, the weights have the form below instead of that in (1.50):

$$a_i = |a_i| \exp(-jk_0 \mathbf{r_i} \cdot \hat{\mathbf{r}}_0) \tag{1.52}$$

with

$$k_0 = 2\pi/\lambda_0$$

for some frequency $f_0 = c/\lambda_0$. In this form, the array pattern has its peak at a location that depends on frequency. Throughout the rest of this section, the phase-steered expression above will be used. The time-delayed expression can be recovered by omitting the subscript.

Among the important parameters of array antennas, those of primary importance to system designers are the gain, beamwidth, sidelobe level, and bandwidth of the array system. These subjects will be dealt with in greater detail in following sections and in Chapter 2, but the definitions and relevant bounding values are given here.

Phase Scanning in One Dimension ($\phi_0 = 0$)

Figure 1.6 show the several geometries used in the analysis of scanning in one dimension. Consider an array of N elements arranged in a line as shown, with element center locations $x_n = nd_x$. The elements can be individual radiators, as shown in 1.6(a), or can themselves be columns of elements, as indicated in 1.6(b). Under the assumption that all element patterns are the same, the normalized array radiation pattern in the far field is given at frequency f_0 by the summation over all N-elements as

$$\mathbf{E}(\theta) = \mathbf{f}(\theta, \phi) \sum a_n \exp(jk_0(nd_x u)) \tag{1.53}$$

for $u = \sin(\theta) \cos(\phi)$.

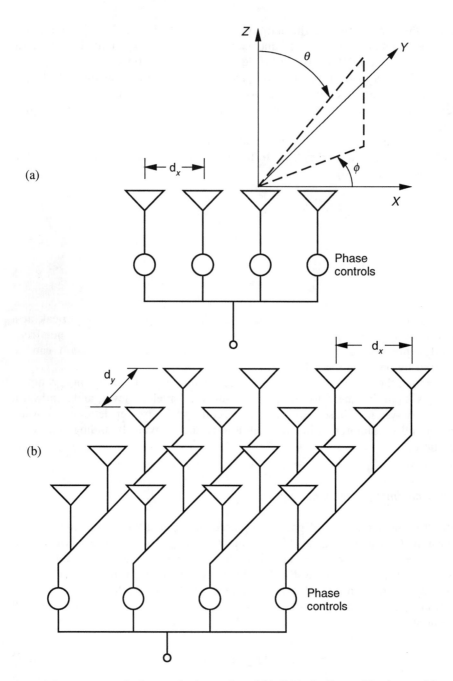

Figure 1.6 Array geometries for scanning in one plane: (a) individual radiators; (b) columns of elements.

The a_n are complex weights assigned to each element, and $\mathbf{f}(\theta, \phi)$ is the radiation pattern (or element pattern) that is assumed the same for all elements. In this case, at a fixed frequency one can create a maximum of $\mathbf{E}(\theta, \phi)$ in the direction $(\theta_0, 0)$ by choosing the weights a_n to be

$$a_n = |a_n| \exp(-jk_0 n d_x u_0) \tag{1.54}$$

and so

$$F(\theta) = \sum |a_n| \exp(jn d_x k_0 (u - u_0)) \tag{1.55}$$

where

$$u_0 = \sin(\theta_0)$$

This expression implies the use of phase shifters to set the complex weights a_n. Equation (1.55) shows that the array factor is a function of $u - u_0$, so that if the array were scanned to any angle, then the pattern would remain unchanged except for a translation. This is the main reason for the use of the variables u and v (often called *sine space* or *direction cosine space*) for plotting generalized array patterns.

For an array with all elements located in the plane $z = 0$, the pattern is symmetric about $\theta = \pi/2$, and the array factor forms a second, mirror-image beam below the plane $z = 0$. Most scanning arrays are required to have only a single main beam, and this is achieved using elements with a ground screen to make the element patterns nearly zero for the region behind the array.

The array factor of an array at frequency f_0 with all equal excitations is shown in Figure 1.7 (solid) and can be derived from (1.55). Normalized to its peak value, this expression is

$$F(u) = \sin[N\pi d_x (u - u_0)/\lambda_0][N \sin(\pi d_x (u - u_0)/\lambda_0)] \tag{1.56}$$

In this figure, $L = N d_x$ is the effective array length, N is 8, and the elements are spaced one-half wavelength apart.

The 3-dB beamwidth (in radians) for this uniformly illuminated array at broadside is $0.886\lambda_0/L$, which is the narrowest beamwidth (and highest directivity) of any illumination, except for certain special *superdirective* illuminations associated with rapid phase fluctuations and closely spaced elements. Except for very small arrays, the superdirective illuminations [1] have proven impractical because they have very large currents and high loss, and require very precise excitation. In most cases, they are also very narrow-band. The level of the first sidelobes for the uniformly illuminated linear array is relatively high (about -13 dB). Figure 1.7

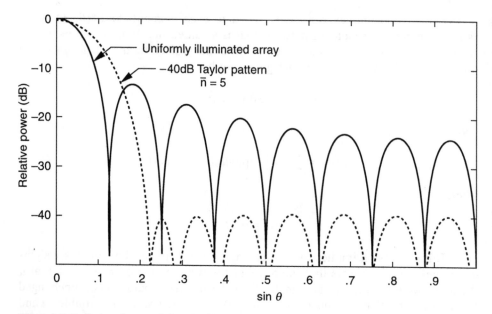

Figure 1.7 Radiation characteristics of uniformly illuminated and low-sidelobe 16-element arrays.

(dashed) shows the same array radiating a low-sidelobe (Chebyshev) pattern, with −40-dB sidelobe levels. This figure illustrates the beam broadening that generally accompanies low sidelobe illuminations.

The beamwidth increases as the array is scanned. For a large array and not near endfire, the beam broadens according to sec θ_0, but the more general case is given later in this section.

Two-Dimensional Scanning of Planar Arrays

The array factor for the two-dimensional array of Figure 1.8(a) with elements at locations

$$\mathbf{r_{m,n}} = \hat{\mathbf{x}}m\,d_x + \hat{\mathbf{y}}n\,d_y \tag{1.57}$$

and using phase steering to place the beam peak at θ_0, ϕ_0 at frequency f_0 is given by the following:

$$F(\theta, \phi) = \sum_{m,n} |a_{m,n}| \exp\{jk_0[md_x(u - u_0) + nd_y(v - v_0)]\} \tag{1.58}$$

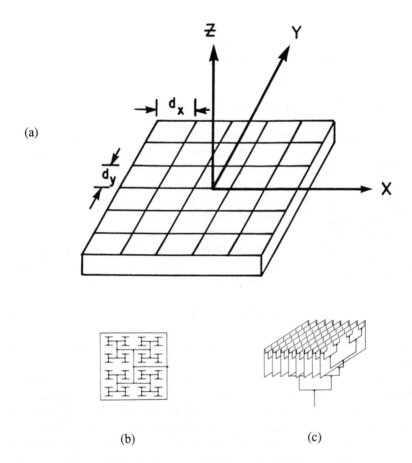

Figure 1.8 Array geometry for two-dimensional scanning: (a) generalized planar array geometry; (b) equal line-length planar feed; (c) equal line-length column feeds.

Often, for a rectangular array aperture, a separable amplitude distribution is chosen so that

$$a_{m,n} = b_m c_n$$

and then the factor can be written as the product of two independent factors of u and v.

$$F(\theta, \phi) = \left\{ \sum b_m \exp[jk_0 m d_x(u - u_0)] \right\} \left\{ \sum c_n \exp[jk_0 n d_y(v - v_0)] \right\} \quad (1.59)$$

Seen in this form, it is clear that the pattern of the linear array (1.55) is of vast importance because of its relevance to planar arrays with separable distributions.

Beamwidth and Directivity of Scanning Arrays

The beamwidth and sidelobe level of an array antenna are governed by the chosen aperture taper. An example of sidelobe reduction is shown by comparing the curves in Figure 1.7. This figure shows antenna patterns for uniform illumination and a low-sidelobe (-40 dB Taylor) illumination of a 16-element array. Antenna sidelobes are reduced by *tapering* the array excitation so that elements at the array center are excited more strongly than those near the edge. Some of the more useful examples of tapering are described in Chapter 2. In addition to sidelobe reduction, however, tapering broadens the array beamwidth. For this more general case, the half-power beamwidth of the radiation pattern for a linear array or in the principal planes of a rectangular array at broadside is

$$\theta_3 = 0.886 B_b \lambda / L \qquad (1.60)$$

where B_b is called the *beam broadening factor* and is obviously chosen as unity for the uniformly illuminated array.

Table 1.1 [15] shows the variation of beamwidth of a continuous line source for several selected illuminations with varying sidelobe levels. The continuous line source pattern is a good approximation of the pattern of a large array with elements spaced a half wavelength or less apart. In this table, the parameter w is equal to Lu/λ. These data indicate a generalized pattern broadening and lowering of the principal sidelobes as the aperture distributions are made smoother. Beyond that, as pointed out by Jasik, the far-sidelobe decay is controlled by the derivatives of the aperture illumination at the edge of the aperture. A uniform illumination, which has a discontinuity in the function and its derivatives, has far sidelobes that vary as $(Lu/\lambda)^{-1}$. For the cosine or gabled distributions, which are continuous but have discontinuous derivatives at the aperture edge, the far sidelobes have a $(Lu/\lambda)^{-2}$ variation. The cosine squared illumination, which is continuous, has a continuous first derivative and a discontinuous second derivative; the far sidelobes vary as $(Lu/\lambda)^{-3}$.

In his original paper on line source synthesis, Taylor [16] documented the relationships between aperture edge behavior, far sidelobes, and array pattern zero locations. His analysis and insights led to a most practical technique for the synthesis of low-sidelobe beams and is described in Chapter 2, Section 2.2.

Table 1.1 also gives the *gain factor* for each illumination, which is the pattern directivity normalized to the maximum directivity of the line source. This parameter

Table 1.1
Line-Source Distributions

Type of Distribution $-1 \leq x \leq 1$	Directivity Pattern E(u)		Half Power Beamwidth (Degrees)	Angular Distance to First Zero	Intensity of First Sidelobe (dB Below Maximum)	Gain Factor
$f(x) = 1$	$l \dfrac{\sin u}{u}$		$50.8 \dfrac{\lambda}{l}$	$57.3 \dfrac{\lambda}{l}$	13.2	1.0
$f(x) = 1 - (1 - \Delta)x^2$	$l(1 + L) \dfrac{\sin u}{u}$ $L = (1 - \Delta)\dfrac{d^2}{du^2}$	$\Delta = 1.0$	$50.8 \dfrac{\lambda}{l}$	$57.3 \dfrac{\lambda}{l}$	13.2	1.0
		$\Delta = .8$	$52.7 \dfrac{\lambda}{l}$	$60.7 \dfrac{\lambda}{l}$	15.8	.994
		$\Delta = .5$	$55.6 \dfrac{\lambda}{l}$	$65.3 \dfrac{\lambda}{l}$	17.1	.970
		$\Delta = 0$	$65.9 \dfrac{\lambda}{l}$	$81.9 \dfrac{\lambda}{l}$	20.6	.833
$\cos \dfrac{\pi x}{2}$	$\dfrac{\pi l}{2} \dfrac{\cos u}{\left(\dfrac{\pi}{2}\right)^2 - u^2}$		$68.8 \dfrac{\lambda}{l}$	$85.9 \dfrac{\lambda}{l}$	23	.810
$\cos^2 \dfrac{\pi x}{2}$	$\dfrac{l}{2} \dfrac{\sin u}{u} \dfrac{\pi^2}{\pi^2 - u^2}$		$83.2 \dfrac{\lambda}{l}$	$114.6 \dfrac{\lambda}{l}$	32	.667
$f(x) = 1 - |x|$	$\dfrac{l}{2} \left(\dfrac{\sin \dfrac{u}{2}}{\dfrac{u}{2}}\right)^2$		$73.4 \dfrac{\lambda}{l}$	$114.6 \dfrac{\lambda}{l}$	26.4	.75

Source: [15].

is analogous to the aperture efficiency of an aperture antenna. If a continuous aperture antenna has the same illumination as the line source in both separable dimensions, then the sidelobe levels quoted in Table 1.1 pertain in the principal planes $(u, v) = (0, v)$ or $(u, 0)$ and the sidelobes are far less in the diagonal planes (and in fact are the product of the principal plane patterns).

Table 1.2 [15] shows the relative gain, beamwidth, and sidelobe level for a circular aperture antenna with various continuous aperture illuminations. In this case, the parameter $w = (2\pi a/\lambda)u$, where a is the aperture radius and $D = 2a$ is the diameter.

Table 1.2
Circular-Aperture Distributions

Type of Distribution $0 \le r \le 1$	Directivity Pattern E(u)	Half Power Beamwidth (Degrees)	Angular Distance to First Zero	Intensity of First Sidelobe (dB Below Maximum)	Gain Factor
$f(r) = (1 - r^2)^0 = 1$	$\pi\sigma^2 \dfrac{J_1(u)}{u}$	$58.9 \dfrac{\lambda}{D}$	$69.8 \dfrac{\lambda}{D}$	17.6	1.00
$f(r) = (1 - r^2)$	$2\pi\sigma^2 \dfrac{J_2(u)}{u^2}$	$72.7 \dfrac{\lambda}{D}$	$93.6 \dfrac{\lambda}{D}$	24.6	0.75
$f(r) = (1 - r^2)^2$	$8\pi\sigma^2 \dfrac{J_3(u)}{u^3}$	$84.3 \dfrac{\lambda}{D}$	$116.2 \dfrac{\lambda}{D}$	30.6	0.56

Source: [15].

The aperture illuminations used in Tables 1.1 and 1.2 are relatively simple and not specifically optimized for low sidelobes.

Figure 1.9 shows the normalized beamwidth for Chebyshev antenna patterns as a function of design sidelobe level. This result uses an approximation due to Drane [17] that is given in Chapter 2. Figure 1.9(b) shows the aperture (or taper) efficiency for a 16-element Chebyshev array pattern as a function of sidelobe level. This result was also computed using an approximation by Drane [17].

Equation (1.50) indicates that the pattern does not change with scan if plotted in terms of the parameter $u = \sin \theta$. When the beam is scanned to the angle θ_0 at frequency f_0, the entire pattern is displaced from the broadside pattern. Though constant in u-space, the beamwidth is not constant in angle space, since it broadens with scan angle according to the equation below, and the directivity changes accordingly.

$$\theta_3 = [\sin^{-1}(u_0 + 0.443B_b \lambda/L) - \sin^{-1}(u_0 - 0.443B_b \lambda/L)] \qquad (1.61)$$

for

$$L = Nd_x$$

This result is for a linear array of N elements or in the principal scan plane of a rectangular array of length L in the plane of scan. Figure 1.10 shows this

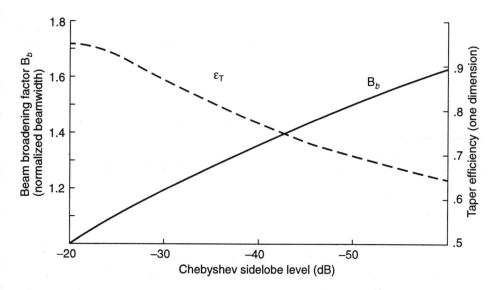

Figure 1.9 Beam broadening (solid line) and taper efficiency (dashed line) versus sidelobe level.

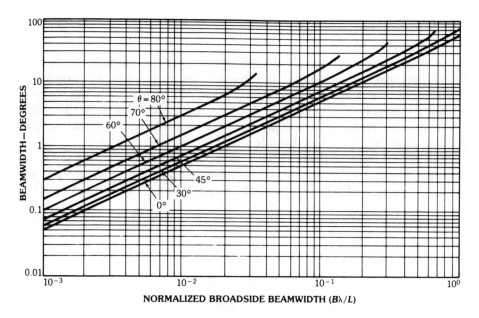

Figure 1.10 Beamwidth variation with scan.

variation with scan for arrays of various sizes. For a large array, the beamwidth computed from the above expression increases approximately as $1/(\cos\theta)$, and so in the large array limit,

$$\theta_3 \approx \theta_3(\text{broadside})/\cos\theta_0 \tag{1.62}$$

This expression is valid for linear and in any scan plane (independent of ϕ) of large planar arrays.

Neither the cosine relationship nor (1.61) is valid for an array scanned within a beamwidth of endfire ($\theta = \pi/2$). Scanning to endfire is discussed in Chapter 2.

Directivity of Linear Arrays

Although the above expressions give the proper beam broadening for linear arrays scanned along their axis and for planar arrays, the gain degradation or scan loss is quite different for aperture and linear arrays. For linear arrays, the scan loss also depends on the directive gain in the plane orthogonal to the scan plane. There is, however, one very simple and important case for linear arrays of isotropic elements with spacings that are any integer number of half-wavelength. In this case, Elliott [18] shows that the directivity is independent of scan angle and is given by (see Chapter 2)

$$D_0 = \frac{|\Sigma a_n|^2}{\Sigma |a_n|^2} \tag{1.63}$$

A note of caution: one should not assume that the constant directivity of (1.63) means that one can design a linear array with no scan loss. Increasing array mismatch due to element mutual coupling negates this possibility, even for omnidirectional elements. In addition, the discussion in Chapter 2 indicates that arrays with element patterns narrowed in the plane orthogonal to scan suffer substantially increased losses when scanned to wide angles.

Since the maximum value of this expression (1.63) is equal to N and occurs when all a_n values are the same, it is convenient to define a *taper efficiency* ϵ_T such that the above result for half-wavelength-spaced isotropic elements is thus [19]

$$D_0 = N\epsilon_T \tag{1.64}$$

where here

$$\epsilon_T = \frac{1}{N}\frac{|\Sigma a_n|^2}{\Sigma |a_n|^2}$$

This taper efficiency is the discrete analog of the *gain factor* used for continuous apertures, as tabulated in Table 1.1.

Equation (1.63) is exact and pertains to omnidirectional elements with integer half-wavelength spacings. A more general but approximate expression that illustrates the linear dependence of directivity and element spacing is due to King [20] and given below [19]. This result applies for isotropic elements spaced less than a wavelength apart and with the beam at broadside so that no grating lobes exist, and for beam shapes that concentrate most of their power in the main beam. In this case, the directivity is given approximately by

$$D_0 = [2d/\lambda][\epsilon_T N] \tag{1.65}$$

a formula that includes (1.63) as a special case (see Chapter 2).

Directivity of Planar Arrays

If the elements of the linear array have significantly narrowed patterns in the orthogonal plane, then, in general, one must perform the integral of (1.4) to evaluate directivity for the scanning array. Section 2.1 gives equations for directivity of more generalized arrays, but for the purposes of this section there is one very convenient form for system applications. The beamwidth and directivity of a relatively large planar array are related by the following approximate equation due to Elliott [21]:

$$D = 32,400 \cos \theta_0 / (\theta_{x3} \theta_{y3}) \tag{1.66}$$

where θ_{x3} and θ_{y3} are the 3-dB beamwidths of the pencil or elliptical beam at broadside. In this formula, the beamwidths are in degrees.

The formula is exact for a uniform matched aperture at broadside. It is a good approximation for most other pencil beam array patterns and shows that the directivity is decreased approximately by the product of the beam broadening factors in each plane for a lower sidelobe array. Stegen [22] points out that the numerator of this expression should be larger for low-sidelobe antennas. This simple formula reveals the well-known cosine dependence of the directivity of large planar arrays, but does not apply at endfire ($\theta = \pi/2$), where it yields zero directivity. The endfire case is described in Section 2.1.

It is possible to test the expression in one limiting case for an aperture with a uniform illumination. Using the uniform array beamwidths (from (1.60)) in the above (at broadside) shows this equation to be consistent with the known relation for the maximum directivity $4\pi A/\lambda^2$ (1.11). The relationship to the number of array elements is obtained in terms of the cell area $A = L_x L_y = N A_{CELL}$, where A_{CELL} is the area of the grid occupied by a single element:

$$D_{max} = 4\pi N A_{CELL}/\lambda^2 \tag{1.67}$$

which is the maximum directivity except in the superdirective limit referred to earlier. Again, introducing the concept of an aperture efficiency ϵ_A and introducing the scan loss for a large array, the actual directivity for a large scanned aperture array is

$$D_0 = D_{max}\epsilon_A \cos\theta \tag{1.68}$$
$$= \frac{4\pi A}{\lambda^2} \epsilon_A \cos\theta$$

This expression can also be derived [19] directly from the integral expression for directivity in the limit of a very large array.

Elliott [23] shows that for a relatively large rectangular array, with a separable distribution and not scanned too close to endfire, the directivity is approximately given by the following expression:

$$D_0 = \pi D_x D_y \tag{1.69}$$

where D_x and D_y are the directivities of the linear arrays of isotropic elements with the separable distributions. The elements in the planar array are assumed to have hemispherical element patterns. This expression is not exact, and the discussion in Section 2.1 indicates that it can be high by up to 2 dB in some cases involving relatively small arrays, but it remains useful for system sizing applications.

Array Gain and Scan Loss

Since the directivity can be related to the beamwidth, and the variation of beam-width with scan is well known, approaching the ideal $1/(\cos\theta)$ dependence (1.61) for large arrays, one might assume that the gain of a scanned array is also simply established. However, the array gain and directivity are related by

$$G = \epsilon_L(1 - |\Gamma|^2)D_0 \tag{1.70}$$

for Γ, the reflection coefficient of the array input terminals. This equation also requires that the array reflection coefficient be known in order to predict the scan loss. In practice, it is very difficult to achieve $\cos\theta$ scan dependence because of the impedance mismatch that results from interelement coupling or *mutual imped-*

ance. Figure 1.11(a) shows the expected scan loss $(1 - |\Gamma|^2)\cos \theta$ for several assumed scan dependences of the form $(\cos \theta)^n$ for $n = 1, 3/2, 2$. These represent reasonable array design goals depending on the array elements and plane of scan. System designers can assume that the $\cos \theta$ can be approached for slot arrays and dipole arrays in one plane of scan, but not often for both. Such benign scan characteristics are the result of extensive research and development to optimize scanned impedance match for these two fundamental elements. More general elements, including most printed circuit and dielectric elements, suffer more severe scan degradation. In addition, without careful design, some element designs can exhibit the catastrophic pattern degradation called *scan blindness*, which results in almost complete cancellation of all radiation for certain beam directions. This phenomenon is depicted in the sketch of Figure 1.11(b). The scan properties of specific array elements are discussed in Chapter 5.

Grating Lobes of a Linear Array

A linear array with its peak at θ_0 can also have other peak values subject to the choice of spacing d_x. This ambiguity is apparent, since the summation also has a

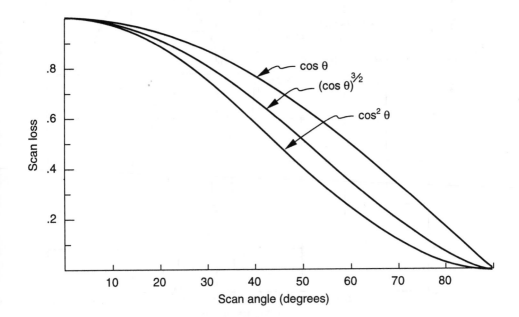

Figure 1.11(a) Typical scan loss curves.

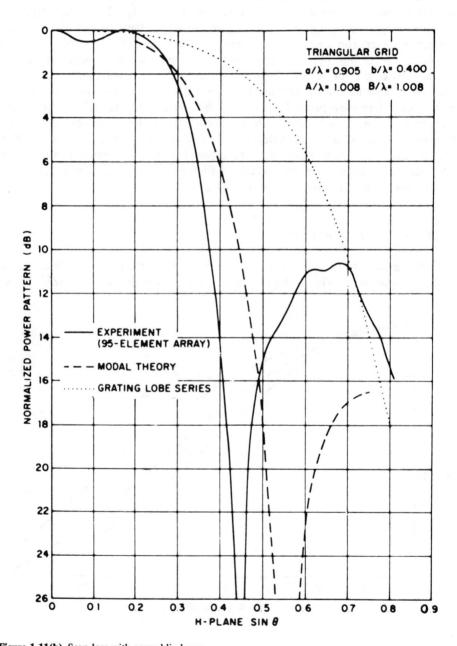

Figure 1.11(b) Scan loss with array blindness.

peak whenever the exponent is some multiple of 2π. At frequency f and wavelength λ, this condition is

$$2\pi \frac{d_x}{\lambda} (\sin \theta - \sin \theta_0) = 2\pi p \qquad (1.71)$$

for all integers p. Such peaks are called *grating lobes* and are shown from the above to occur at angles θ_p such that

$$\sin \theta_p = \sin(\theta_0) + \frac{p\lambda}{d_x} \qquad (1.72)$$

$$p = \pm(1, 2, \ldots)$$

for values of p that define an angle with a real sine ($|\sin \theta_p| \leq 1$).

If the element spacing exceeds a critical dimension, grating lobes occur in the array factor, as indicated in Figure 1.12. This figure shows several patterns of an array of eight elements spaced one wavelength apart, excited by a Chebyshev tapered illumination that would produce -25-dB sidelobes in an array with half-wave spacing. The two sets of patterns are for scan angles of broadside and 30 deg ($u_0 = .5$). The far-field pattern is the product of the element pattern (shown dashed) and the array factor, shown solid in Figures 1.12(a,b). The grating lobe may be suppressed somewhat by the element pattern zero for a broadside array as shown in the figure. However, when the array is scanned (and the element pattern is not), the grating lobe location moves away from the null and can be a substantial source of radiation. In the case shown (Figure 1.12(d)), it is fully as large as the desired main beam. A criterion for determining the maximum element spacing for an array scanned to a given scan angle θ_0 at frequency f is to set the spacing so that the nearest grating lobe is at the horizon. Using (1.72), this leads to the condition

$$\frac{d_x}{\lambda_0} \leq \frac{1}{1 + \sin \theta_0} \qquad (1.73)$$

at the highest operating frequency f_0, which requires spacing not much greater than one-half wavelength for wide angles of scan. In practice, the spacing must be further reduced in order to avoid the effects of array blindness, described in Chapter 5.

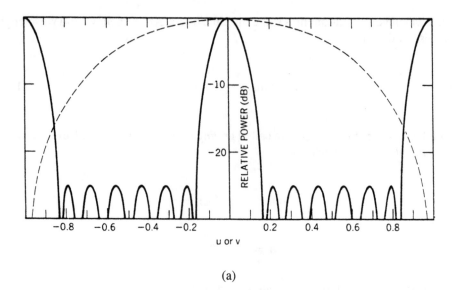

(a)

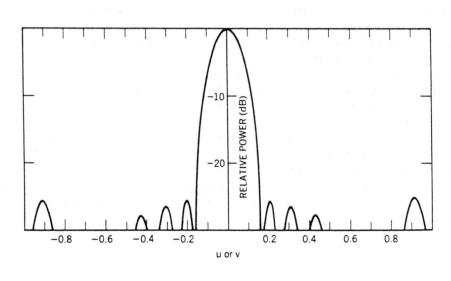

(b)

Figure 1.12 Array factors, element patterns, and grating lobes for a linear array: (a) −25-dB Chebyshev array factor for one wavelength spacing (solid line), assumed element pattern (dashed line); (b) radiation pattern for part (a).

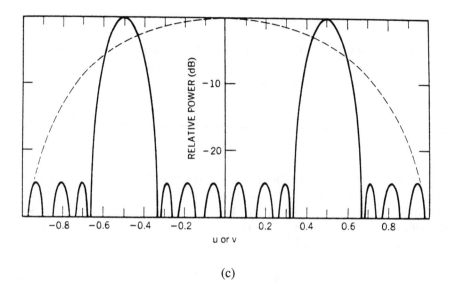

(c)

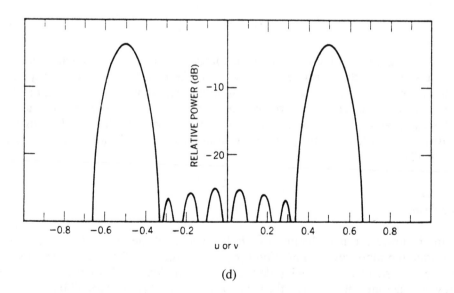

(d)

Figure 1.12 (cont.) Array factors, element patterns, and grating lobes for a linear array: (c) scanned array factor (solid line), element pattern (dashed line); (d) radiation pattern for part (c).

Grating Lobes of a Planar Array

Similar relations hold for a planar array, since the grating lobe phenomenon occurs in these cases also, and one can show for a rectangular grid array (Figure 1.8(c)) with spacings d_x and d_y that lobes occur at

$$u_p = u_0 + p\lambda/d_x \quad p = 0, \pm1, \pm2, \ldots \quad (1.74)$$
$$v_q = v_0 + q\lambda/d_y \quad q = 0, \pm1, \pm2, \ldots$$

This spectrum of grating lobes is shown graphically in the grating lobe lattice of Figure 1.13, which shows the (u_p, v_q) grating lobe locations in u, v space for a rectangular grid array. Not all values of p and q correspond to allowed angles of radiation, however, since the angle (θ_{pq}) associated with grating lobe designated by indices p and q is defined by

$$\cos \theta_{pq} = (1 - u_p^2 - v_q^2)^{1/2} \quad (1.75)$$

There can only be real values of θ_{pq} if the u_p and v_q are constrained to be within the unit circle, or

$$u_p^2 + v_q^2 \leq 1 \quad (1.76)$$

Grating lobes inside the unit circle correspond to real angles θ and radiate, but those outside the unit circle do not. As is the case for the linear array, this limits element spacings to approximately a half-wavelength or slightly more for most applications. The area occupied per element for a 60-deg scan is about $0.29\lambda^2$. In practice, it is necessary to reduce the element spacings further (by 5% to 10%) in order to avoid the pattern deterioration associated with mutual coupling effects [24].

Bandwidth

Array bandwidth can be limited by the bandwidth of the elements in the array, but often the more severe limitation is caused by the use of phase shifters to scan the beam instead of time-delay devices. The complex weights chosen in (1.52) provide true time delay, and so the beam peak occurs at (θ_0, ϕ_0) for all frequencies. If phase shifters are used to scan the beam, the peak is scanned to the desired angle only at center frequency f_0. Otherwise, it is scanned to that angle which makes

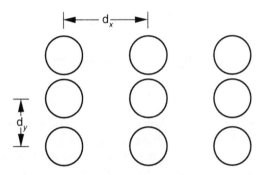

Rectangular grid
array

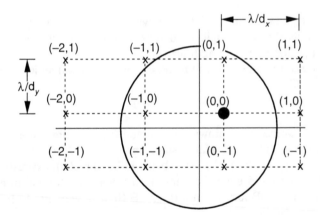

Figure 1.13 Grating lobe spectrum for planar arrays with rectangular grids.

the exponent of (1.50) equal and of opposite sign to the exponent of (1.49). For phase steering in one dimension, the complex weighting has the form

$$a_n = a_0 \exp\left(\frac{-2\pi}{\lambda_0} nd_x u_0\right) \tag{1.77}$$

and the value of u corresponding to beam peak is given by

$$u = u_0 f_0/f \tag{1.78}$$

The result is pattern "squint" like that shown in Figure 1.14(a), in which the beam peak angle is reduced for frequencies above the design frequency and increased for frequencies below the design frequency. If the bandwidth is defined by the frequency limits at which the gain is reduced to half power, the resulting fractional bandwidth is given by

$$\frac{\Delta f}{f} = \frac{\Delta u}{u_0} = \frac{\theta_3}{\sin \theta_0} = 0.886 B_b \left(\frac{\lambda}{L \sin \theta_0}\right) \tag{1.79}$$

for an array with beamwidth θ_3. The bandwidth becomes smaller as the array is made larger or as the scan angle is increased. Figure 1.14(b) shows bandwidth versus scan angle for various-length arrays.

For small scan angles, the following expression is convenient.

$$\Delta f/f_0 = 1/\eta_B \tag{1.80}$$

where η_B is the number of beamwidths scanned (in one dimension).

Another commonly used relationship can be derived from (1.79) for the limit of wide-angle scan (± 60 deg). Using the beamwidth of (1.60), expressed in degrees, and choosing as the band edge the one-quarter beamwidth condition, which corresponds to about 3/4-dB loss and not the 3-dB (half-power) limit used in previous expressions, one obtains [27]

$$\text{Bandwidth (percent)} = \text{beamwidth (degrees)} \tag{1.81}$$

If the 3-dB beamwidth criterion is used, the relation is

$$\text{Bandwidth (percent)} = 2 \cdot \text{beamwidth (degrees)} \tag{1.82}$$

The above relations relate to fractional and percentage bandwidth of an array. However, there is a direct relationship between actual bandwidth and array size

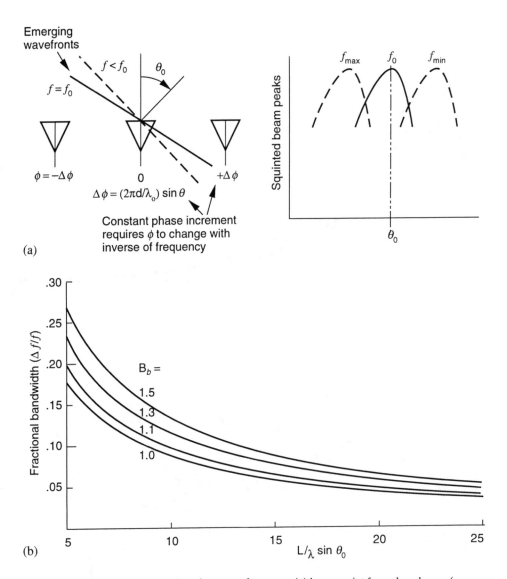

Figure 1.14 Wide-band effects in phased array performance: (a) beam squint for a phased array (wavefronts and beam peak motion; (b) array 3-dB bandwidth versus $(L/\lambda)\sin\theta_0$ (B_b is beam broadening factor).

implied by (1.79), irrespective of whatever the fractional bandwidth may be. From (1.79) one obtains

$$L \sin\theta_0 = 0.886 B_b(300)/\Delta f_M \qquad (1.83)$$

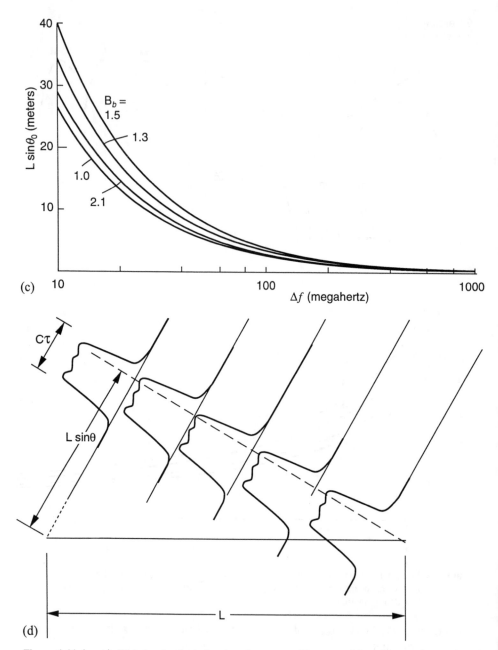

Figure 1.14 (cont.) Wide-band effects in phased array performance: (c) array or subarray length $(L/\lambda)\sin\theta_0$ versus bandwidth (MHz) (B_b is beam broadening factor); (d) narrow pulse incident on array (array fill time).

In this expression, Δf_M is the bandwidth in megahertz and L the array length in meters. Thus, a 300-MHz signal bandwidth operating with a uniformly illuminated array ($B_b \approx 1$) can have a maximum length of about one meter at 60-deg scan. Figure 1.14(c) gives the 3-dB bandwidth of arrays of various lengths and illustrates that there is a maximum array size corresponding to a given array bandwidth.

An alternate perspective on array bandwidth comes from the concept of an array "fill time" T. Figure 1.14(d) illustrates a pulsed waveform modulating a plane wave incident upon the array at an angle θ from the array normal. The sketch shows that a very short pulse will arrive at different edges of the array at entirely different times, and without delaying those signals received by the right side of the array, there is no way to sum the signals at each element and thus benefit from the array gain. The pulse length has to be significantly larger than the fill time, or for a pulse incident from angle θ,

$$\tau > T = \frac{L \sin \theta}{c} \tag{1.84}$$

where c is the velocity of light, τ is the pulse length (duration), and T is the antenna fill time.

Since any measure of pulse bandwidth is inversely proportional to the pulse duration τ, the bandwidth is

$$\Delta f = \frac{K_P}{\tau} < \frac{K_P c}{L \sin \theta} \tag{1.85}$$

for proportionality constant K_P, which is on the order of one.

The fractional bandwidth thus assumes a form similar to (1.79):

$$\frac{\Delta f}{f} < \frac{K_P \lambda}{L \sin \theta} \tag{1.86}$$

Equation (1.79) was written for a continuous-wave (CW) signal and implied an amplitude modulation of 3 dB at the band edges. Equation (1.86) merely states a similar dependence for the pulse case, and is included here for purposes of exposition. It is necessary to perform the more detailed spectral (transform plane) analysis in order to compute a more realistic bandwidth based on tolerable pulse distortion. Detailed treatments of the frequency response of arrays are given by Kinsey and Horvath [25] for a center-fed array, and by Knittel [26] for a phase-scanned array. Frank [27] gives both CW and pulse bandwidth criteria for various series and parallel feeds and shows that for similar criteria of CW signal loss and pulse spectrum loss, the bandwidth of an array passing a pulse with a uniform spectrum is about twice that of the CW signal. Thus, in many cases one can operate a wider bandwidth signal than is given by (1.79) without significant loss of information.

The array bandwidth restriction is, in most cases, a severe limitation. It can be removed only at great cost by replacing phase shifters by time-delay devices. Moreover, present day time-delay units are switched transmission lines, and their bulk and weight make them unsuitable for many array applications. Wide-band array techniques are addressed in Section 1.2.3.

1.2.2 Array Size Determination

Given the specifications required of an array antenna, the first task facing the system designer is to determine the size of the aperture. Gain is one system parameter that defines the size of an array, but when resolution is important, the array beamwidth may be the determining factor. In addition, there are special instances in which the number of elements in the array is governed by the scan volume or the ultimate depth of pattern nulls or null bandwidth. This section enumerates some of the factors that influence the required array size.

EIRP and G/T for Large, Two-Dimensional Passive or Active Arrays

A fundamental consideration is whether the array should be *active*, with solid-state amplifiers at each element, or *passive*, with a single RF power source and a single receiver. In addition, there is an intermediate solution with active devices at various levels within the array (at columns, rows, or groups of elements called *subarrays*). These various levels of organization are indicated in Figure 1.15 (shown for a transmitting array), but the only cases described here are the planar array, with a single power supply P_m (passive), the case with N amplifiers for a two-dimensional array with N elements, and amplifier output P_{mod} at each.

Equations (1.38) and (1.39) indicate that one important feature of the radar or communications transmitter is the product of its gain and input power. This term is called the *effective isotropic radiated power* (EIRP). For a large array with N elements and array aperture area Nd_xd_y, the EIRP for active and passive arrays with uniform illumination are given in the expressions below.

Passive Array

$$\text{EIRP} = N\epsilon_L P_{\text{in}}(D_{\text{CELL}})(1 - |\Gamma|^2) \tag{1.87}$$

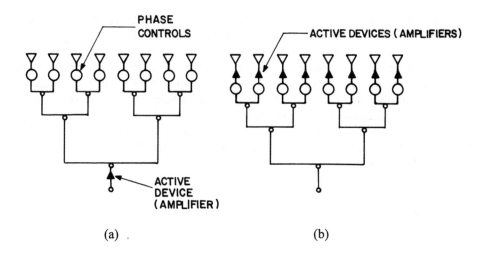

Figure 1.15 Active and passive array configurations: (a) passive array; (b) active array.

Active Array

$$\text{EIRP} = N^2 P_{\text{MOD}}(D_{\text{CELL}})(1 - |\Gamma|^2) \qquad (1.88)$$

where D_{CELL} is the directivity of one cell of the periodic array (or one element) and is defined

$$D_{\text{CELL}} = \frac{4\pi}{\lambda^2} (d_x d_y) \cos \theta \qquad (1.89)$$

For a 0.5λ matched square lattice at broadside, $D_{\text{CELL}} = \pi$.

The ϵ_L (loss efficiency) term used in the above equation for the passive array is a dissipative loss and accounts for power lost in the array feed network and phase shifters. This loss can be several decibels, and can therefore significantly impact required array size, as shown in the next section.

In these expressions, the large-array assumption is used to require that each element of the array sees the same reflection coefficient Γ. This is a good approximation for a large array because most of the elements are far from the edges, and the elements that are near the edges are not excited strongly.

A significant difference in the active and passive arrays is that for the active arrays, EIRP varies like N^2 (increasing the number of elements increases both the input power and the directivity), while the passive array EIRP varies directly with N. If the distribution network were lossless, the ratio of EIRP to net RF power would simply be the directivity and there would be no power balance difference between the active and passive arrays. The remaining difference would lie in the relative efficiency, output power, and cost of the RF amplifiers in the two cases. However, for a lossy distribution network, the advantages of the active array are readily apparent, as can be seen in the next section.

The receiving array in a communication or radar system is characterized by the ratio G/T_S, as given below, with reference to Figures 1.2 and 1.15, assuming uniform illumination.

Passive Array

$$G = D_0 \epsilon_L (1 - |\Gamma|^2)$$
$$T_a = \epsilon_L T_A + T_0(1 - \epsilon_L) \qquad (1.90)$$
$$T_S = T_a + (F_R - 1)T_0$$

where

$$D_0 = ND_{\text{CELL}}$$

Active Array

$$G = D_0(1 - |\Gamma|^2)$$
$$T_a = T_A \qquad (1.91)$$
$$T_S = (F - 1)T_0 + T_a$$

where ϵ_L is the network loss factor for the passive array (the fraction of the power received at the antenna terminals that reaches the receiver), F is the noise figure of the active array receivers, F_R is the noise figure of the passive array receiver, and T_A is the antenna temperature. Here the number of array elements N enters only once in the array gain expression. All else being equal, the active array has the advantage of lower T_a with lossy distribution networks.

Gain Limitations Due to Circuit Losses

Equation (1.11) shows the array directivity increasing linearly with aperture area. If the array is small enough and circuit losses not too large, then gain continues to increase with size, but gain is ultimately limited if line losses are not negligible. Figure 1.16(a) shows the gain of a square array of N elements like those of Figure 1.8, with interelement spacing d in either direction and with each element fed by

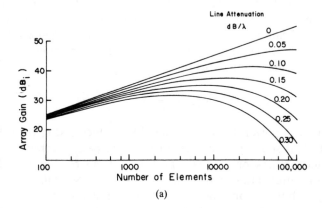

(a)

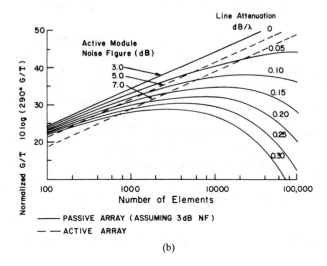

(b)

Figure 1.16 Gain and G/T limitations due to circuit losses: (a) passive array gain; (b) passive and active array G/T.

equal-length transmission lines of length $(N^{1/2} - 1)d$ (as for the array shown in the figure). In this case, the maximum array gain at broadside is just the gain of (1.11) reduced by the loss of the line.

$$\text{Gain} = \frac{4\pi A}{\lambda^2} 10^{-(d/\lambda)(N^{1/2} - 1)(\alpha_{dB/\lambda}/10)} \tag{1.92}$$

where $\alpha_{dB/\lambda}$ is the attenuation loss of the transmission line in decibels per wavelength. In this formula, the array elements are assumed matched. This equation does not include loss due to power dividers in the corporate feed network, a factor which can be significant in some cases. Either of the equal-line-length feeds shown in Figures 1.8(b) or 1.8(c) contains N_{PD} power dividers in series with the element where, for an N-element square array,

$$N_{PD} = \log_2(N) = 3.32 \log_{10}(N) \tag{1.93}$$

and the loss of each power divider may need to be included in the calculation.

Figure 1.16(a) shows gain curves for a square passive array of matched elements separated by 0.5λ on a square grid, and compares the available gain for various values of attenuation. Except for the lossless case, gain does not increase monotonically with the number of elements, and in fact reaches a maximum value and then decreases with further increase in size. The gain of an active array is shown in the figure as the zero loss case because the amplifiers are at the element level. In the active array case, the gain increases linearly without any saturation limit.

Figure 1.16(b) shows the G/T for passive and active receive arrays with the line attenuation parameters used in 1.16(a). In this figure, the passive array curves are shown solid, while the active array curves are shown dashed. The assumed phase shift loss is not shown, but should be included in the system evaluation. The G/T is altered even more than the gain because the temperature is increased by thermal loss in the line.

Transmission line loss is a major factor leading to the integration of solid-state amplifiers into large arrays, and to the fabrication of arrays using several transmission media. It is often convenient to do several layers of power division in low-loss media like waveguides or coaxial lines instead of using a higher loss media like microstrip transmission line throughout the array.

Directivity and Illumination Errors: Random Error and Quantization Error

The net antenna gain is the directive gain reduced by the various system losses. Apart from the loss associated with aperture efficiency (1.12), which is deterministic in nature and built into the choice of aperture illumination as a compromise between gain and sidelobe level, there are usually two other factors that contribute to reduced directivity. These factors are array tolerance errors and errors due to phase, amplitude, or time-delay quantization. They reduce directivity (and gain) by distorting the chosen aperture illumination.

Data describing peak sidelobes and pattern structure due to these effects are given in Chapter 6. Equations for gain reduction and average sidelobe level are given below for arrays with random phase and amplitude errors. The directivity in the presence of amplitude and phase errors is

$$\frac{D}{D_0} = \frac{1}{1 + \overline{\Phi}^2 + \overline{\delta}^2} \tag{1.94}$$

where $\overline{\delta}^2$ is the amplitude ratio variance normalized to unity, $\overline{\Phi}^2$ is the phase error variance in radians squared, and D_0 is the directivity without error.

The average sidelobe level, far from the beam peak and normalized to the peak, is a constant given by

$$SL_{dB} = 10 \log_{10} \overline{\sigma^2}$$

$$\overline{\sigma}^2 = \frac{\overline{\Phi}^2 + \overline{\delta}^2}{N\epsilon_A} \tag{1.95}$$

and ϵ_A is the aperture efficiency.

The above expressions pertain to linear and planar arrays. Since they are normalized to the beam peak, the element pattern gain has been removed from the expressions. Figure 1.17(a) shows the root-mean-square (rms) sidelobe level for a square array as a function of array directivity for various phase errors (and no amplitude errors). The dashed curve of Figure 1.17(a) gives sidelobe levels for an array of the same size, but organized into columns for a one-dimensional scan. In this case, the rms sidelobes cited are in the plane orthogonal to the axes of the columns.

A particularly revealing way to restate the sidelobe results is the expression given below, valid for a planar array with $\lambda/2$ spacing (and element pattern broadside gain π, as in (1.89)). In this expression, the sidelobe level is given relative to the isotropic (zero gain) level as

$$\overline{\sigma}_x^2 = \overline{\sigma}^2 \cdot D_0$$

$$= \overline{\sigma}^2(\pi N\epsilon_A) \tag{1.96}$$

$$= (\overline{\Phi}^2 + \overline{\delta}^2)\pi$$

This level is shown conveniently as a family of circles in Figure 1.17(b) [28].

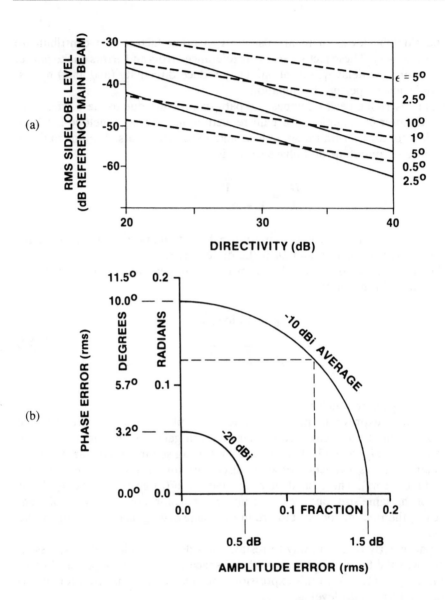

COMBINATION OF RANDOM ERRORS (FROM RUZE)
FOR $\lambda/_2$ SEPARATION

Figure 1.17 Tolerance effects in array antennas: (a) rms sidelobe level for square array with errors at elements (solid lines) or columns (dashed lines) in plane orthogonal to columns (reference to main beam); (b) average (rms) sidelobe level (relative to isotropic) for array with amplitude and phase errors.

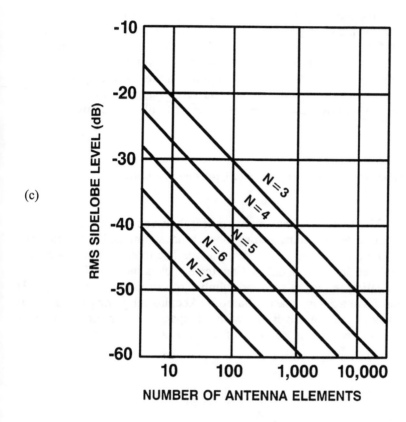

Figure 1.17 (cont.) Tolerance effects in array antennas: (c) rms sidelobes due to *N*-bit phase-shift quantization (*N* = number of phase shifter bits).

A digitally controlled phase shifter with *P* bits has 2^P phase states separated by phase steps of $2\pi/(2^P)$. If the array is made up of such phase shifters, then there is an additional loss due to the staircase approximation of the required phase shift. This loss and the resulting sidelobe level increase are described in much more detail in Chapter 6. The resulting loss in directivity and the average sidelobe level produced by the error are approximated [29] by the equations above using the phase error variance:

$$\overline{\Phi^2} = \frac{1}{3}\frac{\pi^2}{2^{2P}} \tag{1.97}$$

which is evidently the mean square value of the triangular error distribution with height one-half of the phase step. Figure 1.17(c) shows the average sidelobe level for an array with N bits of phase quantization. Chapter 6 gives peak sidelobe levels for such distributions, but standard practice is to break up the periodic error by several means, and so to make the error occur with a more random spatial distribution. In this case, as an approximation, one can assume that discrete peak sidelobes resulting from this error are on the order of 10 dB above this level (see Chapter 6).

*Minimum Number of Elements Versus Scan Coverage: Limited
Field-of-View Arrays*

According to (1.73), there is a maximum spacing between array elements that cannot be exceeded without exciting grating lobes. If not suppressed by the element pattern, these lobes are as large as the main beam. The topic of *limited field-of-view arrays* is treated in more detail in Chapter 8, but is included here for the purposes of evaluating the array size and number of elements. Equation (1.73) gives a condition for maximum spacing based on keeping all grating lobes out of real space throughout the scan coverage. With this spacing, the minimum number of elements in a conventional linear array of length L is

$$N_{min} = L/D_{max} \tag{1.98}$$
$$= \frac{L}{\lambda}(1 + \sin\theta)$$

where D_{max} is the interelement spacing.

Although this expression leads to the use of fewer elements if the scan is limited, this is still a restrictive condition, leading to an absolute minimum number of elements of one per square wavelength even if the array is unscanned, or four per square wavelength if the array is scanned to endfire.

If the array is periodic, however, there is a way to reduce the number of controls by grouping the elements into subarrays that allow one to use extra large spacing between these subarrays while suppressing the resulting grating lobes. This can be done using networks that produce approximate flat-topped element patterns that are nearly constant for $|D/\lambda \sin\theta| \le 0.5$, and zero for $|D/\lambda \sin\theta| \ge 0.5$. With this element spacing (in one dimension), one can scan the array to the maximum scan angle θ_{max}, which is related to the maximum intersubarray spacing D_{max} by

$$(D_{max}/\lambda)\sin\theta_{max} = 0.5 \tag{1.99}$$

and to the condition for the minimum number of controls for a one-dimensional array of length L and beamwidth $\sin(\theta_3) \approx \lambda/L$:

$$N_{\min} = \frac{L}{D_{\max}} = \frac{\sin \theta_{\max}}{(0.5) \sin \theta_3} \qquad (1.100)$$

This minimum number of controls is equal to the number of beams that an orthogonal beam matrix can form over the given scan sector. Networks and circuits for producing such element (or subarray) patterns are described in Chapter 8 and have a variety of characteristics, some approaching this ideal element pattern. The basic flat-topped pattern is produced by a technique called *overlapped subarraying*. Most practical systems need several times the minimum number of elements given in (1.100); but if the scan is restricted, this can be only a small fraction of the elements for an array designed for wide-angle coverage. Array techniques that use these features are called *limited field-of-view* or *limited scan systems*, but are relatively complex compared to conventional arrays.

For a rectangular two-dimensional array, the minimum number of controls is the product of two numbers of the form of (1.100).

$$N_{\min} = \frac{\sin \theta^1_{\max} \sin \theta^2_{\max}}{0.25 \sin \theta_3^{(1)} \sin \theta_3^{(2)}} \qquad (1.101)$$

Since the number $N_{\min}$ is the smallest achievable, it is convenient to define a term called the *element use factor* $N/N_{\min}$, which measures the array against this standard [30]. An array with elements spaced d_x and d_y apart has the element use factor $N/N_{\min}$.

$$\frac{N}{N_{\min}} = \frac{D^{(x)}_{\max} D^{(y)}_{\max}}{d_x d_y}$$

$$= \frac{0.25\lambda^2}{d_x \sin \theta^{(x)}_{\max} d_y \sin \theta^{(y)}_{\max}} \qquad (1.102)$$

Figure 1.18 shows the relative number of elements (controls) for a conventional two-dimensional array with a conical scan sector as compared with the theoretical minimum. This result is due to Stangel and is comparable to the result of using (1.102). The techniques for achieving this reduction in controls and the relative complexity of systems that approach this ideal are detailed in Chapter 8.

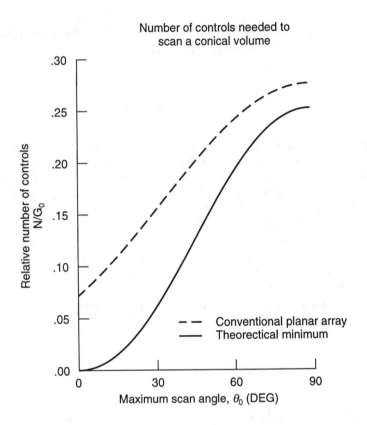

Figure 1.18 Required controls for arrays with limited field of view.

1.2.3 Time-Delay Compensation

The bandwidth limitations imposed by (1.79) severely restrict the use of arrays in many practical radar and communication systems. The use of time delays instead of phase shifts can give enhanced bandwidth, but often at prohibitively large cost and at the cost of other performance goals.

In order to maintain the beam peak at a constant angle θ_0 for all frequencies, one needs time-delayed signals at each element. The excitation coefficients for a linear array are given:

$$a_n = \exp[-j(2\pi/\lambda)nd_x \sin \theta_0] \tag{1.103}$$
$$= \exp[j\Phi_n]$$

In terms of equivalent phases Φ_n at each element, these phase shifts are

$$\Phi_n = -2\pi \frac{nf}{c} d_x \sin \theta_0 \tag{1.104}$$

and thus need to vary linearly with frequency.

The customary way to provide time delay is to insert incremental lengths of transmission line of length $L_n = nd_x \sin \theta_0$ to produce the time delays,

$$\tau_n = \frac{-nd_x \sin \theta_0}{c} = L_n/c \tag{1.105}$$

using actual delay lines by switching sections of transmission lines behind each element or group of elements. Since the phase shift inserted by length of line L_n is

$$\Phi_n = \frac{2\pi L_n}{\lambda} \tag{1.106}$$

each line length (near the ends of the array) has to be variable over the range

$$\frac{-L}{2} \sin \theta_0 \leq L_n \leq \frac{L}{2} \sin \theta_0 \tag{1.107}$$

In this case, the negative value does not indicate a negative line length, since an equal length of line is first added to each path. The required lengths of switched line are extremely bulky and expensive for large arrays, and the large number of discrete time-delay positions requires a highly complex switching network. Furthermore, the relative dispersion in the various transmission line sections may prohibit accurate beam forming. For these reasons, there are few systems that are designed around true time-delay controls.

There are other techniques that may someday be useful for approximating true time delay, at least over relatively broad ranges of frequency. Variable analog microwave acoustic or magnetic delay lines may have adequate dispersion characteristics, and optical-fiber delay lines have promise as well. It is likely, however, that in the near term the digital processing of the signals at each array element will provide the most realistic broadband steering. In such digital beam forming, the signals are down-converted and digitally sampled. The frequency dependence can then be inserted by processing in the frequency domain to produce the linear frequency dependence of (1.50) or by inserting incremental time delays.

Time Delay at the Subarray Level

Two methods have been used to produce time-corrected steering in practical phased arrays. The first method is to group the array into subarrays, use time-delay devices behind each subarray, and use phase shift behind each element, as shown in Figure 1.19(a). To a first approximation, the bandwidth of an array of M subarrays is M times that of the phase-steered array. The primary disadvantage of providing time delay at the subarray level is the resulting periodic distortion in the aperture phase, which causes grating lobes. Detailed evaluations of this bandwidth and the resulting grating lobe power levels are given in Chapter 6 for contiguous subarrays. It is possible, however, to create optimal subarrays that allow good suppression of the grating lobes over very wide bandwidths. These techniques are similar to those used for limited field-of-view antennas and are described in detail in Chapter 8.

Scanning About Time-Delayed Beam Positions

The second method of incorporating time-delay devices into an array combines a complete set of time-delay devices, or a true time-delay multiple-beam network, and a complete set of phase shifters. Shown schematically in Figure 1.19(b), these networks provide exact time delay at only a small number (M) of beam positions, as few as two to four. The scan sector is thus divided into M sections, each centered on the M true time-delayed beam positions. In effect, the phase shifters only need to scan the beam from the time-delayed position half way to the next time-delayed position. The maximum phase scan for any beam position is thus to the angle

$$\theta_{scan} = \frac{\sin \theta_{max}}{M} \qquad (1.108)$$

and with (1.79) one can compute the system squint bandwidth as

$$\frac{\Delta f}{f} = \frac{0.886 B_b \lambda M}{L \sin \theta_{max}} \qquad (1.109)$$

This equation represents a direct bandwidth multiplication by the number of fixed, time-delayed positions. Moreover, unlike the case of subarray level time delay, if analog (not discrete) phase shifters are used, this approach does not introduce any periodic phase error across the array, and so there is no sidelobe degradation.

Figure 1.19(b) shows implementation of this broadband approach using time-delay units (switched lines) with M states and varied across the array. This configuration requires different sets of switched lines for every element of the array

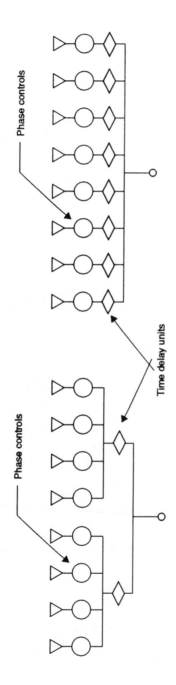

Figure 1.19 Array configurations for wide-band arrays.

and so is inherently more costly than the contiguous subarraying technique. Multiple beam matrices with true time delay can also be used for this application. Their use is described in Chapter 8.

REFERENCES

[1] Hansen, R. C., *Microwave Scanning Antennas*, Vol. 1, New York: Academic Press, 1964, Ch. 1, pp. 82–91.

[2] Kraus, J. D., *Radio Astronomy*, New York: McGraw-Hill, 1966, Ch. 3, p. 100.

[3] Ibid., p. 102.

[4] Skolnik, M. I., *Radar Handbook*, New York: McGraw-Hill, Ch. 2, 1st edition, p. 5.

[5] Balanis, *Antenna Theory: Analysis and Design*, New York: Harper and Row, 1982.

[6] King, R. W. P., *The Theory of Linear Antennas*, Cambridge, MA: Harvard University Press, 1956, p. 15.

[7] Balanis, op. cit., pp. 48–53.

[8] Bayliss, E. T., "Design of Monopulse Antenna Difference Patterns With Low Side Lobes," *Bell Syst. Tech. J.*, Vol. 47, 1968, pp. 623–640.

[9] Kirkpatrick, G. M., "Aperture Illuminations for Radar Angle-of-Arrival Measurements," *IRE Trans. on Aeronautical and Navigational Electronics*, Vol. ANE-9, Sept. 1953, pp. 20–27.

[10] Barton, D. K., and H. R. Ward, *Handbook of Radar Measurement*, New Jersey: Prentice-Hall, 1969, Ch. 2, p. 42.

[11] Sharenson, S., "Angle Estimation Accuracy With a Monopulse Radar in the Search Mode," *IRE Trans.*, Vol. ANE-9, No. 3, Sept. 1962, p. 175–179.

[12] Skolnik, op. cit., Ch. 21, p. 32.

[13] Hacker, P. S., and H. E. Schrank, "Range Distance Requirements for Measuring Low and Ultralow Sidelobe Antenna Patterns," *IEEE Trans.*, Vol. AP-30, No. 5, Sept. 1982, pp. 956–965.

[14] Hansen, R. C., "Measurement Distance Effects on Low Sidelobe Patterns," *IEEE Trans.*, Vol. AP-32, No. 6, June 1984, p. 591–594.

[15] Johnson, R. C., and H. J. Jasik, *Antenna Engineering Handbook*, New York: McGraw-Hill, 2nd edition, 1984, Ch. 2, Table 2.1, p. 16, Table 2.2, p. 20.

[16] Taylor, T. T., "Design of Line Source Antennas for Narrow Beamwidth and Low Sidelobes," *IEEE Trans.*, Vol. AP-3, Jan. 1955, p. 16–28.

[17] Drane, C. J., "Useful Approximations for the Directivity and Beamwidth of Large Scanning Dolph-Chebyshev Arrays," *IEEE Proc.*, Vol. 56, No. 11, Nov. 1968, pp. 1779–1787.

[18] Elliott, R. S., "The Theory of Antenna Arrays," Ch. 1, Vol. 2 in *Microwave Scanning Antennas*, R. C. Hansen, ed., New York: Academic Press, 1966, pp. 29.

[19] Tang, R., and R. W. Burns, Ch. 20 in R. C. Johnson, and H. Jasik, op. cit., p. 15.

[20] King, H. E., "Directivity of a Broadside Array of Isotropic Radiators, *IRE Trans.*, Vol. Ap-7, No. 2, 1959, pp. 187–201.

[21] Elliott, op. cit., pp. 44–45.

[22] Stegen, R. J., "The Gain-Bandwidth Product of an Antenna," *IEEE Trans.*, Vol. AP-12, July 1964, pp. 505–507.

[23] Elliott, op. cit., p. 44.

[24] Knittel, G. H., A. Hessel, and A. A. Oliner, "Element Pattern Nulls in Phased Arrays and Their Relation to Guided Waves," *IEEE Proc.*, Vol. 56, No. 11, Nov. 1968, pp. 1822–1836.

[25] Kinsey, R. R., and A. L. Hovath, "Transient Response of Center-Series Fed Array," *Phased Array Antennas*, A. Oliner and G. Knittel, eds., Dedham, MA: Artech House, 1972, pp. 261–272.

[26] Knittel, G. H., "Relation of Radar Range Resolution and Signal-to-Noise Ratio to Phased-Array Bandwidth," *IEEE Trans.*, Vol. AP-22, No. 3, May 1974, pp. 418–426.

[27] Frank, J., "Bandwidth Criteria for Phased-Array Antennas," *Phased-Array Antennas*, A. Oliner and G. Knittel, eds., Dedham, MA: Artech House, pp. 243–253.

[28] Schrank, H. E., "Low Sidelobe Phased Arrays," *IEEE AP-S Newsletter*, Apr. 1983; see also J. Ruze, "Pattern Distortion in Space Fed Phased Arrays," Lincoln Laboratories Report SBR-1, Dec. 1974.

[29] Miller, C. J., "Minimizing the Effects of Phase Quantization Errors in an Electronically Scanned Array," *Proc. 1964 Symp. on Electronically Scanned Array Techniques and Applications*, RADC-TDR-64-225, Vol. 1, pp. 17–38, RADC, Griffiss AFB, New York, 1964.

[30] Patton, W. T., "Limited Scan Arrays," *Phased Array Antennas, Proc. 1970 Phased Array Symp.*, A. A. Oliner and G. H. Knittel, eds., Dedham, MA: Artech House, 1972, pp. 332–343.

Chapter 2
Pattern Characteristics of Linear and Planar Arrays

2.1 ARRAY ANALYSIS

2.1.1 The Radiation Integrals

As shown in many texts [1], the free-space electromagnetic field can be expressed in terms of integrals over elementary electric and magnetic current sources. The field due to an electric current density $\mathbf{J}$ in a volume $dv' = dx'dy'dz'$ is obtained from the vector potential integral $\mathbf{A}$, where $\mathbf{A}$ is given by

$$\mathbf{A} = \frac{\mu}{4\pi} \int \mathbf{J}(v') \frac{e^{-jk_0R}}{R} \, dv' \tag{2.1}$$

for

$$R = [(x - x')^2 + (y - y')^2 + (z - z')^2]^{1/2}$$

and the associated electric and magnetic fields are given by

$$\mathbf{E_A} = -j\omega\mathbf{A} - \frac{j}{\omega}\nabla(\nabla \cdot \mathbf{A}) \tag{2.2}$$

$$\mathbf{B_A} = \nabla x \mathbf{A} \tag{2.3}$$

and $\omega = 2\pi f$.

The segment of wire shown in Figure 2.1 indicates that the vector potential is routinely used to compute the radiation from wire antenna structures.

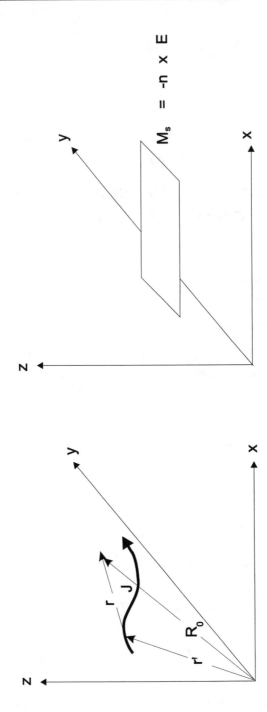

Figure 2.1 Radiation from electric and magnetic current sources.

The field due to a volume density of magnetic current is obtained from a potential function termed the *electric potential* and given by

$$\mathbf{F} = \frac{\epsilon}{4\pi} \int \mathbf{M}(v') \frac{e^{-jk_0 R}}{R} \, dv' \tag{2.4}$$

and the associated fields are

$$\mathbf{E_F} = -\frac{1}{\epsilon} \nabla x \mathbf{F} \tag{2.5}$$

$$\mathbf{B_F} = -j\omega\mu\mathbf{F} - \frac{j}{\omega\epsilon} \nabla(\nabla \cdot \mathbf{F}) \tag{2.6}$$

In classical radiation problems, the magnetic current is understood to be a mathematical artifice, not a realizable current. Its value in antenna analysis is that it is regularly used to represent radiation from apertures described in terms of their known electric fields. In the case of an aperture antenna, the magnetic current is identified with the tangential electric field at the radiating aperture using

$$\mathbf{M_S} = -\hat{\mathbf{n}} \times \mathbf{E_S} \tag{2.7}$$

for $\hat{\mathbf{n}}$, the outward-directed normal at the aperture. The subscripts **S** refer to surface magnetic currents, and in this expression the volume integral has shrunk to a surface integral. The aperture in Figure 2.1 depicts this use of the magnetic current to represent surface electric fields.

The potential functions are integral solutions to Maxwell's equations. At distances far from any source, their radial dependence has the $(1/R)$ form required for energy conservation in (1.1) and the exponential dependence of an outward-traveling spherical wave.

Although both solutions are independent when there are no boundaries, the general electromagnetic field requires the sum of fields from both potentials. In general,

$$\mathbf{E} = \mathbf{E_A} + \mathbf{E_F} \qquad \mathbf{B} = \mathbf{B_A} + \mathbf{B_F} \tag{2.8}$$

is the complete form that may be necessary to satisfy physical boundary conditions.

One boundary condition of vast importance in antenna and array theory is that of an antenna mounted over or in a perfectly conducting ground plane (the term *ground screen* is used interchangeably).

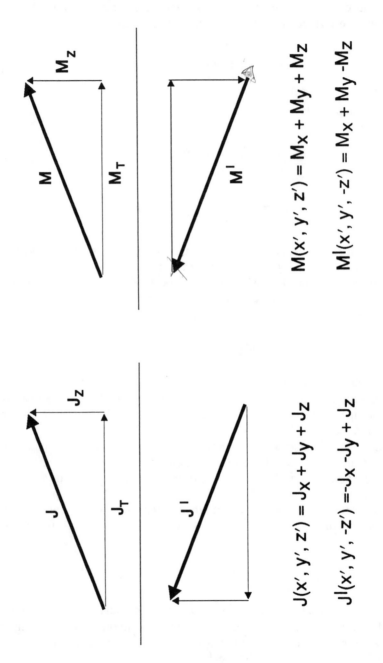

Figure 2.2 Image principle for electric and magnetic currents.

The well-known *image principle*, depicted in Figure 2.2, provides a recipe for superimposing fictitious image sources beneath the ground screen in order to satisfy the required boundary condition that the total tangential electric field be zero at the screen. Potential functions corresponding to these imaged sources are

$$\mathbf{A} = \frac{\mu}{4\pi} \int_v \left\{ \mathbf{J}(v') \frac{e^{-jk_0R}}{R} + \mathbf{J}_I(v_I') \frac{e^{-jk_0R_I}}{R_I} \right\} dv' \tag{2.9}$$

where

$$\mathbf{J}_I = -\hat{\mathbf{x}}J_x - \hat{\mathbf{y}}J_y + \hat{\mathbf{z}}J_z$$

and

$$\mathbf{F} = \frac{\epsilon}{4\pi} \int_v \left\{ \mathbf{M}(v') \frac{e^{-jk_0R}}{R} + \mathbf{M}_I(v_I') \frac{e^{-jk_0R_I}}{R_I} \right\} dv' \tag{2.10}$$

where

$$\mathbf{M}_I(v') = \hat{\mathbf{x}}M_x + \hat{\mathbf{y}}M_y - \hat{\mathbf{z}}M_z$$

and

$$R_I = [(x - x')^2 + (y - y')^2 + (z + z')^2]^{1/2}$$

These equations are used later to describe the radiation from elementary wire and slot elements over a ground screen.

One special case for which the above is used is to express the radiation into the hemisphere from an aperture in a conducting sheet (Figure 2.3). In this case, one uses the electric potential, and the source and image coalesce to double the effective source term. The electric potential for the half-space problem is therefore

$$\mathbf{F} = \frac{\epsilon}{2\pi} \int_s M_S(s') \frac{e^{-jk_0R}}{R} ds'$$
$$= \frac{\epsilon}{2\pi} \int_s -\hat{\mathbf{z}}x\mathbf{E}(s') \frac{e^{-jk_0R}}{R} ds' \tag{2.11}$$

The radiation from more complex structures can also be evaluated using the potential functions, as can the mutual coupling between antenna array elements (Chapter 6). The image principle is one way of constructing solutions to the inhomo-

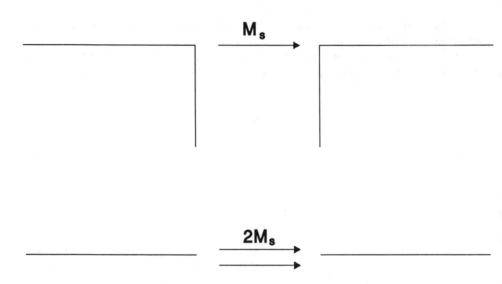

Figure 2.3 Radiation from an aperture in a conducting screen.

geneous vector Helmholtz equations that define the magnetic and electric potentials for half-space radiation over a perfectly conducting ground screen. In the more general case, one can use the inhomogeneous equations

$$(\nabla^2 + k_0^2)\mathbf{F} = -\epsilon\mathbf{M} \tag{2.12}$$

$$(\nabla^2 + k_0^2)\mathbf{A} = -\mu\mathbf{J} \tag{2.13}$$

for magnetic and electric sources, along with the requisite boundary conditions. A description of the use of vector and dyadic Green's functions in the solution of inhomogeneous Helmholtz equations is given in [2,3]. These methods are used to analyze structures in Chapter 6.

Far-Zone Fields in Terms of Radiation Integrals

Figures 2.1 and 2.2 show elements at generalized locations. The integrals of (1.1) and (2.11) are taken over the primed coordinates. In Chapter 1, it is shown that the form of these equations can be simplified if the receiving point is very far from the array. Using vector notation and denoting the source position at the location $\mathbf{r}'$ and the receiving point at $\mathbf{r}$, one can then write the distance R as

$$R = |\mathbf{r} - \mathbf{r}'| \approx R_0 - \mathbf{r}' \cdot \hat{\mathbf{r}} \tag{2.14}$$

where the unit vector $\hat{\mathbf{r}}$ is in the direction of the receiving point $\mathbf{r}$, and the distance R_0 is measured from the center of the coordinate system (usually chosen as the center of the array).

Using the above, one can write the approximate expression (below), which simplifies the potential function integrals considerably, since R_0 is a constant and can be removed from the integrals.

$$\frac{e^{-jk_0 R}}{R} \approx \frac{e^{-jk_0 R} e^{jk_0(\mathbf{r}' \cdot \hat{\mathbf{r}})}}{R_0} \tag{2.15}$$

The radial components of $\mathbf{F}$ and $\mathbf{A}$ are zero (decay faster than $1/R$) in the far zone, and the far-zone fields can be given by [4]

$$\mathbf{E_A} = -j\omega\mathbf{A_T} \tag{2.16}$$

$$\mathbf{H_A} = -\frac{j\omega}{\eta}\hat{\mathbf{r}}x\mathbf{A_T}$$

$$\mathbf{H_F} = -j\omega\mathbf{F_T} \tag{2.17}$$

$$\mathbf{E_F} = j\omega\eta\hat{\mathbf{r}}x\mathbf{F_T}$$

where $\eta = (\mu/\epsilon)^{1/2}$ is the characteristic impedance of the medium, and the subscript T means only transverse components of $\mathbf{A}$ and $\mathbf{F}$ need be considerd.

2.1.2 Element Pattern Effects, Mutual Coupling, Gain Computed From Element Patterns

The array gain is related to the gain of the individual elements in the array, as will be shown layer. However, the gain of an isolated element may be very different from the gain of the same element in the presence of the rest of the array. In addition, the element patterns and gain vary across the array with the elements near the edge behaving quite unlike those near the center. This behavior is due to the electromagnetic coupling between elements and can result in more or less element gain in the array environment than when isolated.

Figure 2.4 illustrates the coupling of a single excited element with all others terminated in matched loads. The actual radiated pattern is formed by the directed radiation from the excited element combined with reradiated fields from all of the elements illuminated by the radiation from the excited element. Depending on element gain and spacing, the radiation pattern of a low-gain element can be substantially narrowed by the interaction, but if a large array is composed of high-gain elements, then the element gain is decreased from the isolated element gain in order to limit the maximum area gain to no more than $4\pi A/\lambda^2$.

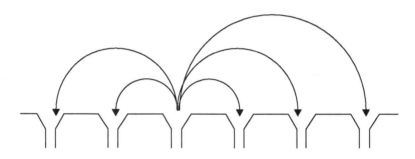

Figure 2.4 Coupling between array elements.

Following this introduction, it should be clear that the actual element gain is usually not known. It is found as the result of a detailed calculation involving the most fundamental electromagnetic analysis. This *mutual coupling* is discussed in Chapter 6. In the following sections, it is assumed that such coupling exists and can be measured or computed to completely describe the array. The sections present an alternative description of the array in terms of *active element patterns*, the patterns of elements embedded in the array environment. This description is fully equivalent to and embodies all of the physics in the array model with mutual coupling.

Element Patterns and Mutual Coupling

The complex subject of mutual coupling and array element patterns should be introduced in the simplest of terms. Consider an array of small waveguide-fed apertures, as shown in Figure 2.5, with apertures located in the plane $z = 0$, but otherwise arbitrarily located. The aperture field of every element will be assumed

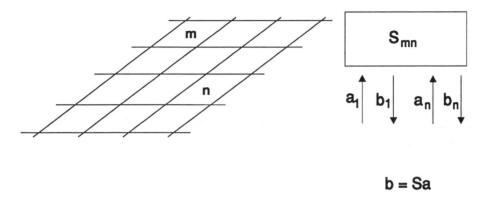

Figure 2.5 Scattering matrix representation for interelement coupling of waveguide apertures.

to have the same distribution, namely that of the exciting waveguide, a linearly polarized TE_{10} mode. For the mth element, the tangential aperture field is

$$\mathbf{E_T}(x_m, y_m, z_m) = \hat{\mathbf{y}} A_m e_{10}(x', y') \qquad (2.18)$$

where the function e_{10} is the spatial distribution of electric field in the aperture with coordinates at $(x', y', 0)$. In the far field of the mth element, the radiated electric field from (1.5) and (1.11) is reduced to the following compact form:

$$\mathbf{E_m} = \frac{jk_0}{2\pi} \frac{e^{-jk_0 R_0}}{R_0} \sum \int dS'_m [\cos\theta\, \mathbf{E_T}(x'_m, y'_m) - \hat{\mathbf{r}}\hat{\mathbf{r}} \cdot \mathbf{E_T}(x'_m, y'_m)] e^{jk_0(\mathbf{r'} - \hat{\mathbf{r}})} \qquad (2.19)$$

The constant A_m is the complex amplitude of the tangential aperture field. This term contains not only the applied field at the antenna aperture, but also the field due to the reflected signal at the aperture and the field induced in the aperture by other array elements. In this case, the entire radiation and interelement coupling behavior for an N-element array is specified in terms of an N-by-N element scattering matrix that relates the various transmitted incident and reflected fields at each element.

When all of the elements of the array are excited by incident signals a_m that one might associate as the voltage of the incident waveguide fields, the reflected signals b_m at each terminal are given in terms of a conventional scattering matrix formalism [5], as indicated schematically in Figure 2.5. For each element of the array,

$$[b] = \{S\}[a] \qquad (2.20)$$

where the column matrix $[a]$ is the incident signal vector and the column matrix $[b]$ is the vector of reflected signals. The tangential field is given by the sum of incident and reflected fields evaluated at the aperture. The constant A_m is therefore the sum of incident and reflected signal amplitudes given by

$$A_m = (a_m + \sum S_{mn} a_n) \qquad (2.21)$$

and the radiated field of the array is

$$\mathbf{E}(\mathbf{r}) = \frac{jk_0}{2\pi} \frac{e^{-jk_0 R_0}}{R_0} [\hat{\mathbf{y}} \cos\theta - \hat{\mathbf{z}} v] c_0 \sum g(m)(a_m + \sum S_{mn} a_n) \qquad (2.22)$$

where

$$g(m) = e^{-jk_0(\mathbf{r}'_m - \hat{\mathbf{r}})}$$

and

$$c_0 = \int e_{10}(x', y') e^{+jk_0(ux' + vy')} dS'_m$$

The factor

$$[\hat{\mathbf{y}} \cos\theta - \hat{\mathbf{z}} v] c_0 = \mathbf{f}_x(\theta, \phi) = \hat{\boldsymbol{\theta}} \sin\phi + \hat{\boldsymbol{\phi}} \cos\theta \cos\phi \qquad (2.23)$$

is the pattern of an isolated element and is polarized transverse to the radial direction. This equation supports two alternative views of array radiation. The following paragraphs illustrate these two perspectives.

The first of these alternatives sees each element from a circuit point of view, with incident signals coupling to all array elements as indicated in (2.22). From this *mutual impedance* perspective, each element is considered to radiate separately, based on its aperture field $\mathbf{E}_T$. In order to maintain a desired radiation pattern, one must control all the aperture fields as a function of scan. As the array is scanned, the array mismatch increases (assuming it is matched at broadside), and the aperture fields at any given element do not change in proportion to the incident signal, because the reflection coefficient is scan-dependent. The array control task is here seen as that of specifying the correct incident fields to produce the desired aperture fields in the mutually coupled environment.

Rearranging this expression emphasizes the nature of the element pattern in a scanned array and illustrates the alternative point of view which describes array scan phenomena. From the perspective of the *active element pattern* (or just *element pattern*), each element is excited with all other elements terminated in matched loads. The resulting radiation pattern $f_m(\theta, \phi)$ is the active element pattern of that element. The element pattern does not change with scan, but includes all inter-element coupling for all scan angles. For elements in a finite array, the radiated field is given by

$$
\begin{aligned}
\mathbf{E}(\mathbf{r}) &= \frac{jk_0}{2\pi} \frac{e^{-jk_0 R_0}}{R_0} \mathbf{f_i} \sum a_m g_m \left[1 + \sum S_{mn} \frac{g_n}{g_m}\right] \\
&= \frac{jk_0}{2\pi} \frac{e^{-jk_0}}{R_0} \sum a_m g_m \mathbf{f_m}(\theta, \phi)
\end{aligned}
\tag{2.24}
$$

where

$$
\mathbf{f_m}(\theta, \phi) = \mathbf{f_i}(\theta, \phi)\left[1 + \sum S_{mn} \frac{g_n}{g_m}\right]
$$

This expression shows the far field written as the sum of element excitation coefficients a_m multiplied by the time-delay factor g_m and an element pattern $\mathbf{f_m}(\theta, \phi)$, which is now different for each element. The $\mathbf{f_m}(\theta, \phi)$ has a term representing radiation from the excited element and a sum of terms to account for radiation from all of the other elements with phase centers at positions across the array, hence the term g_n/g_m, multiplying the scattering coefficients S_{nm}. The basic array element field pattern is thus the product of the isolated active element pattern and a space factor, which accounts for all of the other coupled elements. Some of the mutually coupled terms can produce very angle-sensitive changes to the element patterns, resulting in rippled and distorted patterns with strong frequency dependence. The element patterns for centrally located elements of a large array tend to be very similar, while the ones near the array edges are distorted and asymmetrical. This distortion limits the sidelobe level that can be maintained if the various elements are excited with some predetermined illumination. Figure 2.6 shows element patterns and reflection coefficients of the center element in several small arrays of parallel plate waveguides. These data, due to Wu [6], illustrate substantial changes due to mutual coupling as a function of the number of array elements N.

Historically, the most significant use of element patterns has been to experimentally verify the scan behavior of particular elements in test arrays. This is done

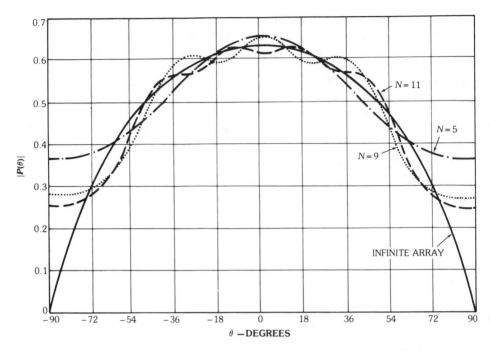

Figure 2.6(a) Element pattern $P(\theta)$ and reflection coefficient R of center element in unloaded waveguide array $[b/\lambda = a/\lambda = 0.4]$ after Wu [6]: radiation patterns.

[7] by building an array of sufficient length (10λ to 20λ or more on a side) and to terminate all but one element in matched loads. The resulting measured radiated pattern of a central element is the approximate element pattern of the scanned array, and the patterns of edge elements likewise approximate the patterns near the edge of a larger array.

Throughout the rest of Chapter 2, it is assumed that all element patterns in the array are identical. However, in Chapter 3 it is shown that by using the calculated or measured element patterns it is possible to synthesize low-sidelobe patterns, even in the presence of mutual coupling. Alternatively, if the elements can be assumed to each support the same current distribution (single-mode assumption), then one can always perform the synthesis and solve for the required source voltages using the mutual impedance matrix.

Gain Computed From Element Patterns (for Large Array)

Although the element gain may vary across the array, many of the central elements of a large array have the same gain and element patterns. For such a large array,

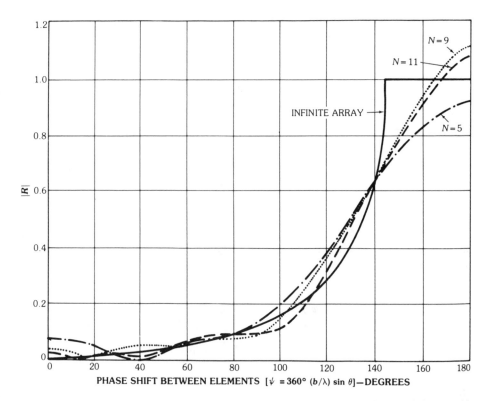

Figure 2.6(b) Element pattern $P(\theta)$ and reflection coefficient R of center element in unloaded waveguide array $[b/\lambda = a/\lambda = 0.4]$ after Wu [6]: reflection coefficients.

one can obtain a good approximation of the array gain by assuming that all element patterns are the same. In this case, the gain for each element is

$$g_E^n(\theta, \phi) = \frac{4\pi R^2}{P_E^n} S_E^n(\theta, \phi) \qquad (2.25)$$

where $S_E^n(\theta, \phi)$ is the radiated power density of the nth element at the distance R from the array, and P_E^n is the power input to the nth element (note that this power also includes that which is lost in the feed network). If the element is matched, this normalized power input is proportional to the square of the input signals, or (in a normalized form)

$$P_E^n = |a_n|^2 \qquad (2.26)$$

and the input power for the whole array is given by the sum of the excitation coefficients at each element.

$$P_{\text{in}} = \sum_n P_E^n$$
$$= \sum |a_n|^2 \qquad (2.27)$$

where the coefficients a_n represent voltages, currents, or incident wave amplitude. The far field for any input signal a_n is proportional to

$$[S_E^n(\theta, \phi)]^{1/2} = \frac{g_E^{1/2}(\theta, \phi)}{[4\pi R^2]^{1/2}} a_n \qquad (2.28)$$

Assuming that the excitation is chosen with a progressive phase to scan the beam, the fields add directly at the peak. The array far-field power pattern at the beam peak (θ_0, ϕ_0) is

$$S(\theta_0, \phi_0) = \left[\sum (S_E^n(\theta_0, \phi_0))^{1/2}\right]^2$$
$$= \frac{1}{4\pi R^2}\left\{\sum [g_E^n(\theta, \phi)]^{1/2}|a_n|\right\}^2 \qquad (2.29)$$

and so the array gain is

$$G = \frac{4\pi R^2 S(\theta_0, \phi_0)}{P_{\text{in}}}$$
$$= g_E(\theta_0, \phi_0)\frac{\{\sum|a_n|\}^2}{\sum|a_n|^2} \qquad (2.30)$$

This expression, due to Allen [8], can be extremely useful for any large array, whether linear or planar, because it allows gain to be computed directly from the array excitation coefficients. It is strictly correct only if the embedded element power pattern is known and the array is large enough for most element patterns to be the same. Care must always be taken to use the embedded element pattern gain, not that of the isolated element pattern. Since the use of this expression implies that all element patterns are the same, it is more correct for elements whose pattern shape does not change much when embedded in an array (like dipoles or slots spaced $\lambda/2$ apart), and less correct for high-gain elements, whose gain is significantly altered in the array environment and so changes across the array, or for small arrays in which edge effects dominate.

An approximate expression for taper efficiency is also derivable from (2.30), since it shows the maximum array gain as N times the element gain, and so the array gain can be written as

$$G = Ng_E(\theta_0, \phi_0)\epsilon_T \tag{2.31}$$

where the taper efficiency ϵ_T is thus

$$\epsilon_T = \frac{|\Sigma a_n|^2}{N\Sigma|a_n|^2} \tag{2.32}$$

for N, the total number of elements in the array.

This definition of taper efficiency extends the definition of column array gain for omnidirectional elements spaced $\lambda/2$ apart, as given in (1.64) in Chapter 1, to full two-dimensional arrays with arbitrary elements (subject to the large-array approximation). The expression is written in terms of gain rather than directivity because it is usually used with measured element gain patterns that include losses. The array aperture efficiency ϵ_A is the taper efficiency for the large-array case, as is readily seen by comparing (2.32) above with (1.67) of Chapter 1 at broadside, using $g_e(0, 0) = 4\pi A_e/\lambda^2$, where A_e is the area of the unit cell occupied by the element. The terms *taper efficiency* and *aperture efficiency* are often used interchangeably, but throughout this text aperture efficiency will be related to the fundamental definition of area gain in (1.12), while taper efficiency will be assumed calculated by (2.32) for a two-dimensional array and by (1.64) for linear arrays with multiple of $\lambda/2$ spacing.

The relationship of (2.30) also leads to an expression for the scan dependence of the active element pattern. Using (1.68) and the relationship between directivity and gain, one obtains (with Γ the network reflection coefficient)

$$G = D_0\epsilon_L(1 - |\Gamma|^2)$$
$$G = \frac{4\pi A_e N}{\lambda^2}(1 - |\Gamma|^2)\epsilon_A\epsilon_L \cos(\theta) = Ng_e(\theta_0, \phi_0)\epsilon_A \tag{2.33}$$

So the element gain is given by

$$g_e(\theta_0, \phi_0) = \frac{4\pi}{\lambda^2}A_e\epsilon_L(1 - |\Gamma|^2) \cos\theta_0 \tag{2.34}$$

It is important to bear in mind that this definition assumes a very large array with a periodic lattice, so that essentially all of the array element patterns are the same, the taper efficiency is the aperture efficiency, and the array spacing is such that no grating lobes radiate.

Unlike most of the definitions of gain used in this chapter, the element gain above is an aperture gain and assumes that the aperture radiates into a half space. The gain formulas of Section 2.2.1 assume that the radiation occurs into both half spaces, and so for any beam at angle θ there is another symmetrical beam below the horizontal plane. Other definitions of array directivity are introduced in the following sections.

2.2 CHARACTERISTICS OF LINEAR AND PLANAR ARRAYS

2.2.1 Linear Array Characteristics

Comparison With Continuous Illumination

It is often convenient to model the discrete array as the limiting case of a continuous aperture illumination. This is a convenient model because some of the most useful synthesis procedures are those developed for continuous apertures, where the analysis is more readily tractable. The normalized broadside radiation patterns of both a uniformly illuminated N-element array and a line source of length L are given below.

Linear Array

$$f(\theta) = \frac{\sin [N\pi d_x u/\lambda]}{N \sin (\pi d_x u/\lambda)} \tag{2.35}$$

Line Source

$$f(\theta) = \frac{\sin (L\pi u/\lambda)}{L\pi u/\lambda} \tag{2.36}$$

For arrays of more than a few elements, these two patterns are very similar for small values of the argument. The array length is taken as (Nd_x). Figure 2.7(a) [9] shows radiation patterns for a continuous line source of length 4λ and an eight-element array of $\lambda/2$-spaced elements with uniform illumination. The line source pattern differs very little from the array up to the second sidelobe, and the null positions are unchanged.

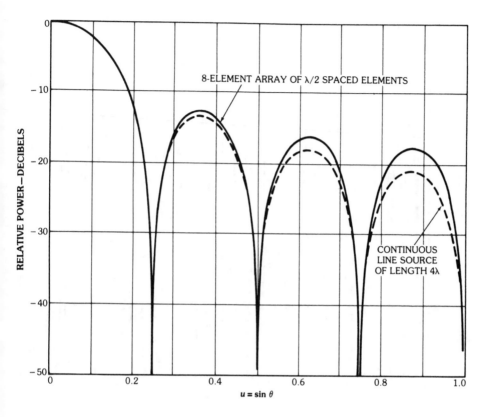

Figure 2.7(a) Line source patterns and array patterns: patterns of 4λ line source and 8-element array with λ/2 spacing.

Figure 2.7(b) [9] shows the patterns of a continuous line source of length 32λ, an array of 64 elements spaced λ/2 apart, and an array of 8 elements spaced 4λ apart. The patterns have nearly identical beamwidths and are very similar through the first few sidelobes. Comparison with the 4λ-spaced array shows that the similarity pertains about half way to the grating lobe, and the deviations begin to occur because the pattern repeats with period $\lambda/d_x = 0.5$ in the sin θ parameter.

Pattern Characteristics and Directivity Formulas for Linear Arrays

A broadside linear array of isotropic elements has a very wide pattern in the plane orthogonal to the array axis and a narrow pattern in the plane that includes the

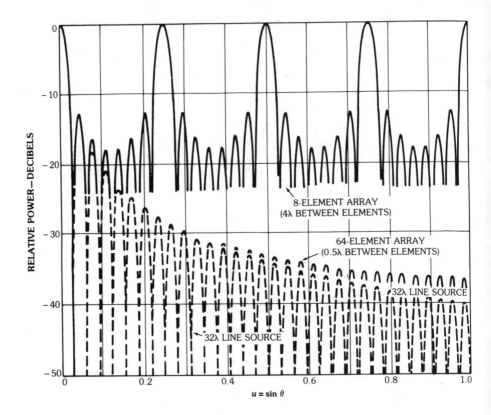

Figure 2.7(b) Line source patterns and array patterns: patterns of uniformly illuminated 64-element array with 0.5λ spacing, 32λ line source, and 8-element array with 4λ spacing.

array axis. This type of pattern is termed a *fan beam*, with reference to its appearance in Figure 2.8(a), which shows the broadside and scanned patterns.

As the array is scanned, the linear array fan beam pattern takes on the conical shape shown, which can lead to significant ambiguity if the pattern were used for radar tracking.

The ϕ dependence of the elevation angle θ for a beam at frequency f_0, scanned to $(\theta_0, 0)$, is readily obtained from (1.55), in which the beam peak is evidently at

$$u = \sin \theta \cos \phi = \sin \theta_0$$

so

$$\sin \theta = \frac{\sin \theta_0}{\cos \phi} \qquad (2.37)$$

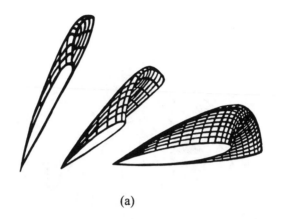

(a)

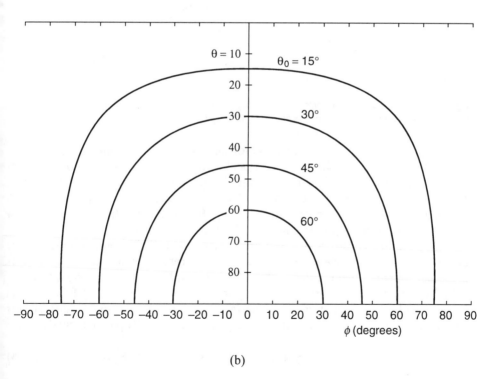

(b)

Figure 2.8 Beam shape for scanned fan beam and pencil beam arrays: (a) beam shape versus scan angle for fan beam (linear array) antenna; (b) beam peak contours near endfire.

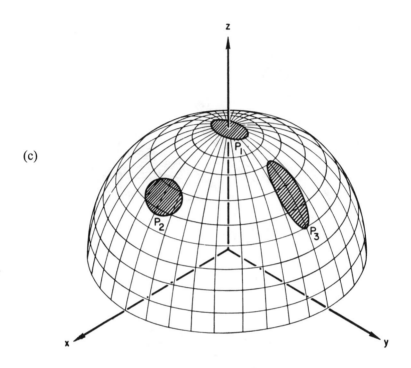

Figure 2.8 (cont.) Beam shape for scanned fan beam and pencil beam arrays: (c) beam shape versus scan position for a pencil beam (after Elliott [10]).

Figure 2.8(b) is a plot of this relationship for an array scanned to the various θ angles to 60 deg, showing the beam peak contour curving as a function of scan angle θ_0. An array with a narrow beam in ϕ does not have a significant curvature, but a broad beam will have its peak extending over a significant conical region as shown in the lowest curve of Figure 2.8(b). Figure 2.8(c) illustrates the way a slightly elliptical beam projects in several directions of scan.

In general, the directivity of a linear array of realistic element patterns can only be obtained by integration. However, for the case of omnidirectional and certain other simple element patterns, the directivity can be integrated in closed form.

To perform the integration to compute directivity, the array of Figure 2.9 is oriented with element centers at $z = nd$ (so the ϕ integrals are uncoupled). In this coordinate system, the array pattern of equally spaced isotropic elements is

$$E(\theta, \phi) = \sum |a_n| \exp \left[jk(nd(\cos \theta - \cos \theta_0)) \right] \tag{2.38}$$

The array is scanned to some angle θ_0, measured from endfire, as indicated in the equation, but because of the array orientation, θ_0 is the complement of the usual scan angle measured from broadside.

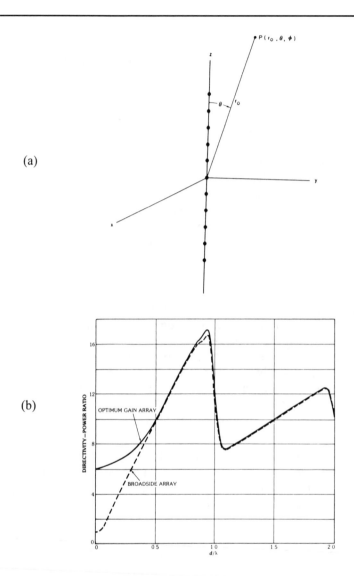

Figure 2.9 Directivity of an array of omnidirectional elements: (a) array geometry; (b) array directivity for a 10-element array (after Tai [11]).

For omnidirectional elements, directivity is readily integrated and reduced to

$$D = \frac{\{\Sigma_n |a_n|\}^2}{\Sigma_n \Sigma_m |a_n||a_m| \exp[-jkd(n-m)\cos\theta_0]\,\text{sinc}[kd(n-m)]} \quad (2.39)$$

where sinc $(x) = \sin x/x$.

Several special cases of the above are particularly revealing. At broadside, the directivity of this tapered array of isotropic elements reduces to the expression:

$$D = \frac{|\Sigma \, a_n|^2}{\Sigma\Sigma \, |a_m||a_n| \, \text{sinc}[(n - m)kd]} \tag{2.40}$$

Figure 2.9(b) shows the dependence of directivity on the spacing d for a uniformly illuminated array (dashed curve) and an array with excitation coefficients chosen to optimize directivity. For spacings larger than about $\lambda/2$, the optimum and uniform array directivities are nearly identical. The reduced directivity near $d/\lambda = 1$ is a result of grating lobes entering real space. These curves also reveal that the pattern has the same value for d any multiple of $\lambda/2$. For such spacings, the directivity becomes

$$D = \{\Sigma \, |a_n|\}^2/\Sigma \, |a_n|^2 \tag{2.41}$$

This relationship is fully general as long as the elements radiate isotropically and does not imply any particular distribution. A given, well-tapered illumination with controlled sidelobes may have directivity D, but rearranging the element excitations in any order would leave the directivity unchanged, even though the sidelobe structure is severely distorted. The linear relation evident for spacing less than $\lambda/2$ leads to the simple relationship given in (1.65) and due to King [12]:

$$D = [2d/\lambda]\epsilon_T N \tag{2.42}$$

In the case of a scanned array, the double summation of (2.39) is reduced when all the elements are excited equally. To understand this, let n and m run from 1 to N and substitute P for $n - m$ in (1.39). Tabulating these terms p in the matrix below shows a diagonal symmetry.

				m			
		1	2	3	$\cdot$	$\cdot$	$\cdot$	N
	1	0	-1	-2	$\cdot$	$\cdot$	$\cdot$	$-(N - 1)$
	2	1	0	-1	$\cdot$	$\cdot$	$\cdot$	$-(N - 2)$
n	3	2	1	0	$\cdot$	$\cdot$	$\cdot$	$-(N - 3)$
	4	3	2	1	$\cdot$	$\cdot$	$\cdot$	$-(N - 4)$
	$\cdot$	$\cdot$	$\cdot$	$\cdot$	$\cdot$	$\cdot$	$\cdot$	$-(N - 5)$
	$\cdot$	$\cdot$	$\cdot$	$\cdot$	$\cdot$	$\cdot$	$\cdot$	$-(N - 6)$
	N	$(N - 1)$	$(N - 2)$	$\cdot$	$\cdot$	$\cdot$	$\cdot$	0

The previous double summation adds terms with the above values of p by summing N rows of N columns. However, the matrix has odd symmetry about the diagonal and all terms equal in any minor diagonal. Thus, one can combine terms using this symmetry. The resulting summation (as long as all amplitudes are equal) is given [13]:

$$D = \frac{N^2}{N + 2 \sum_{n=1}^{N-1} (N - n) \, \text{sinc}(nkd)\cos(nkd \cos \theta_0)} \qquad (2.43)$$

This result shows that if the array spacing $d/\lambda = 0.5, 1.0, 1.5, \ldots$, the directivity is equal to the number of elements N, independent of the angle of scan. This result, which promises a directivity invariance with scan, is the result of assuming omnidirectional element patterns. The constant directivity is due to the real-space imaginary space boundary ($u^2 + v^2 \leq 1$). This causes a narrowing of the pattern in the plane orthogonal to scan as the array scan angle approaches endfire. The use of elements with narrower beams in the plane orthogonal to scan would thus lead to directivity that falls off more severely, as will be described in a later section. Furthermore, although the directivity may be constant, the gain varies with the array reflection coefficient and so generally tends to decrease with scan if the array is matched at broadside.

The directivity formulas given above are for omnidirectional elements. Hansen [14] also gives convenient formulas for several fundamental elements, including the broadside directivity of short dipoles and half-wave dipoles (or slots). These equations are not included here because of their availability and because they are ultimately based on isolated element patterns.

One can obtain more general formulas for directivity in a manner similar to that done for generalized element patterns, but based on the self- and mutual-resistance of the array elements. In general, using the peak far fields E_0 and H_0 and the average power radiated at some distance R_0 the directivity is written

$$D = \frac{2\pi R^2 E_0 H_0^*}{P_{\text{rad}}} = \frac{R^2 E_0^2}{60 \, P_{\text{rad}}} \qquad (2.44)$$

since $|H_0| = |E_0|/(120\pi)$.

When the coupling can be described in terms of single mutual impedance terms between elements (i.e., when higher order effects can be neglected) the denominator term can be evaluated by circuit relations that include all mutual coupling terms in the N-by-N matrix.

$$P_{\text{rad}} = \frac{1}{2} \sum_n \text{Re}[I_n V_n^*]$$

$$= \frac{1}{2} \sum_n I_n \sum_m I_m^* R_{nm}$$

(2.45)

This expression is fully general, and what remains is to evaluate the peak far field E_0 in terms of the element current. Hansen [13] uses the relationship for an array of half-wave dipoles at broadside:

$$E_0 = \frac{60}{R_0} \sum_{n=1}^{N} I_n$$

(2.46)

and in this case the directivity becomes

$$D = \frac{120[\Sigma I_n]^2}{\Sigma_n \Sigma_m I_n I_m R_{nm}}$$

(2.47)

For an array of half-wave dipoles [14,15] with uniform illumination:

$$D = \frac{120\ N^2}{\displaystyle\sum_{n=1}^{N} \sum_{m=1}^{N} R_{nm}} = \frac{120\ N}{R_{00} + (2/N) \displaystyle\sum_{n=1}^{N-1} (N - n) R_n}$$

(2.48)

In this expression, R_{00} is the element self-resistance and R_{nm} is the mutual resistance between the mth and nth elements. The reduction from double to single summation noted in the above is accomplished as explained for (2.43). This result, like the others in this section, pertains to arrays in free space. In the case of slots in a half space, the directivity is doubled.

Optimum Directivity and Superdirectivity for Linear Arrays

The uniformly illuminated, constant phase excitation of linear array antennas gives near-optimum directivity for most arrays. However, higher values of directivity can be obtained for certain nonuniform phase distributions. This phenomenon, called *supergain*, or more properly *superdirectivity*, has been well understood for many years and is clearly explained in Hansen [15,16]. Superdirectivity is produced using rapid phase variations across an array of closely spaced elements. Unfortunately, the higher directivity results from an interference process, and only the sidelobes are in real space, with the pattern main beam in or partly in "invisible space" (sin $\theta > 1$). The resulting ratio of stored-to-radiated energy (Q) is extremely high, and so the circuit bandwidth is very small. Furthermore, since the radiation resistance is very low, the efficiency is poor and the antenna noise temperature is high in the presence of losses due to finite antenna and matching network conductivity. Since the high directivity depends on cancellation of the contributions from all the array currents, superdirective array behavior is dependent on highly accurate current determination, and small errors in array excitation can destroy the properties of superdirective arrays.

The above comments were qualitative, not quantitative, but it is the degree of superdirectivity that determines the ultimate practicality of the synthesis. C. T. Tai [11], in his paper on optimum directivity of linear arrays, shows the onset of superdirectivity to occur when the element spacings are less than $\lambda/2$. When the element spacing is greater than $\lambda/2$, broadside arrays have their maximum gain approximately equal to the gain for the uniformly illuminated array. As the element spacings are further decreased and the optimum directivity sought, the degree of superdirectivity is increased. Small degrees of superdirectivity are achievable and practical in single small elements or endfire arrays (the Hansen-Woodyard [17] condition is an example), or for small, closely spaced arrays [11]. There have, in fact, been very practical applications of superdirectivity combined with superconductive antenna matching networks to improve circuit efficiency.

As the degree of superdirectivity is increased, so is the degree of difficulty in practically implementing the synthesis. Hansen quotes the data of Yaru [18], who studied a nine-element Chebyshev array with $\lambda/32$ spacing between elements. The required tolerance for maintaining the designed -26-dB sidelobes was one part in 10^{10}. Hansen [14] lists other examples, including the extensive results of Bloch, Medhurst, and Pool [19].

In all, it appears that superdirectivity is an interesting phenomenon, which can be exploited to a small degree. There is new interest and excitement in using high-temperature superconductivity to decrease the losses in superdirective arrays, and that may open further possible uses, especially for small to medium gain arrays. However, there remain the issues of high Q (limited bandwidth), difficult impedance matching, and very high required precision for superdirective arrays that will continue to limit the general use of this phenomenon. The synthesis topics discussed in Chapter 3 will assume that spacings are approximately $\lambda/2$ and will therefore exclude superdirective geometries.

2.2.2 Planar Array Characteristics

Pattern Characteristics and Grating Lobes/Array Grid Selection

A moderate- or large-size planar array of dimensions L_x and L_y, with uniform illumination, has beamwidths of 0.886 λ/L_x and $0.886\lambda/L_y$. In the principal planes

($\phi = 0$ and $\phi = \pi/2$), the patterns are the same as for a linear array aligned with the scan plane. If $L_x = L_y$ for a beam at broadside, the beam shape at the -3-dB contour is approximately circular, and this is often termed a *pencil beam*. For L_x not equal to L_y, the -3-db contour becomes an approximate ellipse, as shown in Figure 2.8(c).

The scanned planar array pattern also exhibits some distortion with scan, as indicated in Figure 2.8(c), but if both beamwidths are kept narrow, the angle ambiguity is much smaller than for the linear array.

Equation (1.58) gives the pattern of a planar array of equally spaced elements arrayed in a rectangular grid. The grating lobe structure for this array is given in that section also.

It is often advantageous to choose an alternate grid location with elements arranged in a triangular lattice, as shown in Figure 2.10. In this case, the elements are located at positions (x_m, y_n), where

$$y_n = nd_y \quad \text{and} \quad x_m = md_x \qquad \text{for } n \text{ even}$$
$$x_m = (m + 0.5)d_x \qquad \text{for } n \text{ odd}$$

The grating lobe lattice for this triangular grid is shown in Figure 2.10, and the lobe positions are given by

$$u_p = u_0 + p\lambda/d_x : v_q = v_0 + q\lambda/d_y \qquad \text{for } p = 0, \pm 2, \pm 4, \ldots \tag{2.49}$$
$$= v_0 + (q - 0.5)\lambda/d_y \qquad \text{for } p = \pm 1, \pm 3, \pm 5, \ldots$$

Other grid selections can lead to reduction of specific grating lobes within the scan sector. One extreme of this is indicated in Figure 2.11, where all the rows of the array are displaced by different distances Δ_n. In this case, the array factor is given by

$$E(\theta, \phi) = \sum_m \sum_n |a_{mn}| \exp\{j[(md_x + \Delta_n)k(u - u_0) + nd_y k(v - v_0)]\} \tag{2.50}$$

If the amplitude distribution a_{mn} is chosen as being separable, then the array factor is

$$E(\theta, \phi) = \left\{ \sum b_m \exp(jk[md_x(u - u_0)]) \right\} \left\{ \sum c_n \exp(jk[(v - v_0)nd_y \right.$$
$$\left. + \Delta_n(u - u_0)]) \right\} = f(u)g(u, v) \tag{2.51}$$

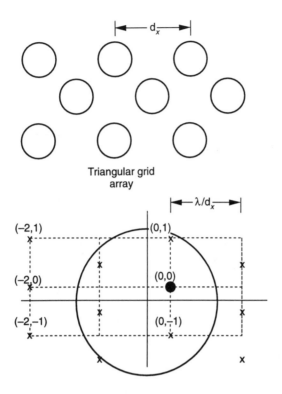

Figure 2.10 Geometry and grating lobe lattice of a triangular grid array.

In this form, it is clear that the sum over b_m is unchanged by the row displacements, but the sum over the rows c_n is significantly altered by the exponential factor that includes the displacements Δ_n, and the array factor is not separable. The triangular grid, which is discussed above, has the displacements

$$\Delta_n = (0, d_x/2, 0, d_x/2, \ldots)$$

For a uniform array, the array factor is different for the various u_p locations. For $p = \pm 1, \pm 3, \pm 5$, and so on, the pattern shape is

$$g(u_p, v) = \frac{1}{N_y} \frac{\sin[N_y \pi (v - v_0) d_y/\lambda]}{\cos[\pi (v - v_0) d_y/\lambda]} \tag{2.52}$$

This pattern has a zero at $v = v_0$ and an asymmetrical distribution in $v - v_0$, with

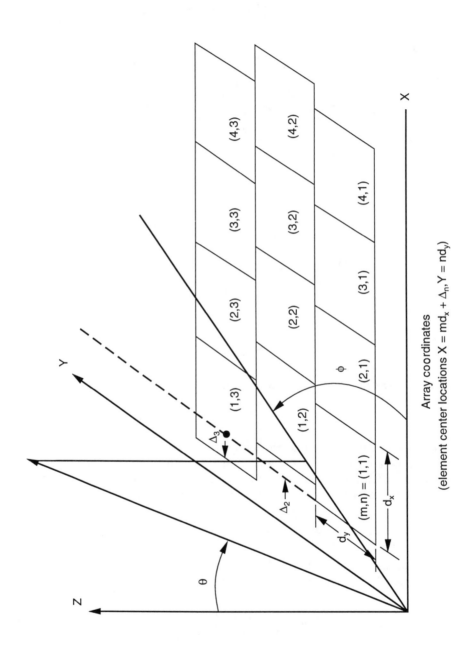

Figure 2.11 Array grating with displaced rows.

principal maxima of unity (grating lobes) at $(v - v_0) = 0.5 + q\lambda/d_y$, and so produces the grating lobes at locations indicated above and in Figure 2.10.

At the grating lobes $p = \pm 2, \pm 4, \ldots$, the summation becomes

$$g(u_p, v) = \frac{1}{N_y} \frac{\sin[N_y\pi(v - v_0)d_y/\lambda]}{\sin[\pi(v - v_0)d_y/\lambda]} \tag{2.53}$$

which again is the same distribution (see (1.65)) as for uniform with $\Delta_n = 0$ and offers no grating lobe suppression.

The triangular grid distribution thus suppresses the grating lobes with p odd in one sector of space by splitting them each into two lobes and moving each out to a relatively wide angle, where they are reduced by the element pattern. The distribution does not alter the even grating lobes at all.

It is possible to choose other displacements that suppress grating lobes in various regions of space, and this may be important for certain applications. Several examples of such a choice are given in [20]. The best example of such selective suppression is the triangular grid considered earlier, which suppresses those grating lobes along the ridge (u_p, v) for p odd, but does not suppress those for p even. This structure is advantageous because in most conventional arrays the elements are spaced between 0.5λ and λ apart, so the grating lobes adjacent to the main beam $(p = \pm 1)$ are most significant. However, if the array element spacings are much larger, so that many grating lobes are allowed to radiate, then by using a random displacement Δ_n, one can still obtain good grating lobe suppression everywhere, except along the ridge that includes the main beam $(u = u_0)$. One can show [20] that, in general, although the peak grating lobes can be reduced, the average power in the grating lobes is a constant. Consider the integral of the power within the region $-0.5 \leq (v - v_0)d_y/\lambda \leq 0.5$. After normalizing the total power to the power at the peak of the main beam, one obtains for the normalized power per unit length in $(d_y/\lambda)(v - v_0)$ space:

$$P_{avg} = \frac{\sum_{n=1}^{N_y} |c_n|^2}{\left| \sum_{n=1}^{N_y} c_n \right|^2} \tag{2.54}$$

independent of the Δ_n. For uniform illumination in the y-direction, this suppression is the factor $1/N_y$. Although it may be possible to choose the Δ_n displacements so as to reduce the peak value of the grating lobe throughout the region specified, the average value will remain constant at that level for an array with N_y rows. For an array with uniform distribution in the y-direction $(|c_n| = 1)$, one can thus obtain the maximum of about 9-dB suppression of the peak lobes for an array of 8 rows, 12 dB for an array of 16 rows, and so on. The choice of a low-sidelobe illumination in the y-direction reduces this suppression by the amount of the taper efficiency.

This technique can be a significant advantage for certain types of limited scan antennas, as will be described in Chapter 8.

Directivity Formulas for Planar Arrays

If the array average element pattern directivity is known, the directivity of a planar array is given by (2.33):

$$D = Nd_E(\theta_0, \phi_0)\epsilon_T \tag{2.55}$$

where d_E is the average element directivity (gain divided by loss efficiency) and ϵ_T is the taper efficiency.

A second expression pertains if elements are spaced to avoid grating lobes and if the aperture efficiency is known. In this case, for a pencil beam antenna radiating into a hemisphere (thus assuming a ground screen), one can use the area formula ((1.68) repeated)

$$D = \frac{4\pi A}{\lambda^2}\epsilon_A \cos\theta \tag{2.56}$$

to obtain the directivity. Then, if gain is desired, one can approximate the scan loss for the average element using calculated mutual coupling parameters or measured element patterns, or replace the $\cos\theta$ by scan loss according to $\cos\theta$ to some power (see Figure 1.11).

Similarly, as in Chapter 1, one can use the half-power beamwidths for a pencil beam antenna at any scan angle to estimate directivity using

$$D = \frac{4\pi(0.886)^2}{\theta_{x3}\theta_{y3}} \tag{2.57}$$

where the beamwidths are othogonal and here given in radians. This expression is equivalent to (1.66), where the angles are in degrees. This relation is approximate and implies a degree of control over array average sidelobes. It has been found accurate [21] for most pencil beam array distributions, including uniform, cosine on a pedestal, and even Chebyshev distributions with sidelobes down to the level where gain limitation sets in (see Chapter 3). In another convenient approximate form, the directivity of a planar two-dimensional array with separable illuminations can be written in terms of the directivities D_x, D_y of the illuminations that excite its orthogonal planes ((1.69) repeated):

$$D = K \, D_x D_y \tag{2.58}$$

In this expression, the linear array directivities D_x and D_y are the values for omnidirectional elements.

Elliott [21] gives the constant $K = \pi$ for the case of the maximum directivity of an array over a ground plane (i.e., with hemispheric element patterns). This constant is evidently too large, since an array of $N_x N_y$ elements spaced $\lambda/2$ apart and radiating into a half space has directivity $\pi N_x N_y$, which is π times the product of the directivities of two-linear arrays of *isotropic* elements.

By comparing the results of (2.58) with several other approximate results and the exact integration for a 4-by-4 and a 16-by-16 array of short dipoles, Hansen [22] has shown that (2.58) does yield too large a value of directivity. In this case, the elements radiated into a half space, and the constant $K = \pi/2$ was used in the comparison. Hansen showed that the above and several other approximate methods of calculation give directivities consistently high (by up to about 2 dB) for element spacings less than one wavelength at broadside. In this regime, the correct factor should be about $K = 2$ for arrays over a ground screen.

An expression in terms of self- and mutual resistance is given by Hansen [23]:

$$D = \frac{120 \left\{ \sum\limits_m \sum\limits_n I_{mn} \right\}^2}{\Sigma_m \Sigma_n \Sigma_p \Sigma_q \, I_{mn} I_{pq} R_{mnpq}} \tag{2.59}$$

In this equation, each element has the double index mn, and R_{mnpq} is the mutual resistance between the mnth and pqth elements. The relationship is valid at broadside for arrays of small elements (slots or dipoles) spaced to eliminate grating lobes.

Beyond the above expressions, a number of synthesis procedures for large arrays are based on the near equivalence of the patterns of discrete arrays and continuous aperture illuminations. In these cases, it is sometimes possible to obtain a closed-form expression for the directivity or aperture efficiency. Among others, this method has been used to derive aperture efficiency expressions for the Taylor line source illuminations given later.

2.3 SCANNING TO ENDFIRE

Equation (2.43) gives the directivity of a uniformly illuminated linear array of isotropic elements for all scan angles, even scanned to endfire $(\theta, \phi) = (\pi/2, 0)$.

In order to scan to endfire, the element spacing should be less than $\lambda/2$ so that no grating lobe will enter real space at $(\pi/2, \pi)$. However, if the array is composed of elements or subarrays (rows or columns) that are directive in the plane orthogonal to scan, as in Figure 2.12, then the directivity of the two-dimensional array falls off more severely with scan, and varies approximately like $\cos \theta$. Since the array is finite, the directivity is not zero at the horizon, but approaches a constant times the square root of the array length.

An extremely convenient general (though approximate) formula can be obtained from (2.57) relating beamwidth and directivity of pencil beam antennas. The beamwidth of an array of length $L = Nd$ in free space, with a perfectly conducting ground screen and scanned to endfire, is obtained directly from (1.61) by expanding the direction cosine $u = \sin \theta$ in a power series near the angle $\theta = \pi/2$. Setting $\theta = \pi/2 - \Delta\theta$ and $u = 1 - \Delta u$ gives an expression for the beamwidth $\Delta\theta$ in terms of Δu as:

$$\Delta\theta = [2\Delta u]^{1/2} \tag{2.60}$$

For an array over a ground screen, $\Delta u = 0.443 B_b \lambda/L$, and so one obtains the endfire beamwidth

$$\theta_3 = [0.886 B_b/(L/\lambda)]^{1/2} \tag{2.61}$$

Without the ground screen, the beamwidth is doubled.

For a planar array over a ground screen, the directivity can now be written directly using the relationship between directivity and beamwidth (2.57) using the broadside beamwidth for the length L_T of the array in the plane orthogonal to scan.

$$D = \frac{4\pi(0.886 B_b)^{1/2}}{B_{bT}} (L_T/\lambda)(L/\lambda)^{1/2} \tag{2.62}$$

In this expression, B_{bT} is the beam broadening factor in the transverse plane. Though approximate, this result gives a value only 0.5 dB less than that obtained from a direct integration [24].

One can obtain further narrowing of the beam and increased directivity by scanning the array "beyond" endfire to values of the $\sin \theta$ parameter greater than unity. Figure 2.12(b) shows a progression of scanned array factors as the array is scanned toward endfire, at endfire, and beyond endfire. Only the main beam is shown to avoid confusion. The array factor is bidirectional, with even symmetry about $\theta = \pi/2$. The solid curve shows the pattern scanned several beamwidths from endfire, where the beamwidth is well defined and given by (1.61). The dashed

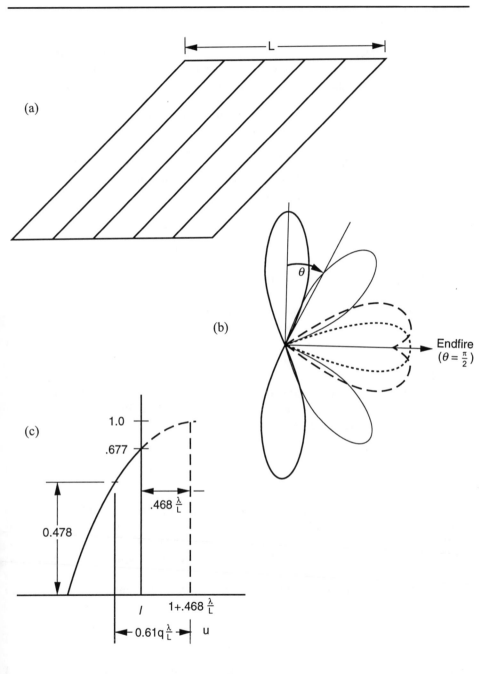

Figure 2.12 Scanning to endfire: (a) array of directive elements; (b) beam shape near horizon; (c) conditions for Hansen-Woodyard endfire gain.

curve shows the beam scanned to less than one-half beamwidth from endfire, where the beam for $\theta_0 < \pi/2$ and that for $\theta_0 > \pi/2$ have begun to merge, and the definition of beamwidth is ambiguous. At $\theta_0 = \pi/2$, both beams coincide and the beamwidth is given by the equation above. If $\sin \theta_0$ is increased beyond unity, the peak of the beam does not radiate and is said to be in "invisible" space, but what is left of the main beam is narrowed and the directivity can increase beyond the normal endfire value. The dotted curve represents this condition.

An early example of obtaining increased directivity by scanning beyond endfire is known as the *Hansen-Woodyard* [17] condition. In this case, the array is scanned beyond endfire to the angle

$$u_0 = 1 + \frac{2.94\lambda}{2\pi(N-1)d} \tag{2.63}$$

or by adding the additional phase lag $\delta = 2.94/(n-1)$ to the interelement phase $2\pi u_0 d/\lambda$.

The beam peak for a large array is at approximately

$$u_0 = 1 + 0.468\lambda/L \tag{2.64}$$

The Hansen-Woodyard condition is depicted in Figure 2.12(c), where the dashed part of the beam indicates that the beam is in imaginary space ($\sin \theta > 1$). One can estimate the 3-dB beamwidth for the uniformly illuminated case, since the beam shape is then given by

$$F(u) = \frac{\sin[\pi(u - u_0)L/\lambda]}{\pi(u - u_0)L/\lambda} \tag{2.65}$$

and at the actual peak $u = 1$ and $u - u_0 = -0.468\lambda/L$, $F(1) = 0.677$. At the 3-dB point, $F(u)$ is 0.478, and one can show that $u - u_0 = -0.619\lambda/L$, so the half beamwidth in u-space is $\Delta u = 0.151\lambda/L$ instead of $0.443\lambda/L$. This narrowed beamwidth produces increased directivity and is a practical example of the superdirectivity discussed earlier.

Using (2.60), the beamwidth for the uniformly illuminated case (with no ground screen) is

$$\Delta\theta = 2[0.30\lambda/L]^{1/2} \tag{2.66}$$

which corresponds, upon using (2.57), to an increase in directivity of about 2.3 dB relative to the endfire case.

The Hansen-Woodyard relation, which was derived for large arrays, does not actually produce the optimum directivity, but in most cases has improved directivity relative to that for ordinary endfire arrays. A useful comparison for a number of uniformly excited arrays scanned beyond endfire is given by Ma [25].

The above expressions describe the available endfire directivity. However, the actual array gain is much less than the directivity, because most of the array elements become substantially mismatched when the array is scanned to wide angles. This mismatch is due to the cumulative effects of mutual coupling, which are very severe at or near endfire. The definitive paper by King and Sandler [26] shows examples of this phenomenon and reveals why scanning to endfire is extremely inefficient. Studies have shown that it is necessary to tailor the feedline impedance to optimally match an endfire antenna. Alternative techniques for exciting efficient endfire radiation have been developed, but these are not phased array approaches; rather they are surface wave antenna approaches and involve exciting a passive slow wave structure with a single source [27,28].

2.4 THINNED ARRAYS

A number of applications require a narrow scanned beam, but not commensurably high antenna gain. Since the array beamwidth is related to the largest dimension of the aperture, it is possible to remove many of the elements (or to "thin") an array without significantly changing its beamwidth. The array gain will be reduced in approximate proportion to the fraction of elements removed, because the gain is related directly to the area of the illuminated aperture. This procedure can make it possible to build a highly directive array with reduced gain for a fraction of the cost of a filled array. The cost is further reduced by exciting the array with a uniform illumination, thus saving the cost of a complex power divider network.

Typical applications for thinned arrays include satellite receiving antennas that operate against a jamming environment, where the uplink power is adequate in terms of signal-to-noise ratio in the absence of jamming. For this case, antenna gain is of secondary value; only sidelobe supression or adaptive nulling can counter the jammer noise, and a narrow main beam can discriminate against jammers very near to the main beam. A second application often satisfied by thinned arrays is ground-based high-frequency radars, in which the received signal is dominated by clutter and atmospheric noise. Here again, the emphasis is on processing and array gain is of secondary value to the system. A third application, and one of the most significant, is the design of interferometer arrays for radio astronomy. Here the resolution is paramount, while gain is compensated by increased integration time. For applications such as these, the goal of the antenna system is to produce high resolution, so the array should be large, but not necessarily high gain.

Conventional closely spaced arrays have pattern characteristics that approach those of continuous apertures as closely as desired and have directivity commensurate with their area gain and aperture efficiency $(4\pi A/\lambda^2)\epsilon_A$. Thining the array is always accompanied by pattern deterioration, although the characteristics of this deterioration can be controlled by the method of thinning employed. Figure 1.12 shows an example of array thinning by using very wide spacings in a periodic array and indicates very little beam broadening, but extremely high grating lobes. Periodic thinning is thus seen to produce discrete high sidelobes. Sidelobe levels are also increased for nonperiodic thinning algorithms, but in this case the peak sidelobe level can be minimized.

An excellent summary of developments in the theory of thinned arrays is given by Lo [29]. In this reference, Lo reviews past works and points out that there is no practical synthesis method for obtaining optimized solutions for large nonperiodically or statistically thinned arrays. For small or moderate arrays, it can be convenient to formulate the thinning procedure as a sidelobe minimization problem (see [30–32]). These procedures do control both peak and average sidelobe levels, but are numerically difficult to implement for large arrays.

The variety of statistical procedures for array thinning exert direct control primarily on the average sidelobe level and can produce peak sidelobes for larger than the average level. A paper by Steinberg [33] compares the peak sidelobes of 70 algorithmically designed aperiodic arrays with those of 170 random arrays. The study showed that most techniques led to very similar average levels, although for relatively small arrays the method of dynamic programming [34] was the most successful procedure for control of peak sidelobe levels.

Many thinning algorithms have been developed and applied to the design of arrays. However, the bias of this text is to seek methods applicable to the design of large arrays. For this purpose, the method of Skolnik et al. [35] is presented because it is straightforward to implement for large arrays. In addition, studies by Lo are summarized to state bounds on the operating parameters of arrays subject to statistical thinning.

2.4.1 Average Patterns of Density-Tapered Arrays

Skolnik et al. [35] investigated a statistical thinning technique in which the density of elements is made proportional to the amplitude of the aperture illumination of a conventional filled array. The selection of element locations is done statistically by choosing element weights as unity or zero with probabilities proportional to the filled-array taper. The assumption made here is that the elements are regularly (periodically) spaced, but whether they are excited or not depends on the results of the statistical test. The filled-array pattern $E_0(\theta, \phi)$ is given by

$$E_0(\theta, \phi) = \sum A_n \exp(j\Phi_n) \qquad (2.67)$$

where A_n is the amplitude weight for the filled array.

The pattern of the thinned array is given as

$$E(\theta, \phi) = \sum F_n \exp(j\Phi_n) \qquad (2.68)$$

where F_n takes on the value zero or one, according to whether the nth element is excited.

The probability of exciting a given element with unity excitation in any area of the array is

$$P(F_n = 1) = K \frac{A_n}{A_0} \qquad (2.69)$$

where A_0 is the largest amplitude in the array.

The thinning constant K is defined by Skolnik in the following way. If K is set to unity and the above rule is used to approximate the average pattern of an array with a given sidelobe level, then the array is said to be thinned by the "natural" degree of thinning. The average number of elements of the original N-element array that remain excited are given by N_E. If the array is further thinned so that the total number of elements excited N_r is less than N_E,

$$N_r = KN_E \quad \text{for} \quad K \le 1 \qquad (2.70)$$

and the probability rule (2.69) is used, then the resulting pattern is still an approximation of the desired pattern, but with the maximum probability density K instead of unity and with higher sidelobes, as will be shown.

The resulting average field intensity (an ensemble average over many array selections) is a constant times the pattern $E_0(\theta, \phi)$ of the filled array:

$$\overline{E(\theta, \phi)} = KE_0(\theta, \phi) \qquad (2.71)$$

Skolnik showed that the average radiated power pattern is the sum of two patterns, the first is the pattern of the filled array and the second is an average pattern that is a constant value with no angle dependence.

$$|\overline{E(\theta, \phi)}|^2 = K^2 |E_0(\theta, \phi)|^2 + K \sum A_n(1 - KA_n) \qquad (2.72)$$

Since the far sidelobes of the filled array tend to be very low for most chosen distributions, the average pattern dominates the sidelobe pattern at wide angles. This average sidelobe level is given below, shown normalized to the pattern peak:

$$\overline{SL} = \frac{K\Sigma A_n(1 - KA_n)}{\Sigma |F_n|^2} \tag{2.73}$$

In the limit of a highly thinned array, the average sidelobe level is approximately $1/N_r$.

The average array directivity for a large array is approximately equal to the number of remaining elements times an element pattern directivity D_e, or

$$D = D_e \sum F_n = D_e N_r \tag{2.74}$$

Figure 2.13(a) shows an array with elements arranged on a rectangular grid but thinned to produce a low-sidelobe (-50 dB) pattern. Figure 2.13(b) shows the desired Taylor ($\overline{n} = 8$) [36] pattern for the filled array, and Figure 2.13(c) shows a computed pattern for the statistically thinned pattern. The dashes in Figure 2.14(a) indicate elements that have been removed. The array chosen has elements with $\lambda/2$ grid locations occupying a circle with radius 25λ and consisting of 7,845 elements if filled. The average sidelobe level shown in Figure 2.13(c) exceeds the design sidelobe, so clearly, in this example, the chosen sidelobe level was too low for the array to synthesize. Section 2.4.3 gives data on directivity, EIRP, and sidelobe level for density-tapered arrays with one or a number of different quantized amplitude levels.

The statistical procedure introduced above is readily applicable to the design of large arrays, but it is only one of a number of approaches that have been investigated. It is not optimum in that it does not ensure that peak sidelobes are maintained below a given level.

2.4.2 Probabilistic Studies of Thinned Arrays

The studies of Lo [37] addressed the peak-sidelobe issue and showed that a statistical description of these sidelobes is possible and yields useful bounds for array design. Following Lo's notation, a linear array of length a is excited by signals of equal amplitude. The probability density function $g(X)$ is the probability of placing an element at X, with $|X| \leq a/2$.

$$\int_{-a/2}^{a/2} g(X)\, dX = 1 \tag{2.75}$$

If there are N equally excited elements within the aperture that are placed

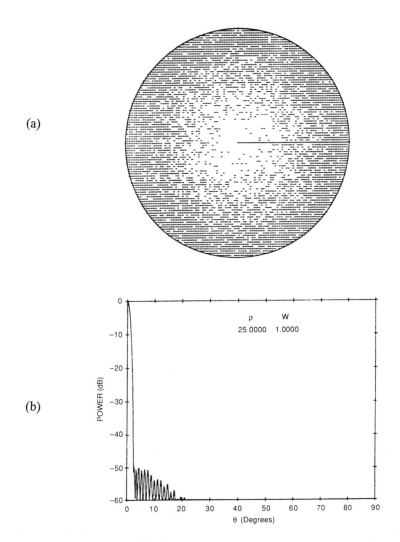

Figure 2.13 Circular array with elements removed: (a) geometry (dashes show elements removed); (b) desired Taylor pattern (filled array).

according to the probability density $g(X)$, then for each set of random samples $[X_1, X_2, \ldots, X_N]$ there is a pattern function

$$F(u) = \frac{1}{N} \sum_{n=1}^{N} \exp(jux_n) \qquad (2.76)$$

where we have normalized the dimension x, so that

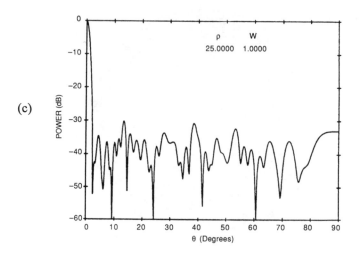

Figure 2.13 (cont.) Circular array with elements removed: (c) thinned array pattern.

$$x_n = 2X_n/a \tag{2.77}$$

and u as defined by Lo is different from that used throughout this text, and is

$$u = a\pi(\sin\theta - \sin\theta_0) \tag{2.78}$$

for the main beam at the observation angle θ_0.

In terms of this length normalization, the aperture extends from -1 to 1, and

$$g(x) = 0 \qquad \text{for } |x| > 1$$
$$\int_{-1}^{1} g(x)dx = 1 \tag{2.79}$$

The major conclusions of Lo's study will only be summarized here. The text by Lo [29] contains many of the details in the original paper and is recommended as a thorough and scholarly review of this material. Among other important points, Lo showed the following:

Mean Pattern

The mean of the pattern function $F(u)$ is given as the Fourier transform of $g(x)$:

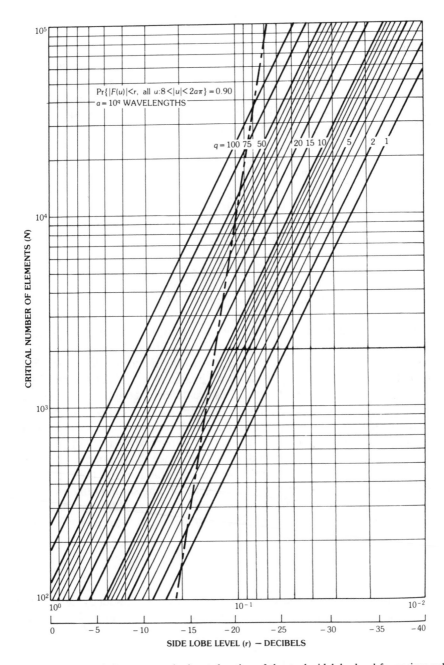

Figure 2.14 Number of elements required as a function of the peak sidelobe level for various values of $a = 10^q$ wavelengths with a 90% probability (after Lo [37]).

$$\phi(u) = E\{F(u)\} = \int_{-\infty}^{\infty} g(x)e^{jux}dx \qquad (2.80)$$

where the $g(x)$ is a continuous function and $E\{. . .\}$ is a probability average operator. Note that this mean value ϕ is equivalent to the average pattern of Skolnik et al., except that Skolnik sampled a discrete set of positions. Moreover, in this summary of Lo's work, the total number of elements is N, and this corresponds to N, in the above description of the Skolnik et al. study.

Variances Between Mean and Sample Patterns

Defining variances σ_1^2 and σ_2^2 as the mean of the squared difference between the mean pattern and the pattern computed from (2.76) for both real (F_1) and imaginary (F_2) parts, one obtains (since the mean pattern is real)

$$\sigma_1^2 = \text{Var } F_1(u) = E\{[F_1(u) - \phi(u)]^2\}$$
$$\sigma_2^2 = \text{Var } F_2(u) = E\{[F_2(u)]^2\}$$

Lo shows that outside of the main beam region, the variances of the real and imaginary parts of the pattern are equal and approximately given by $1/2N$, independent of the probability density function. This significant conclusion implies that although the pattern behavior in the main beam region is determined by $g(x)$, outside of the main beam area the variances are determined only by N, the number of elements, not the probability density function $g(x)$. Therefore, in many cases (unless the near-in sidelobe level is of interest), it may be advantageous to use the uniform density function for $g(x)$ to maintain a narrow beam. As N increases, however, the variances decrease, and $F(u)$ approaches the mean pattern when the variances are significantly less than the design sidelobe level. In these cases it may be appropriate to use a nonuniform $g(x)$. In general, one should only use a tapered function $g(x)$ if the value of the variances ($1/2N$) is less than the desired mean pattern sidelobes, or if only the first several sidelobes are of primary importance.

Peak Sidelobe

Another significant conclusion due to Lo has to do with specifying the highest sidelobe in the visible pattern range. In this case, for a uniform probability density function,

$$g(x) = 1/2 \qquad \text{for } |x| < 1 \qquad (2.81)$$

which thus satisfies the normalization criterion of (2.79). Lo obtained the probability for a sidelobe level less than r. Outside of the main beam region, this is

$$P_r\{|F(u)| < r\} = [1 - \exp(-Nr^2)]\exp\{[-4\pi N^{1/2}r\exp(-Nr^2)](a^2/12\pi)^{1/2}\} \quad (2.82)$$

Computer simulation by Agrawal and Lo [38] has verified this formula for an array as small as 11 elements over an aperture of 5 to 10 λ.

For large numbers of elements, this reduces to

$$P\{|F(u)| < r\} = [1 - 10^{-0.4343Nr^2}]^{[4a]} \quad (2.83)$$

where the bracket [4a] is the integer part of $4a$.

This equation shows that unless the number of elements N is numerically on the order of the sidelobe power r^2, the probability of achieving a given sidelobe level is very low. This similar dependence can be inferred from the variance data previously mentioned. Figure 2.14 (from Lo [37]) is a plot of the above equation and gives this critical number of elements versus the sidelobe level 20 log r for the 90% probability case. The data to the left of the dot-dashed line are less accurate, and the more accurate expression (2.82) above should be used in this region. Figure 2.14 indicates that one needs very large arrays to achieve low sidelobes, especially when considered in the light of decreasing directivity achieved with such highly thinned one-dimensional structures.

Beamwidth

Lo shows [37] that for large arrays the beamwidth of the statistical array converges to that of the mean pattern.

Directivity

The directivity D of a sample pattern function for a large array is related to the directivity of the mean pattern D_0 as

$$(D_0 - D)\,\mathrm{dB} \leq 20 \log_{10}(1 + d_{avg}^{1/2}/\|g(x)\|) \quad (2.84)$$

where

$$\|g(x)\|^2 = \int_{-1}^{1} |g(x)|^2\, dx \quad (2.85)$$

and

$$d_{avg} = (\text{average spacing}) \sim a/N$$

This expression says that the sample pattern for directivity D is less than D_0 by a quantity no greater than the term shown at the right above. As a corollary, two arrays with identical distribution functions but different numbers of elements have their directivities related by

$$(D_1 - D_2)\,\text{dB} = 10 \log N_1/N_2 \qquad (2.86)$$

or D is, with high probability, proportional to N.

Two-Dimensional Arrays

Lo's results are extendable to two-dimensional arrays. If a rectangular array dimension is ab with probability density function $g(x, y)$, (2.82) and (2.83) still give the relation between sidelobe level and total number of elements N, except that $[4a]$ in (2.83) is replaced by $[16\,ab]$. Figure 2.14 is also directly useful by writing $a = 10^q$ and $b = 10^p$, and then the q in Figure 2.14 should be replaced by $(p + q)$ and the 90% probability replaced by $(0.9)^2$, or approximately 80%. Or, indeed, one could redraw Figure 2.14 using (2.83) for the 90% probability.

2.4.3 Thinned Arrays With Quantized Amplitude Distributions

There may be advantages in the use of several discrete, quantized output power levels for the array instead of a continuous taper. This discretization may be appropriate, for example, in arrays of solid-state modules with output amplifiers operated in a saturated state. In such a situation, it is appropriate to arrange the array into regions illuminated by each of the quantized weights and then to use thinning to reduce the sidelobes that would be introduced if the quantization were used alone. This array organization was addressed in two recent papers [39,40]. The values of average parameters in the several included figures are due to Mailloux and Cohen [40].

Figure 2.15(a–c) show the quantization of a circular array amplitude taper and the array geometry in general. Added to the quantization is one of several discretizing algorithms, indicated pictorially in Figure 2.16. The array is divided into rings of radii $\rho_1, \rho_2, \rho_3. \ldots$, with quantized voltage levels V_1, V_2, V_3, and so on. The levels V_n were chosen to minimize the first few sidelobes of the pattern of a quantized continuous aperture (Figure 2.16(b)).

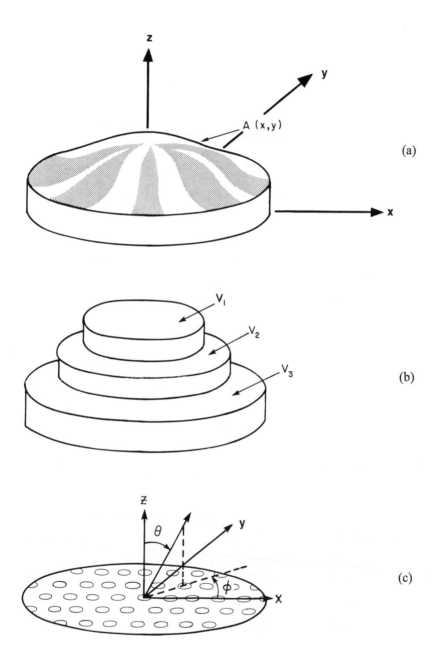

Figure 2.15 Array with quantized amplitude taper (after Mailloux and Cohen [40]): (a) array amplitude taper $A(x, y)$; (b) quantized amplitude taper; (c) array aperture and coordinates.

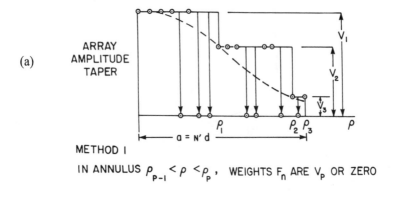

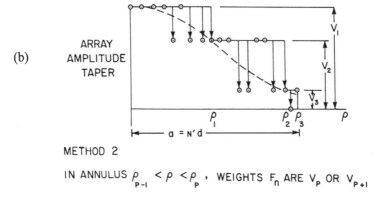

Figure 2.16 Thinning and quantizing geometries (after Mailloux [40]): (a) ideal taper (dashed) and method 1 source weight options; (b) ideal taper (dashed) and method 2 source weight options.

With the algorithm called *method 1*, in any annulus $\rho_{p-1} < \rho < \rho_p$, the array amplitude weights F_n are either V_p or reduced to zero according to the following rule.

The probability of assigning the weight $F_n = V_p$ to an element at location ρ_n in the radial annulus $\rho_{p-1} \leq \rho \leq \rho_p$ is given by

$$P(F_n = V_p) = KA_n/V_p \tag{2.87}$$

where A_n is the amplitude of the ideal illumination at the nth element. Figure 2.17

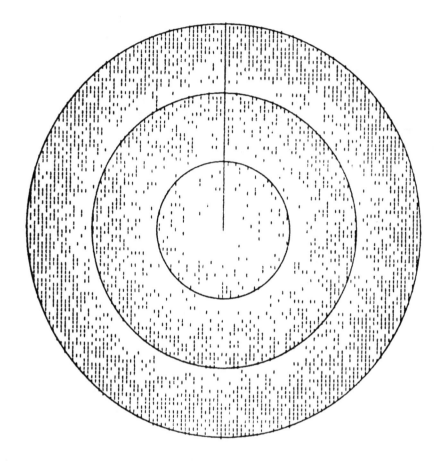

Figure 2.17 Distribution of non-excited (thinned) elements for an array with three quantized steps (after Mailloux and Cohen [40]).

shows an array with some of the elements left at the value V_p and others set to zero. This "thinning" rule reduces to Skolnik's when a single quantized level is used.

With the algorithm called *method 2*, the array is not actually thinned (unless K is less than unity). For $K = 1$, every element is excited, but the level of signal in the annulus $\rho_{p-1} < \rho < \rho_p$ is chosen to be either V_p or V_{p+1} according to the probability rule below:

$$P(F_n = V_p) = \frac{K[A_n - V_{p+1}]}{V_p - V_{p+1}} \qquad P(F_n = V_{p+1}) = \frac{K[V_p - A_n]}{V_p - V_{p+1}} \qquad (2.88)$$

The average power patterns for arrays built according to these algorithms are readily shown to consist of a term given by K^2 times the ideal power pattern plus an error term that is the average sidelobe level. Figure 2.17 shows the geometry of an array filled according to the algorithm of method 1. The figure illustrates that the probability rule forces a symmetrical quantization pattern denoted by dashes that indicate use of the V_p level in an annulus $\rho_{p-1} < \rho < \rho_p$.

Figures 2.18 and 2.19 show the result of using these multiple-step discretization rules. In these figures, the array input power is normalized to the total number of elements N as

$$P_{in} = \frac{\Sigma F_n^2}{N} \tag{2.89}$$

The average sidelobe level, normalized to the peak of the beam, is given by

$$SL = \frac{P_{SL}}{(\Sigma F_n)^2} \tag{2.90}$$

where the values of sidelobe power P_{SL} are given by method 1:

$$P_{SL} = \sum_p V_p \sum_{n(p)} KA_n[1 - kA_n/V_p] \tag{2.91}$$

and method 2:

$$P_{SL} = \sum_p \sum_{n(p)} [KA_n(V_p + V_{p+1}) - V_p V_{p+1}] - \sum_n K^2(A_n)^2 \tag{2.92}$$

The directivity for a thinned array can be computed in several ways, depending on whether the element pattern directivity is known. If the array were not thinned, if elements were placed $\lambda/2$ apart, matched at broadside, tailored to have nearly cosine scan dependences, and if the array were large so that an average element directivity could be assumed, then (2.30) would properly describe the directivity using $D_e = \pi$:

$$D = D_e \frac{[\Sigma F_n]^2}{\Sigma F_n^2} \tag{2.93}$$

This expression is also valid if the array were thinned by simply not exciting but properly terminating some elements of a periodic $\lambda/2$ lattice to accomplish the thinning. Such thinning leaves the element patterns unchanged.

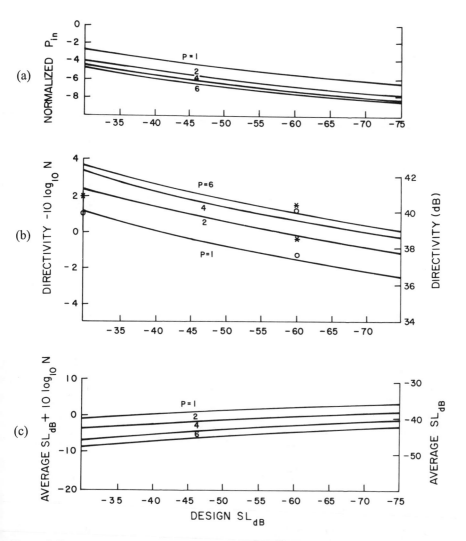

Figure 2.18 Input power (a), directivity (b), and average sidelobe level (c) for method 1 (thinning quantization) (after Mailloux and Cohen [40]).

If, however, the aperture is truly thinned by omitting elements (not just match terminating them), then the element pattern directivity can be less than π and may approach the result for nearly hemispherical element patterns with the directivity of 2, depending on the isolated pattern directivity of the element in question.

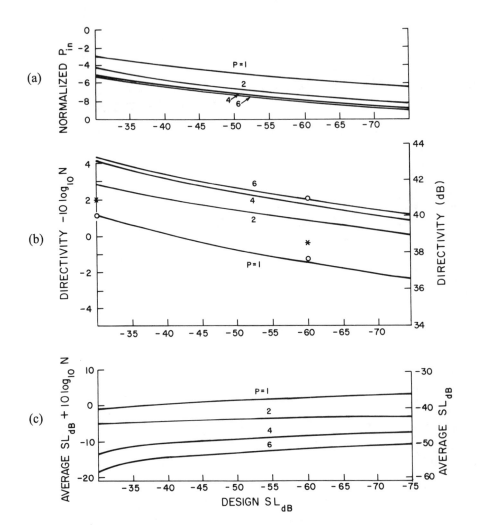

Figure 2.19 Input power (a), directivity (b), and average sidelobe level (c) for method 2 (quantization) (after Mailloux and Cohen [40]).

One can also compute an average directivity for the thinned array radiating into a half space using the basic definition of directivity and the power pattern. The result is given below under the assumption of a constant sidelobe level *SL* (implying a hemispherical element pattern)

$$D = \frac{D_0}{1 + \frac{1}{2K^2} D_0 \, SL} \tag{2.94}$$

where $SL = P_{SL}/P_{max}$, and D_0 is the directivity of the ideal pattern. If a cosine element pattern were to be used, then the 2 in the denominator of (2.94) above should be replaced by 4, and the result of using (2.94) or (2.93) converge. In the above form, (2.94) is most appropriate for highly thinned arrays ($K < 1$) or for use with method 1 with elements removed.

Figures 2.18 and 2.19 give the directivity, normalized input power, and average sidelobe level for a circular planar array of (if filled) 7,845 elements and occupying an area with radius 50λ. The curves are given for one, two, four, and six quantization levels. The axis at the right of the sidelobe and directivity figures is computed directly, but the axes at left are normalized to the number of elements in the array, and so the results are applicable to different-size arrays. The asterisk at several places gives the results using (2.93), while the circle near the same point is the directivity evaluated from a direct numerical pattern integration.

These figures show a general increase in average sidelobe level as the design sidelobe is lowered. Since there is little use in synthesizing a very-low-sidelobe pattern with a thinned array that would have a higher average sidelobe level, Figure 2.20 gives the number of elements for which the design and average sidelobe levels are equal. These curves are readily generalized to maintain average sidelobes some margin (of say 10 or 20 dB) below the design sidelobes by increasing the 10 log N by the chosen margin. For an array with a single quantized level ($p = 1$), the number of elements is seen as equal to the sidelobe level ($r^2 \approx 1/N_r$).

The element numbers for a single quantized level ($p = 1$) on these curves should display some similarity to the peak sidelobe data plotted by Lo [37] and given in Figure 2.14, although Figure 2.14 is given for a linear array, and the two-dimensional equivalent is for a rectangular aperture. For example, taking the rectangular aperture limit, with $a = b = 44\lambda$, then $q = \log_{10} a$ and $p + q = 3.29$, for N approximately 10,000 elements. For this case, Figure 2.14 gives a peak sidelobe level of about -30 dB. Figure 2.20 gives the average sidelobe level (SL) of approximately -40 dB, which is equal to $1/N_e$. A brief look at Figure 2.14 confirms that for arrays of up to thousands of elements, whether linear or planar, almost all of the sidelobes are less than about 10 dB higher than the average sidelobe level ($1/N_e$).

In general, comparing all data for methods 1 and 2 shows that the technique of method 2 results in significantly lower average sidelobes and higher directivity for given design sidelobe levels than can be achieved by the quantized thinning algorithm, method 1.

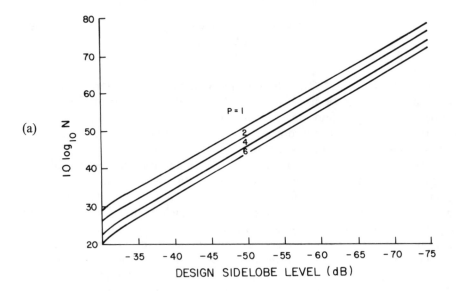

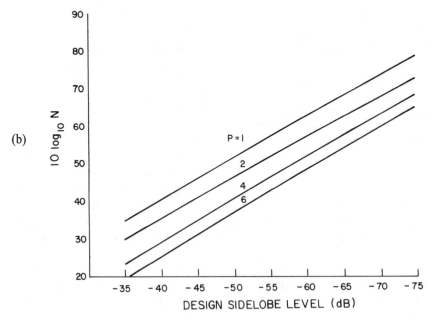

Figure 2.20 Number of array elements for average sidelobe level equal to design sidelobe level: (a) method 1; (b) method 2 (after Mailloux and Cohen [40]).

REFERENCES

[1] Balanis, C., Ch. 3 in *Antenna Theory: Analysis and Design*, New York: Harper and Row, 1982.

[2] Levine, H., J. Schwinger, "On the Theory of Electromagnetic Wave Diffraction by an Aperture in an Infinite Conducting Screen," *Commun. Pure and Applied Math*, Vol. 44, 1950–51, pp. 355–391.

[3] Tai, C. T., *Dyadic Green's Functions in Electromagnetic Theory*, Scranton, PA: Intext Educational Publishers, 1971.

[4] Balanis, op. cit., pp. 92–93.

[5] Collin, R. E., Chap. 4 in *Foundations for Microwave Engineering*, New York: McGraw-Hill, 1966.

[6] Wu, C. P., "Analysis of Finite Parallel-Plate Waveguide Arrays," *IEEE Trans.*, Vol. AP-18, No. 3, May 1970, pp. 328–334.

[7] Diamond, B. L., "Small Arrays—Their Analysis and Their Use for the Design of Array Elements," *Phased Array Antennas, Proc. 1970 Phased Array Antenna Symp.*, Dedham, MA: Artech House, 1972, pp. 1278–1281.

[8] Allen, J. L., et al. "Phased Array Radar Studies, July 1960 to July 1961," Technical Report No. 236(U), Lincoln Laboratory, MIT, (13 Nov. 1961), Part 3, Ch. 1, DDC 271724.

[9] Mailloux, R. J., Chapter 13 in *Antenna Handbook, Theory, Applications and Design*, Y. T. Lo and S. W. Lee, eds., New York: Van Nostrand Reinhold, 1988.

[10] Elliott, R. E., Chap. 1 in *Microwave Scanning Antennas*, Vol. 2, R. C. Hansen, ed., Peninsula Publishing, 1985, p. 43.

[11] Tai, C. T., "The Optimum Directivity of Uniformly Spaced Broadside Arrays of Dipoles," *IEEE Trans.*, Vol. P-12, 1964, pp. 447–454.

[12] King, H. E., "Directivity of a Broadside Array of Isotropic Radiators," *IRE Trans.*, Vol. AP-7, No. 2, 1959, p. 187–201.

[13] Bach, H. and J. E. Hansen, Chapter 5 in *Antenna Theory, Part 1*, New York: McGraw-Hill, R. E. Collin and F. J. Zucker, eds., p. 153.

[14] Hansen, R. C., Chap. 1, in *The Handbook of Antenna Design*, Vol. 2, Rudge, Milne, Olver, Knight, eds., Peter Peregrinus, London, 1987.

[15] Hansen, R. C., Chap. 1 in *Microwave Scanning Antennas*, Vol. 1, Los Altos, CA: Peninsula Publishing, 1985, pp. 82–92.

[16] Hansen, R. C., Chap. 1 in *The Handbook of Antenna Design*, op. cit., pp. 56–62.

[17] Hansen, W. W., and J. R. Woodyard, "A New Principle in Directional Antenna Design," *Proc. IRE*, Vol. 26, 1938, pp. 333–345.

[18] Yaru, N., "A Note on Super-Gain Arrays," *Proc. IRE*, Vol. 39, pp. 1081–1085.

[19] Bloch, A., R. G. Medhurst, and S. D. Pool, "A New Approach to the Design of Super-Directive Aerial Arrays," *Proc. IEE*, Vol. 100, Part 111, pp. 303–314.

[20] Mailloux, R. J., et al., Tech. Rept. 76-307, Sept. 1976.

[21] Elliott, R. E., op. cit., p. 44.

[22] Hansen, R. C., "Comparison of Square Array Directivity Formulas," *IEEE Trans.*, Vol. AP-20, Jan. 1972, pp. 100–102.

[23] Hansen, R. C., *The Handbook of Antenna Design*, op. cit., p. 154.

[24] Ibid., p. 155.

[25] Ma, M. T., Chap. 3 in *Antenna Engineering Handbook*, R. C. Johnson, H. Jasik, eds., New York: McGraw-Hill, pp. 3–18.

[26] King, R. W. P., and S. S. Sandler, "The Theory of Endfire Arrays, *IEEE Trans.*, Vol. AP-12, May 1964, pp. 276–280. Correction Nov. 1968, p. 778.

[27] Walter, C. H., *Traveling Wave Antennas*, New York: McGraw-Hill, 1965, pp. 121–122, 322–325.

[28] Mavroides, W. G., and R. J. Mailloux, "Experimental Evaluation of an Array Technique for Zenith to Horizon Coverage," *IEEE Trans.*, Vol. AP-26, May 1978, pp. 403–406.

[29] Lo, Y. T., Chap. 14 in *Antenna Handbook Theory, Applications and Design*, Y. T. Lo and S. W. Lee, eds., New York: Van Nostrand Reinhold, 1959.

[30] Unz, H., "Linear Arrays with Arbitrarily Distributed Elements," *IRE Trans.*, Vol. AP-8, March 1960, pp. 222–223.

[31] Sandler, S. S., "Some Equivalences Between Equally and Unequally Spaced Elements," *IRE Trans.*, Vol. AP-8, Sept. 1960, pp. 496–500.

[32] Harrington, R. F., "Sidelobe Reduction by Nonuniform Element Spacing," *IRE Trans.*, Vol. AP-9, March 1961, pp 187–192.

[33] Steinberg, B. D., "Comparison Between the Peak Sidelobe of the Random Array and Algorithmically Designed Aperiodic Arrays," *IEEE Trans.*, Vol. AP-21, May 1973, 366–369.

[34] Skolnik, M. I., G. Newhauser, and J. W. Sherman III, "Dynamic Programming Applied to Unequally Spaced Arrays," *IEEE Trans.*, Vol. AP-12, Jan. 1964, pp. 35–43.

[35] Skolnik, M., J. W. Shermon III, F. C. Ogg, Jr., "Statistically Designed Density-Tapered Arrays," *IEEE Trans.*, Vol. AP-12, July 1964, pp. 408–417.

[36] Taylor, T. T., "Design of Line Source Antennas for Narrow Beamwidth and Low Sidelobes," *IEEE Trans.*, Vol. AP-3, Jan. 1955, pp. 16–28.

[37] Lo, Y. T., "A Mathematical Theory of Antenna Arrays With Randomly Spaced Elements," *IEEE Trans.*, Vol. AP-12, May 1964, pp. 257–268.

[38] Agrawal, V. D., Y. T. Lo, "Mutual Coupling in Phased Arrays of Randomly Spaced Antennas," *IEEE Trans.*, Vol. AP-20, May 1972, pp. 288–295.

[39] Numazaki, T., S. Mano, T. Kategi, and M. Mizusawa, "An Improved Thinning Method for Density Tapering of Planar Array Antennas," *IEEE Trans.*, Vol. AP-35, No. 9, Sept. 1987, pp. 1066–1069.

[40] Mailloux, R. J., and Cohen, Edward, "Statistically Thinned Arrays with Quantized Element Weights," *IEEE Trans.*, Vol. AP-39, no. 4., April 1991, pp. 436–447.

Chapter 3
Pattern Synthesis for Linear and Planar Arrays

One of the major advantages of array antennas is that the array excitation can be closely controlled to produce extremely-low-sidelobe patterns or very accurate approximations of chosen radiation patterns. Many intricate procedures have been developed for synthesizing useful array factors. These methods fit into three main classes of synthesis: synthesis of various sector patterns that are usually many beamwidths wide, synthesis of low-sidelobe, narrow-beam patterns, and procedures that optimize some (usually receiving) array parameter, such as gain and signal-to-noise ratio, subject to some constraint on the sidelobe level or the existence of outside noise sources.

Most of the synthesis procedures described in the chapter are for narrow-beam, low-sidelobe array factors. However, the Fourier transform method, the Woodward synthesis technique, and power pattern synthesis methods are very appropriate for the synthesis of shaped-beam patterns. The chapter by Schell and Ishimaru [1], Ma [2], and Rhodes [3] present detailed treatments of the synthesis problem. Such details are beyond the scope of this text, which is devoted to the task of presenting specific results for practical design.

Throughout this chapter, the synthesis is carried out for arrays with broadside beams, without loss of generality, because the scanned performance is obtained from the broadside pattern by multiplying the excitation coefficients by the exponential factor

$$\exp(-jk[u_0 n d_x + v_0 m d_y]) \tag{3.1}$$

for a two-dimensional array. Thus, replacing u by $(u - u_0)$ and v by $(v - v_0)$ will produce the correct equations for the synthesis of beams scanned away from broadside. This translation property ensures that array factors are unchanged with scan, but does not necessarily ensure invariance of average pattern functions such as signal-to-noise ratio or directivity.

3.1 LINEAR ARRAYS AND PLANAR ARRAYS WITH SEPARABLE DISTRIBUTIONS

3.1.1 Fourier Transform Method

Fourier series methods [4] can be applied to array synthesis problems. The pattern function

$$F(u) = \sum a_n e^{jkund_x} \tag{3.2}$$

where the summation is performed over the range

$$-(N - 1)/2 \leq n \leq (N - 1)/2$$

and

$$
\begin{aligned}
n &= \pm 1/2, \pm 3/2, \pm 5/2, \ldots & &\text{for } N \text{ even} \\
&= 0, \pm 1, \pm 2, \pm 3, \ldots & &\text{for } N \text{ odd}
\end{aligned}
$$

is a finite Fourier series and is periodic in u-space with the interval of the grating lobe distance λ/d_x. Thus, given a desired pattern distribution $F(u)$, one can obtain an expression for the excitation coefficients a_n from orthogonality:

$$a_n = \frac{d_x}{\lambda} \int_{-\lambda/(2d_x)}^{\lambda/(2d_x)} F(u) e^{-j(2\pi/\lambda)und_x} du \tag{3.3}$$

This method provides the least mean squared error approximation of the desired pattern for $d_x \geq 0.5\lambda$. If the spacing is closer, the domain of integration exceeds the visible region and the definition of the pattern is not unique.

The Fourier series method is usually applied to the synthesis of shaped-beam patterns that are wide compared to the minimum array beamwidth (λ/L). The example in Figure 3.1 shows two Fourier series approximations of the square-top pattern also shown in the figure. The two curves are for 8- and 16-element arrays with $\lambda/2$ spacing and show that the larger array provides a more accurate approximation of the desired pattern by reproducing steeper slopes to match the ideal pattern.

3.1.2 Schelkunov's (Schelkunoff's) Form

A synthesis procedure developed by Schelkunov [5] makes use of the polynomial form of the array factor and presents an insightful technique for pencil-beam pattern

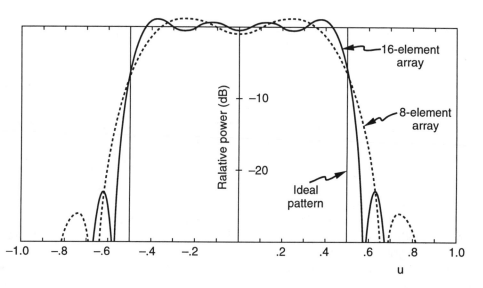

Figure 3.1 Fourier series synthesized representation of pulse-shaped pattern (solid curves for ideal and pattern of 16-element array, dashed curve for 8-element array).

synthesis. The array factor of (3.2) for a one-dimensional array can be written in the form of a polynomial in the complex variable z, where

$$z(u) = \exp(jkud_x) \tag{3.4}$$

and the array polynomial is written

$$F(u) = \sum_{n=0}^{N-1} a_n z^n \tag{3.5}$$

for excitation coefficients a_n at each element.

This form is a polynomial of degree $N - 1$, where N is the number of elements in the array. The summation index range has been changed in the above to run from zero to $N - 1$ in order to simplify the polynomial form. This does not change the form of the array factor, but assumes the zero phase reference at an end element instead of at the array center.

Since the polynomial is of degree $(N - 1)$, it has $(N - 1)$ zeros and may be factored as

$$F(u) = a_{N-1}(z - z_1)(z - z_2)(z - z_3) \cdots (z - z_{N-1}) \tag{3.6}$$

where the terms z_n are the complex roots of the polynomial (as yet unspecified). The magnitude of the array factor is thus

$$|F(u)| = |a_{n-1}||z - z_1||z - z_2| \cdots |z - z_{N-1}| \qquad (3.7)$$

Although the zero locations z_n are unknown in general, those that correspond to real roots in the u-plane must all have magnitude unity, and so if plotted in complex z-space ($z = x + jy$), they all occur on the unit circle shown in Figure 3.2. The magnitude of the array factor, as observed from any point on the unit circle, is the product of the lengths of the straight segments joining that point to the zeros of the array factor.

The uniformly illuminated array has equally spaced zeros located at $u_n = (n/N)(\lambda/d_x)$, and so at $z_n = \exp[j(2\pi/N)n]$, with the zero at $n = 0$ (corresponding to the beam peak at $u = 0$) omitted, and in this special case the polynomial can be written in the compact form

$$F(u) = \frac{z^N - 1}{z - 1} \qquad (3.8)$$

The polynomial representation provides a convenient tool for visualizing the way pattern zeros and grating lobes occur. In the example above, the unit circle has fixed zeros at the z_n given above, equally spaced around the unit circle and indicated by the circle points. For positive values of scan angle, u varies from 0 to 1 and z varies between 1 (at $u = 0$) and $\exp[jkd_x]$. If that path crosses the zeros at location z_n, 3.8 shows that the pattern function $F(u)$ has a null. Figure 3.2(a) shows that for a uniform array of 8 elements with $\lambda/4$ separation, the value of z for $0 < u < 1$ traces through only one quarter of the unit circle (from $z = 1$ at $u = 0$ to $z = \exp[j\pi/2]$ at $u = 1$). This path is indicated by the arrow. The x at $z = 1$ indicates removal of the zero at that point. For $3\lambda/4$ spacing (Figure 3.2(b)), the value of z traces through $\frac{3}{4}$ of the unit circle for positive u. Although the case of one wavelength is not shown, for λ spacing, the z value traces through the full 2π for positive or negative u and in fact reaches the grating lobe at $u = \pm 1$.

If the array is scanned, one replaces u with $u - u_0$, z becomes $\exp[jkd_x(u - u_0)]$, and the unit circle is unchanged. The range of $(u - u_0)$ is not confined by the limits of u ($-1 \le u \le 1$), but instead, for positive scan, the range of $u - u_0$ is $-1 - u_0 \le u - u_0 \le 1 - u_0$, and so the extent of the locus of $z(u - u_0)$ is reduced in the positive direction and increased in the negative direction. Figure 3.2(c) depicts this situation for an array with $3\lambda/4$ spacing. The counterclockwise arrow indicates u varying from u_0 to 1, while the clockwise arrow indicates the locus of $z(u - u_0)$ as u varies from u_0 to -1, passing through a grating lobe at

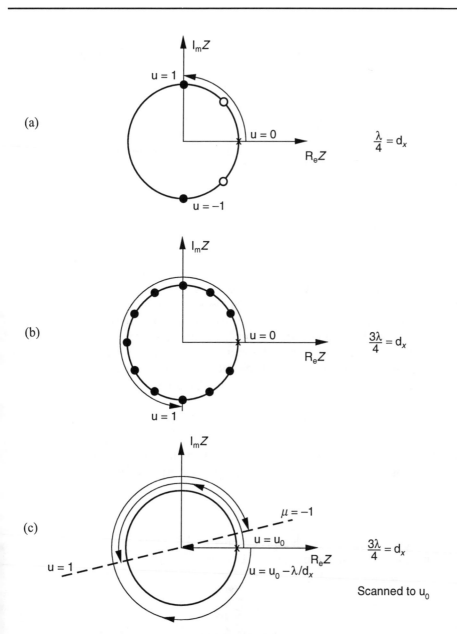

Figure 3.2 Schelkunov unit circle representation of complex root locations for an 8-element uniformly illuminated array: (a) root locations (circles) and path limits (arrow) for positive $0 \leq u \leq 1$ with $\lambda/4$ spaced array; (b) path limit (arrow) for positive u ($0 \leq u \leq 1$) of array with $3\lambda/4$ spacing; (c) path limits (arrows) for array with $3\lambda/4$ spacing scanned to u_0; clockwise arrow for $u_0 \leq u \leq 1$, counterclockwise arrow for $-1 \leq u \leq u_0$.

$u = u_0 - \lambda/d_x$ or $z = -2\pi$ (zero locations have been omitted in Figure 3.2(c) for clarity).

These elements of intuition have led to detailed synthesis procedures as well as iterative approaches to synthesize nearly arbitrary patterns. The first step toward array synthesis is to recognize that if the first zeros were moved further away from the $z = 1$ position, then the zeros would be crowded together and the resulting sidelobes reduced at the expense of a broader main beam. The procedures described by Taylor [6] and Bayliss [7] and summarized later in this chapter are based on manipulation of the pattern zeros.

3.1.3 Woodward Synthesis

The pattern of a uniformly illuminated array, shown in Figure 1.7, has the form $\sin(N\pi z)/(N \sin \pi z)$ for $z = (d/\lambda) \sin \theta$ and is the narrowest pattern that can be formed with an array (superdirectivity excepted). The uniform pattern has another feature that makes it an ideal tool for synthesis: it is a member of an orthogonal set of beams, and therefore one can devise lossless networks to superimpose groups of beams and synthesize desired patterns [8, 9]. For an array of length $L = Nd_x$, there are N such beams that will fill a sector of width $(N - 1)\lambda/L$ in u-space, as shown in Figure 3.3(a). The beam peaks at locations u_i are separated by λ/L in u-space, and their locations are given by the expressions below.

$$u_1 = (\lambda/L)i = [\lambda/(Nd_x)]i \tag{3.9}$$

for $i = \pm 1/2, \pm 3/2, \ldots \pm (N - 1)/2$ for N even or $i = 0, \pm 1, \pm 2, \ldots \pm (N - 1)/2$ for N odd. The ith beam is excited by the phase progression

$$a_n^i = e^{-jkd_x u_i n} \tag{3.10}$$

where n takes on the same values as i (above).

The normalized beam pattern is given by

$$
\begin{aligned}
f_i(u) &= \frac{1}{N} \sum_{n=-(N-1)/2}^{(N-1)/2} e^{jknd_x(u - u_i)} \\
&= \frac{\sin[(N\pi d_x/\lambda)(u - u_i)]}{N \sin[(\pi d_x/\lambda)(u - u_i)]}
\end{aligned}
\tag{3.11}
$$

A given pattern $E(u)$ is thus approximated by sampling it at N-points denoted by the u_i values. As shown in Figure 3.3(a,b) [10], only one of the beam patterns

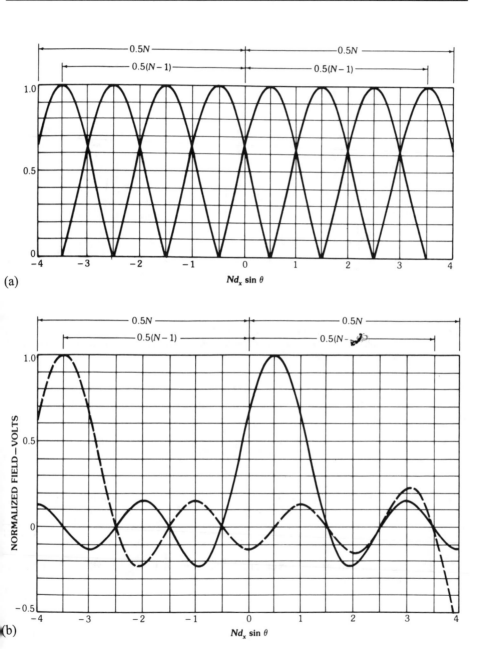

Figure 3.3 Synthesis method of Woodward and Lawson: (a) orthogonal Woodward beams for array of *N* elements (plotted to first zeros) (after [10]); (b) two orthogonal beams (plotted over domain of orthogonality) with $i = 1/2$ (solid) and $i = -7/2$ (dashed) (after [10]).

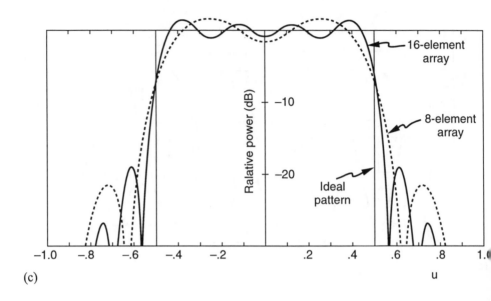

Figure 3.3 (cont.) Synthesis method of Woodward and Lawson: (c) pulse-shaped pattern of Figure 3.1 synthesized by Woodward procedure (solid curve for ideal and pattern of 16-element array, dashed curve for 8-element array).

f_i has a nonzero value at each point, so one can write the approximate pattern as the sum

$$E(u) \approx \sum_i A_i f_i(u) \qquad (3.12)$$

Since the patterns $f_i(u)$ have peak values of unity, A_i is the sampled value

$$A_i = E(u_i) \qquad (3.13)$$

The total current at each element is the sum of those for all the beams. At the nth element,

$$a_n = \sum_i A_i a_n^i$$

$$a_n = \sum_i A_i e^{-jkd_x u_i n} \qquad (3.14)$$

Figure 3.3(c) shows the Woodward synthesis of the same flat-top pattern function approximated in Figure 3.1 by the Fourier series technique. Comparing these two figures shows that the Fourier series provides a lower ripple level and lower sidelobes than the Woodward method. One of the disadvantages of Woodward synthesis is that it does not control the sidelobe level in the unshaped region of the pattern, since only the constituent beams within the shaped region are used in the synthesis. The primary advantage of the Woodward synthesis technique is that it can be implemented using lossless orthogonal beam networks, described in Chapter 7, and so is a relatively simple distribution to approximate with virtually no loss.

The Woodward technique is also the basis for a convenient iterative synthesis procedure due to Stutzman [11]. In that procedure, which is not discussed further in this text, Stutzman adds a correction term to a convenient original pattern whose beamwidth is near to that of the desired pattern. The iterative procedure adds Woodward-type beams, centered at sampling points, to bring the level of the total pattern to the desired level. The procedure is repeated until the desired pattern is matched at all sampled points.

Although Woodward synthesis is often thought of as a procedure for synthesizing shaped beams, Chapter 8 illustrates the synthesis of very-low-sidelobe patterns with Woodward-type beams using so-called dual transform feeds.

3.1.4 Dolph-Chebyshev Synthesis

The procedure commonly referred to as *Dolph-Chebyshev synthesis* [12] equates the array polynomial to a Chebyshev polynomial and produces the narrowest beamwidth subject to a given (constant) sidelobe level. The synthesized pattern for an array of N_T elements spaced d_x apart for $\lambda/2 \le d_x \le \lambda$ at broadside is

$$F(z) = T_M(z) \tag{3.15}$$

for

$$M = N_T - 1$$

where $T_M(z)$ is the Chebyshev polynomial of order M:

$$
\begin{aligned}
T_M(z) &= \cos(M \cos^{-1}z) && \text{for } |z| \le 1 \\
&= \cosh(M \cosh^{-1}z) && \text{for } |z| \ge 1
\end{aligned}
\tag{3.16}
$$

and

$$z = z_0 \cos[(\pi d_x/\lambda) \sin \theta]$$

and

$$z_0 = \cosh(1/M \cosh^{-1}r)$$

for voltage main beam to sidelobe ratio $r > 1$ such that $SL_{dB} = 20 \log_{10} r$ is a positive number. (Note that some of the figures due to other authors use upper case R for the voltage sidelobe ratio, so R and r should be considered interchangeable. Although SL_{dB} is always positive, it is sometimes convenient to refer to sidelobes as negative with respect to the main beam. This should be considered as $-SL_{dB}$.)

If the array polynomial is forced to match the Chebyshev polynomial in such a way that the array sidelobe region occupies the range $|z| \leq 1$ and the beam peak (at $\theta = 0$) is in the region $Z_0 > 1$, then

$$T_M(z_0) = r$$

Figure 3.4 shows the pattern of an 8-element array with Chebyshev illumination and sidelobe levels of -20, -30, and -40 dB. The currents required to produce the synthesized pattern are given by Stegen [13] for spacing $d_x/\lambda \geq 0.5$ as

$$I_m = \frac{2}{N_T}\left[r + 2 \sum_{s=1}^{(N_T-1)/2} T_m\{z_0 \cos(s\pi/N_T)\} \cos[2s\pi m/N_T]\right]$$

$$m = 0, 1, 2, \ldots, (N_T - 1)/2 \tag{3.17a}$$

for N_T odd and as:

$$I_m = \frac{2}{N_T}\left[r + 2 \sum_{s=1}^{N_T/2-1} T_M\{z_0 \cos(s\pi/N_T)\} \cos[(2m - 1)s\pi/N_T]\right]$$

$$m = 1, 2, 3, \ldots, (N_T/2 - 1) \tag{3.17b}$$

for N_T even, where N_T is the number of elements; $2N + 1$ is N_T odd; and $2N$ is N_T even; and again $M = N_T - 1$.

Other authors have given formulas valid for $d_x/\lambda < 0.5$ for arrays with odd numbers of elements. Stegen's formulas are obtained by expanding the Chebyshev radiation pattern in a Fourier series and are more convenient and stable to compute than the original equation of Dolph or those derived prior to Stegen's work. The Chebyshev pattern synthesis procedure has received much attention in the litera-

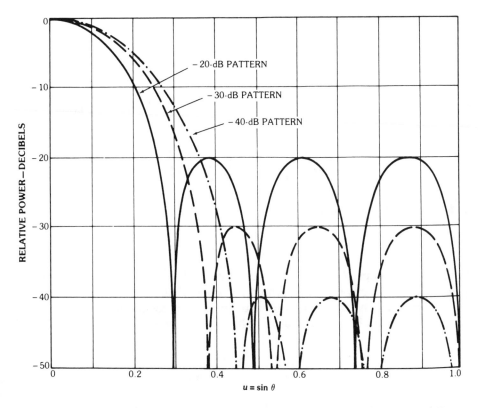

Figure 3.4 Patterns of Dolph-Chebyshev arrays with eight elements (-20-, -30-, -40-dB sidelobes) (after [10]).

ture. Brown and Scharp [14] give extensive tabulations of current distributions computed from the above formulas (although Hansen [15] has pointed out that the numerical accuracy of the tabulated data does not meet current standards). Stegen and others give equations for beamwidth, and there are several convenient expressions for array gain valid for large arrays.

Stegen [16] gives the following expression for directivity

$$D = \frac{N_T}{1 + \dfrac{2}{r^2} \displaystyle\sum_{s=1}^{w} \left[T_M\!\left(z_0 \cos \frac{s\pi}{N_T} \right) \right]^2} \tag{3.18}$$

where

$$W = \frac{N_T}{2} - 1 \quad \text{for } N_T \text{ even}$$

$$= \frac{N_T - 1}{2} \quad \text{for } N_T \text{ odd}$$

For spacings greater than $\lambda/2$, Drane [17] gives the following equation for the directivity of a large array:

$$D = \frac{2r^2}{1 + (\lambda/L')r^2\{\ln[2r]/\pi\}^{1/2}} \tag{3.19}$$

and the beamwidth in radians:

$$\theta_{CH} = 0.18(\lambda/L')(SL_{dB} + 4.52)^{1/2} \tag{3.20}$$

In these expressions, L' is the physical array length $L' = (N_T - 1)d_x$. Drane also gives similar relations for arrays with spacing less than $\lambda/2$.

Elliott [18] gives the following approximate expression for the directivity in terms of the beam broadening factor. This expression is valid for large arrays:

$$D = \frac{2r^2}{1 + (r^2 - 1)(\lambda/L)B_b} \tag{3.21}$$

where the beam broadening factor B_b for a large Chebyshev array is

$$B_b = 1 + 0.636\{(2/r)\cosh[(\cosh^{-1}r)^2 - \pi^2]^{1/2}\} \tag{3.22}$$

and the beamwidth is, as in Chapter 1, $\theta_3 = 0.886(\lambda/L)B_b$. These two equations are used in Figure 1.9.

Figure 3.5(a) compares directivity as computed by Drane [17], using Elliott's formulas [18] with the exact calculation. Good agreement is shown over a wide range of array lengths. The figure also shows that the directivity does not increase indefinitely with L, but reaches a maximum value $2r^2$, or 3 dB greater than the numerical value of the specified sidelobe level. This effect is demonstrated in Figure 3.5(b), due to Elliott [18], which shows the computed directivity versus array length for isotropic elements. The figure shows a linear increase in directivity with array length for relatively small arrays, but each curve reaches a maximum directivity related to its sidelobe level. This effect is due to the forced constant sidelobes that take a progressively large part of the power as the array size increases and beamwidth narrows.

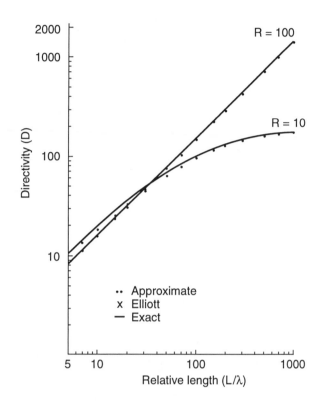

Figure 3.5(a) Characteristics of Chebyshev patterns: array directivity vs. length for −20-dB ($R = 10$) and −40-dB ($R = 100$) sidelobe arrays (after [17]): comparison between approximation of Drane (....), Elliott (xxxx), and exact values.

Figure 3.5(c) shows the Chebyshev beamwidth as computed from (3.20) and the exact value, and Figure 3.5(d) shows the normalized directivity D/N_T or taper efficiency ϵ_T as defined in Chapter 1 as a function of sidelobe level SL_{dB}, computed from (3.19). The general trend of the curves (for $SL_{dB} > 40$) is a result of beam broadening and is almost independent of array size once the array is large enough. For higher sidelobe levels at the left of the figure, the lowered efficiency ratio is a result of the saturation effect mentioned earlier. The larger arrays need lower sidelobes to be efficient.

Although the Chebyshev pattern is a classic synthesis procedure and is well documented and conveniently tabulated, it is not useful for large arrays because of the gain limitation mentioned earlier. The stipulation that the sidelobes remain constant for large angles leads to a maximum in the directivity and then reduced

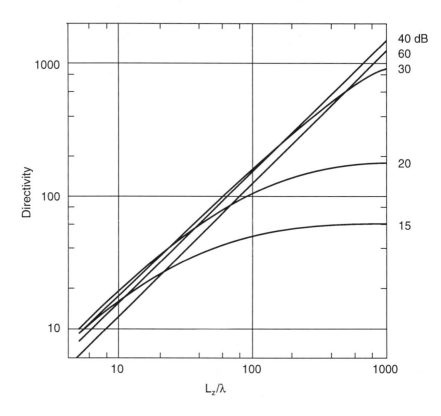

Figure 3.5(b) Characteristics of Chebyshev patterns: array directivity vs. length (after [18]). Note: text uses L in place of Elliott's Lz; sidelobe levels are -15 to -60dB.

directivity with further increases in array length, as shown in Figure 3.5(a,b,d). In addition, for increasingly large arrays, this requires a nonmonotonic aperture illumination with peaks at the array edges and cannot be excited efficiently. These details of aperture illumination are discussed in the next section, since they pertain to Taylor pattern synthesis.

3.1.5 Taylor Line Source Synthesis

In a landmark paper, T. T. Taylor [6] analyzed the deficiencies of the Chebyshev pattern and formulated a pattern function that has good efficiency for large arrays. Taylor examined the limit of a continuous line source and drew conclusions about allowed illuminations and pattern far-sidelobe levels. He compared the pattern of

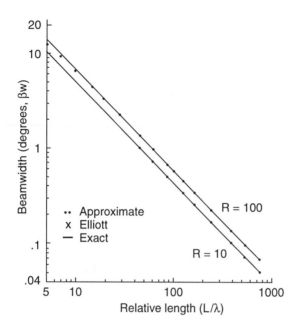

Figure 3.5(c) Characteristics of Chebyshev patterns: array beamwidth versus length for −20- and −40-dB sidelobe arrays (after [17]): comparison between approximation of Drane, Elliot, and exact values.

the Chebyshev illumination with that of a constant illumination [$\sin(\pi z)/\pi z$] for $z = uL/\lambda$, which has the highest efficiency in the large-array limit.

As pointed out by Taylor, the loss in efficiency of the Chebyshev pattern results from the requirement that sidelobe heights are constant. For large arrays, this implies that increasingly more of the energy is in the sidelobe region. In the limit of a very large array, maintaining the Chebyshev sidelobe structure requires an unrealizable aperture illumination. He showed that the far sidelobes of a given line source are a function only of the line source edge illumination. In particular, for a line source of length $2a$, and if the edge illumination has the behavior

$$(a - |x|)^\alpha \tag{3.23}$$

for x measured from the center of the source, then for $\alpha \geq 0$, the far-sidelobe level has the behavior indicated in Table 3.1. The values for $\alpha < 0$ are not given because the illuminations are unrealizable.

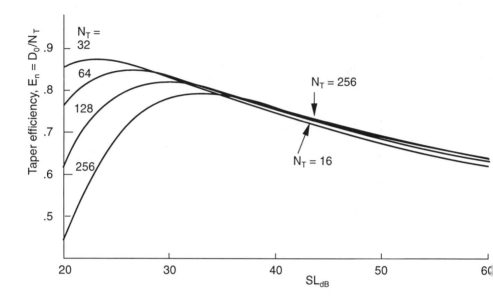

Figure 3.5(d) Characteristics of Chebyshev patterns: taper efficiency $\epsilon_T = D/N_T$ versus sidelobe level.

Table 3.1
Array Far Sidelobe Level versus Edge
Illumination Parameter α

α	Asymptotic $F(z)$
0	$\sin \pi z / (\pi)$
1	$\cos \pi z / (\pi z^2)$
2	$\sin \pi z / (\pi z^3)$
3	$\cos \pi z / (\pi z^4)$

Note: $z = uL/\lambda$.

The above data show that selecting an aperture illumination with zero derivative ($\alpha = 0$) or a pedestal at the array edge leads to far sidelobes with angular dependence $\sin \pi z / \pi z$, like those of the uniform illumination. This pattern distribution maintains its efficiency as the array is made larger. Choice of larger values of α make the far sidelobes decay faster, as indicated in the table, but have generally lower efficiency.

Taylor also showed that the location of the far zeros of the pattern are determined by the edge illumination. The nth pair of pattern zeros (for n large) occur at locations

$$z_n = \pm(n + \alpha/2)$$

as n tends to infinity.

Clearly, this too is consistent with the uniform illumination case for $\alpha = 0$. However, when compared with the actual location of the nth pair of zeros for the Chebyshev pattern, it is found that these occur asymptotically at $\pm(n - 1/2)$. These zero locations correspond to $\alpha = -1$, an unrealizable illumination for the continuous aperture case.

Taylor expanded upon these mathematical insights to suggest a pattern function with zeros far from the main beam at locations that correspond to the uniform illumination, while the zeros closer to the main beam are chosen similar to those of the Chebyshev pattern.

Since Taylor chose to simulate and then modify not the Chebyshev array pattern, but that of a continuous source with similar features to the Chebyshev pattern, he used the following ideal line source pattern as substitute:

$$
\begin{aligned}
F_0(z, A) &= \cos[\pi(z^2 - A^2)^{1/2}] \qquad \text{for } z^2 > A^2 \\
&= \cosh[\pi(A^2 - z^2)^{1/2}] \qquad \text{for } z^2 < A^2
\end{aligned}
\tag{3.24}
$$

where

$$z = uL/\lambda$$

and the sidelobe ratio is evidently given as the value of F_0 at $z = 0$, or

$$r = \cosh(\pi A) \tag{3.25}$$

so A is defined as

$$A = \frac{1}{\pi} \cosh^{-1} r \tag{3.26}$$

As shown by Van der Mass [19], this pattern corresponds to the limiting case of the Chebyshev array as the number of elements is indefinitely increased, and has zeros at the locations

$$Z_N = \pm[A^2 + (N - 1/2)^2]^{1/2} \qquad N = 1, 2, 3, \ldots, \infty \tag{3.27}$$

The pattern has the Chebyshev characteristics with all sidelobes equal, but is physically unrealizable for the reasons described earlier, since the far nulls have asymptotic locations corresponding to $\alpha = -1$.

An expression for the beamwidth of this idealized pattern is readily obtained from the pattern function, since, in the main beam region,

$$\cosh^{-1} F_0(z, A) = \pi\{[(\cosh^{-1} r)/\pi]^2 - z^2\}^{1/2} \tag{3.28}$$

and at the half-power point

$$F_0(z_3, A) = \cosh^{-1}(r/2^{1/2}) \tag{3.29}$$

Combining these relations gives the half-power beamwidth (in u-space) as

$$\begin{aligned} u &= \frac{\lambda}{L} \frac{2}{\pi} [(\cosh^{-1} r)^2 - (\cosh^{-1}(r/2^{1/2}))^2]^{1/2} \\ &\approx \frac{\lambda}{L} \beta_0 \end{aligned} \tag{3.30}$$

The beamwidth is thus a constant β_0 times the "standard beamwidth" λ/L.

Although the idealized pattern is unrealizable, Taylor recognized that by selecting a new function with near zeros very close to those of the ideal pattern (3.27), but with far zeros corresponding to those of the $\sin \pi z/(\pi z)$ function at integer values of z, he could satisfy the requirements on both near and far sidelobes. Taylor chose to keep all nulls at the integer location for $|u| \geq \bar{n}$, and to move those for $|u| < \bar{n} - 1$ near the locations (3.27) that would produce the nearly constant sidelobes near the main beam.

To match these two sets of zeros, Taylor introduced a *dilation factor* σ that is slightly greater than unity to stretch the ideal space factor horizontally by moving the ideal zero locations z_n, such that eventually one of the zeros becomes equal to the corresponding integer $\bar{n}$.

The synthesized pattern normalized to unity is

$$F(z, A, \bar{n}) = \frac{\sin \pi z}{\pi z} \prod_{n=1}^{\bar{n}-1} \frac{1 - z^2/z_n^2}{1 - z^2/n^2} \tag{3.31}$$

for

$$z = uL/\lambda$$

The numbers z_n are the zero locations of the synthesized pattern and are given by

$$z_n = \pm\sigma(A^2 + (n - 1/2)^2)^{1/2} \qquad \text{for } 1 \le n \le \bar{n} \qquad (3.32)$$
$$= \pm n \qquad\qquad\qquad\qquad \text{for } \bar{n} \le n \le \infty$$

where

$$\sigma = \frac{\bar{n}}{[A^2 + (\bar{n} - 1/2)^2]^{1/2}}$$

Note that at $n = \bar{n}$, $z_n = \bar{n}$.

Since the dilation factor σ stretches or dilates the "ideal" space factor to move its zeros away from the main beam, then the beamwidth is increased to a first approximation by that same factor. A good approximation for the beamwidth is therefore given by

$$\theta_3 \approx \sigma\beta_0\lambda/L \text{ radians} \qquad (3.33)$$

for $\beta_0\lambda/L$, the beamwidth of the idealized pattern (3.30). Table 3.2 [20] gives values of the parameter β_0 in degrees and the dilation factor σ used in computation of the approximate beamwidth (3.33).

The aperture distribution required to produce Taylor patterns is expanded as a finite Fourier series of terms with zero derivatives at the aperture edges.

$$g(x) = F(0, A, \bar{n}) + 2 \sum_{m=1}^{\bar{n}-1} F(m, A, \bar{n}) \cos\left(\frac{2m\pi x}{L}\right) \qquad (3.34)$$

for

$$-L/2 \le x \le L/2$$

The coefficients $F(z, A, \bar{n})$ are evaluated to be

$$F(m, A, \bar{n}) = \frac{[(\bar{n} - 1)!]^2}{(\bar{n} - 1 + m)!(\bar{n} - 1 - m)!} \prod_{n=1}^{\bar{n}-1} [1 - m^2/z_n^2] \qquad (3.35)$$

Figure 3.6(a,b) shows 40-dB Taylor patterns computed from the function (3.31) using $\bar{n} = 2$ and 11. These patterns show that choosing $\bar{n}$ too small leads to some pattern distortion. In this case, the distortion is evident because for $\bar{n} = 2$ only one sidelobe is controlled, while the other zero locations are the same as for the uniform illumination case. Only one sidelobe is suppressed in this case, and

Table 3.2
Design Sidelobe Level and Beamwidth for Taylor Distributions

Design Sidelobe Level (dB)	R (Sidelobe Voltage Ratio)	$180\beta_N/\pi$ (Degrees)	A^2	Values of the Parameter (σ)								
				$\bar{n}=2$	$\bar{n}=3$	$\bar{n}=4$	$\bar{n}=5$	$\bar{n}=6$	$\bar{n}=7$	$\bar{n}=8$	$\bar{n}=9$	$\bar{n}=10$
15	5.62341	45.93	0.58950	1.18689	1.14712	1.11631	1.09528	1.08043	1.06969	1.06112	1.05453	1.04921
16	6.30957	47.01	0.64798	1.17486	1.14225	1.11378	1.09375	1.07491	1.06876	1.06058	1.05411	1.04887
17	7.07946	48.07	1.6267	1.13723	1.11115	1.11115	1.09216	1.07835	1.06800	1.06001	1.05367	1.04852
18	7.94328	49.12	0.77266	1.15036	1.13206	1.10843	1.09050	1.07724	1.06721	1.05942	1.05328	1.04815
19	8.91251	50.15	0.83891	1.13796	1.12676	1.10563	1.08879	1.07609	1.06639	1.05880	1.05273	1.04777
20	10.00000	51.17	0.90777	1.12549	1.12133	1.10273	1.08701	1.07490	1.06554	1.05816	1.05223	1.04738
21	11.2202	52.17	0.97927		1.11577	1.09974	1.08518	1.07367	1.06465	1.05750	1.05172	1.04697
22	12.5893	53.16	1.05341		1.11009	1.09668	1.08329	1.07240	1.06374	1.05682	1.05119	1.04654
23	14.1254	54.13	1.13020		1.10430	1.09352	1.08135	1.07108	1.06280	1.05611	1.05064	1.04610
24	15.8489	55.09	1.20965		1.09840	1.09029	1.07934	1.06973	1.06183	1.05538	1.05007	1.04565
25	17.7828	56.04	1.29177		1.09241	1.08598	1.07728	1.06834	1.06083	1.05463	1.04948	1.04518
26	19.9526	56.97	1.37654		1.08632	1.08360	1.07517	1.06690	1.05980	1.05385	1.04888	1.04669
27	22.3872	57.88	1.46395		1.08015	1.08014	1.07300	1.06543	1.05874	1.05305	1.04826	1.04420
28	25.1189	58.78	1.55406			1.07661	1.07078	1.06392	1.05765	1.05223	1.04762	1.04368
29	28.1838	59.67	1.64683			1.07300	1.06851	1.06237	1.05653	1.05139	1.04696	1.04316
30	31.6228	60.55	1.74229			1.06934	1.06619	1.06079	1.05538	1.05052	1.04628	1.04262
31	35.4813	61.42	1.84044			1.06561	1.06382	1.05916	1.05421	1.04963	1.04559	1.04206
32	39.8107	62.28	1.94126			1.06182	1.06140	1.05751	1.05300	1.04872	1.04488	1.04149
33	44.6684	63.12	2.04472				1.05893	1.05581	1.05177	1.04779	1.04415	1.04091
34	50.1187	63.96	2.15092				1.05642	1.05408	1.05051	1.04684	1.04341	1.04031
35	56.2341	64.78	2.25976				1.05386	1.05231	1.04923	1.04587	1.04264	1.03970
36	63.0957	65.60	2.37129				1.05126	1.05051	1.04792	1.04487	1.04186	1.03907
37	70.7946	66.40	2.48551					1.04868	1.04658	1.04385	1.04107	1.03843
38	79.4328	67.19	2.60241					1.04681	1.04521	1.04282	1.04025	1.03777
39	89.1251	67.98	2.72201					1.04491	1.04382	1.04176	1.03942	1.03711
40	100.0000	68.76	2.84428					1.04298	1.04241	1.04068	1.03808	1.03643

Source: [20].

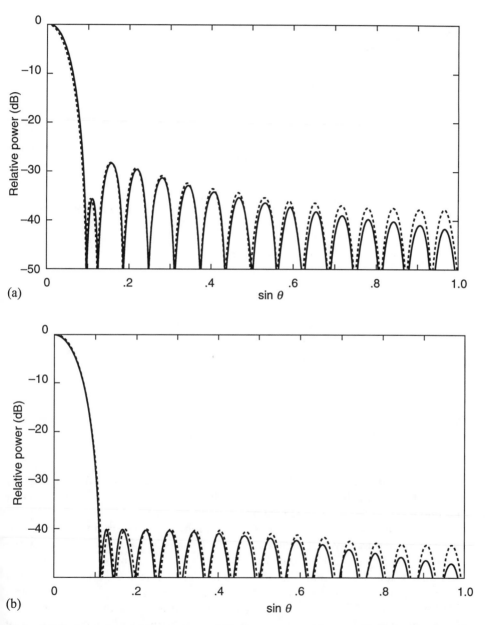

Figure 3.6 Taylor patterns of line sources and 32-element arrays: (a) array with Taylor $\bar{n} = 2$, -40-dB sidelobe pattern (solid curve from function, dashed curve from array currents); (b) array with Taylor $\bar{n} = 11$, -40-dB sidelobe pattern (solid curve from function, dashed curve from array currents).

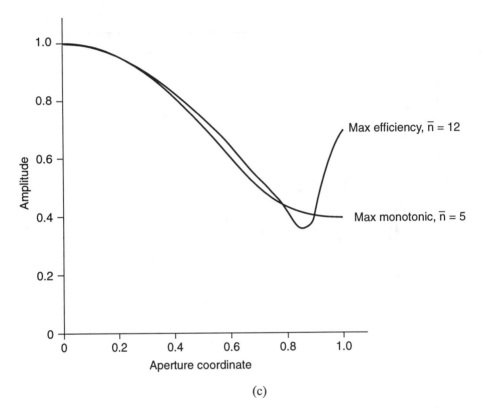

Figure 3.6 (cont.) Taylor patterns of line sources and 32-element arrays: (c) Taylor $\bar{n}$ aperture distributions for -25-dB sidelobe level (after Hansen [15]).

the rest tend to return to the levels of the sin $\pi z/\pi z$ function, which is greater than -30 dB, even though the -40-dB Taylor taper is selected. Again, it is clear that one must increase $\bar{n}$ as the sidelobe level is lowered.

Both figures show the pattern of a continuous source computed using (3.31). The pattern of an array of 16 elements is plotted on the same figure to show the result of sampling the continuous distribution (3.34). It is important that the distribution be sampled at points one-half element spacing from the end of the Taylor distribution function, and so the aperture illumination is sampled at the points $(L/N\lambda)i$ for $\pm i = 1/2, 3/2, 5/2, \ldots, (N-1)/2$ for arrays with an even number of elements, and $\pm i = 0, 1, 2, \ldots, (N-1)/2$ for arrays with an odd number of elements. The sampling procedure maintains good control of the first sidelobe level.

The efficiency of this distribution is given by Hansen [21] as

$$\eta = \frac{1}{1 + 2 \sum_{m=1}^{\bar{n}-1} F^2(m, A, \bar{n})} \tag{3.36}$$

for factors $F(m, A, \bar{n})$, given by the previous expression. This efficiency η pertains to the continuous distribution, but is analogous to the taper efficiency for the discrete array.

The choice of the parameter $\bar{n}$ is not arbitrary, since increasing $\bar{n}$ retains more of the sidelobes at the design sidelobe level and thus makes the Taylor pattern closer to the Chebyshev pattern. Increasing $\bar{n}$ thus leads to narrower main beam patterns and higher aperture efficiency, but eventually to aperture illuminations that are not monotonic and have increased illumination near the aperture edges. A rough guide to the selection of $\bar{n}$ is given in Table 3.3 below, which is due to Hansen [21]. This table shows the efficiency η for Taylor patterns of various sidelobe levels from -20 to -40 dB for two selections of $\bar{n}$. One choice leads to maximum efficiency, which is also accompanied by a peak in the aperture illumination near the array edge, and one choice corresponds to the maximum efficiency η obtainable with a monotonic illumination.

Increasing $\bar{n}$ to the limit of maximum efficiency may not result in realizable current excitation. Figure 3.6(c) shows results due to Hansen that compare the Taylor aperture distributions for maximum efficiency $\bar{n} = 12$ and maximum efficiency with a monotonic illumination ($\bar{n} = 5$) for a 25-dB pattern. The figure shows severe inverse tapering near the edge of the array with maximum efficiency. This rapid variation in current is difficult to approximate with a discrete array and may be unrealizable in a practical size. Moreover, the data cited in Table 3.3 indicate

Table 3.3
Aperture Efficiency for Taylor Patterns

	Max η Values		Monotonic $\bar{n}$	
SL_{dB}	$\bar{n}$	η	$\bar{n}$	η
20	6	0.9667	3	0.9535
25	12	0.9252	5	0.9105
30	23	0.8787	7	0.8619
35	44	0.8326	9	0.8151
40	81	0.7899	11	0.7729

Source: After [21].

that the efficiency penalty in going from maximum efficiency to maximum efficiency with monotonic illumination is only 1% for the case of Figure 3.6(c).

3.1.6 Modified sin $\pi z/\pi z$ Patterns

Taylor [22] also developed a procedure for synthesizing pattern functions with arbitrary first sidelobe levels and a far sidelobe level similar to that of a uniformly illuminated source. This distribution is known as the modified sin $\pi z/\pi z$ distribution [20] or the Taylor one-parameter distribution [21]. The pattern is given by the expressions below.

$$
\begin{aligned}
E(z) &= \frac{\sinh[\pi(B^2 - z^2)^{1/2}]}{\pi[B^2 - z^2]^{1/2}} \qquad z \le B \\
&= \frac{\sin[\pi(z^2 - B^2)^{1/2}]}{\pi[z^2 - B^2]^{1/2}} \qquad z > B
\end{aligned}
\tag{3.37}
$$

where $z = Lu/\lambda$.

The value of B is chosen as indicated below to set the first sidelobe to some given level r, where again $SL_{dB} = 20 \log_{10} r$. Since the first sidelobe occurs in the region $z > B$, where the function has assumed the second form given above, the level of that sidelobe is about 13.26 dB (or the factor $E(z)$ is equal to 1/4.603). However, at the beam peak, $E(0)$ is equal to $\sinh(\pi B)/(\pi B)$, so the ratio of beam peak to sidelobe level is

$$
r = 4.60333 \frac{\sinh \pi B}{\pi B}
\tag{3.38}
$$

The values of parameter B required to obtain a given sidelobe level are obtained from the solution of the above equation. Table 3.4 from Hansen [21] gives the appropriate values of B to produce the required sidelobe levels.

Inspection of (3.37) shows the far sidelobes to be clearly asymptotic to those of the uniform array ($\sin \pi z/\pi z$), since the far zeros are left at $z_n = \pm n$. The near sidelobes are reduced by the placement of the pattern zeros, which have been set at locations $z_n = [n^2 + B^2]^{1/2}$.

The normalized aperture illumination for maintaining this distribution is given as the following.

$$
a(x) = \frac{1}{I_0(\pi B)} I_0\{\pi B[1 - (2x/L)^2]^{1/2}\}
\tag{3.39}
$$

Table 3.4
Modified sin(πz)/(πz) Line Source Characteristics

SL (dB)	B	Z_3 (radians)	η	Edge Taper (dB)
13.26	0	0.4429	1	0
15	0.3558	0.4615	0.993	2.5
20	0.7386	0.5119	0.933	9.2
25	1.0229	0.5580	0.863	15.3
30	1.2762	0.6002	0.801	21.1
35	1.5136	0.6391	0.751	26.8
40	1.7415	0.6752	0.709	32.4
45	1.9628	0.7091	0.674	37.9
50	2.1793	0.7411	0.645	43.3

Source: After [21].

where

x = distance from the center of the aperture,
L = aperture length,
I_0 = modified Bessel function of the first kind (of order zero), and
B = parameter that determines the sidelobe level and is defined below by its relation to the sidelobe level r.

Sampling this aperture illumination results in a set of array excitation coefficients that give an approximation to the pattern (equation). The normalized aperture illumination is seen from the above to have the maximum value unity at the aperture center, and the value $1/I_0(\pi B)$ at the edge. Table 3.4 also gives the value of this edge taper $[-20 \log_{10} I_0(\pi B)]$ in decibels.

The beamwidth θ_3 is given in terms of the parameter Z_3 by

$$\theta_3 = 2 \sin^{-1}[z_3/(L/\lambda)] \tag{3.40}$$

For all but a very small aperture, the beamwidth expression above is accurately given by

$$\theta_3 = 2z_3/(L/\lambda) \tag{3.41}$$

The parameter z_3 is obtained from the solution of

$$\sinh[\pi B/\sqrt{2}\pi B] = \sin\{\pi[z_3^2 - B^2]^{1/2}\}/\{\pi[z_3^2 - B^2]^{1/2}\} \tag{3.42}$$

The aperture efficiency of this line source illumination is given by Hansen as

$$\eta = \frac{2 \sinh^2 \pi B}{\pi B I_0'(2\pi B)} \tag{3.43}$$

where I_0' is the integral of I_0 from 0 to $(2\pi B)$ and is a tabulated integral. The table below [21] gives values of the parameter B, half-power beamwidth, efficiency, and edge taper for sidelobe levels from 13 to 50 dB.

Figures 3.7(a, b) show patterns of line sources of length 4λ and 16λ, with modified $\sin \pi z/(\pi z)$ patterns designed for -40-dB sidelobes, and compare the patterns computed from (3.37) with those computed using arrays of 8 (Figure 3.7(a)) and 32 (Figure 3.7(b)) elements that sample the aperture illuminations (3.39) at half-wavelength increments. The line source patterns are well approximated by the discretized patterns, especially for the 32-element array.

The modified $\sin \pi z/\pi z$ pattern is an excellent low-sidelobe distribution and has good efficiency. A comparison of Tables 3.3 and 3.4 reveals, however, that the Taylor patterns can have higher efficiency if $\bar{n}$ is chosen appropriately.

3.1.7 Bayliss Line Source Difference Patterns

A useful synthesis procedure for the asymmetrical patterns required of monopulse systems was developed by Bayliss [7]. Like the Taylor procedure, Bayliss patterns are fully described in terms of the two parameters A and $\bar{n}$, which again control the sidelobe level and decay behavior. The synthesized pattern is given by

$$F(z) = \pi z \cos(\pi z) \frac{\prod\limits_{n=1}^{\bar{n}-1} \{1 - (z/\sigma z_n)^2\}}{\prod\limits_{n=0}^{\bar{n}-1} \{1 - (z/(n + \frac{1}{2}))^2\}} \tag{3.44}$$

for

$$z = uL/\lambda \qquad \sigma = \frac{\bar{n} + \frac{1}{2}}{z_{\bar{n}}}$$

and

$$z_{\bar{n}} = (A^2 + \bar{n}^2)^{1/2}$$

The zeros of this function are at σz_n and will be specified later.

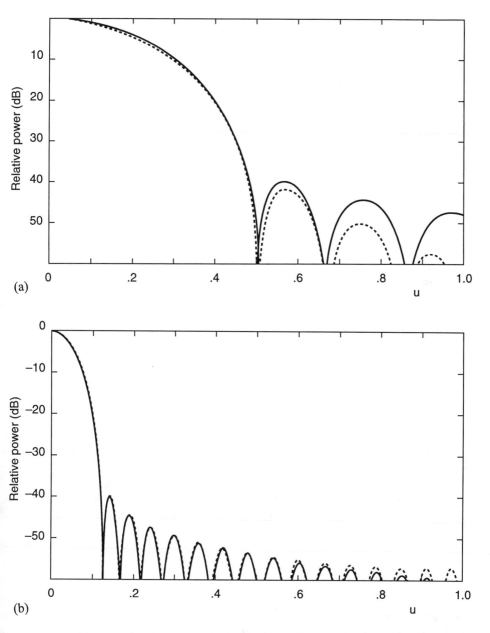

Figure 3.7 Modified $\sin(\pi z)/(\pi z)$ line source and array patterns (-40-dB sidelobes) (solid curves from function, dashed curve from array currents); (a) four-wavelength line sources (solid) and 8-element array patterns (dashed); (b) sixteen-wavelength line source (solid) and 32-element array patterns (dashed).

The line source excitation is given by the Fourier series

$$g(x) = \sum_{n=0}^{\bar{n}-1} B_n \sin[(2\pi x/L)(n + \tfrac{1}{2})] \qquad -L/2 \leq x \leq L/2 \qquad (3.45)$$

and the Fourier coefficients are

$$B_m = \begin{cases} \dfrac{1}{2j}(-1)^m(m + \tfrac{1}{2})^2 \dfrac{\displaystyle\prod_{n=1}^{\bar{n}-1}\left\{1 - \dfrac{[m + \tfrac{1}{2}]^2}{[\sigma z_n]^2}\right\}}{\displaystyle\prod_{\substack{n=0 \\ n\neq m}}^{\bar{n}-1}\left\{1 - \dfrac{[m + \tfrac{1}{2}]^2}{[n + \tfrac{1}{2}]^2}\right\}} & m = 0, 1, 2, \ldots, \bar{n} - 1 \\[20pt] 0 & \text{for } m \geq \bar{n} \end{cases} \qquad (3.46)$$

The null locations σz_n are given with z_n defined as

$$z_n = \begin{cases} 0 & n = 0 \\ \pm \Omega_n & n = 1, 2, 3, 4 \\ \pm(A^2 + n^2)^{1/2} & n = 5, 6, \ldots \end{cases} \qquad (3.47)$$

The coefficients A and the Ω_n are not available in closed form, but Bayliss presented a table of coefficients for fourth-order polynomials to represent these five coefficients as a function of sidelobe level (SL_{dB}). In addition to A and Ω_n, the table lists values for the polynomial approximation of p_0, which is the location of the difference peak. Recall that in u-space the peak locations are given by $u = (\lambda/L)p$. In this case, the polynomial is represented by

$$\text{Polynomial} = \sum_{n=0}^{4} C_n[-SL_{dB}]^n \qquad (3.48)$$

with coefficients c_0 through c_4 given by Table 3.5. In addition, Elliott [23] gives table of the coefficients themselves for sidelobe levels from -15 to -40 dB in increments of 5 dB (Table 3.6). Figure 3.8(a) compares the patterns of the continuous distribution (3.44) and that of a 16-element array sampling the continuous aperture distribution of (3.45) for $\bar{n} = 4$.

Table 3.5
Polynomial Coefficients

Polynomial	C_0	C_1	C_2	C_3	C_4
A	0.30387530	−0.05042922	−0.00027989	−0.00000343	−0.00000002
Ω_1	0.98583020	−0.03338850	0.00014064	0.00000190	0.00000001
Ω_2	2.00337487	−0.01141548	0.00041590	0.00000373	0.00000001
Ω_3	3.00636321	−0.00683394	0.00029281	0.00000161	0.00000000
Ω_4	4.00518423	−0.00501795	0.00021735	0.00000088	0.00000000
p_0	0.47972120	−0.01456692	−0.00018739	−0.00000218	−0.00000001

Source: After [7].

Table 3.6
Parameters A, Ω_1, Ω_2, Ω_3, Ω_4 for Bayliss Patterns

Polynomial	Sidelobe Level (dB)					
	15	20	25	30	35	40
A	1.0079	1.2247	1.4355	1.6413	1.8431	2.0415
Ω_1	1.5124	1.6962	1.8826	2.0708	2.2602	2.4504
Ω_2	2.2561	2.3698	2.4943	2.6275	2.7675	2.9123
Ω_3	3.1693	3.2473	3.3351	3.4314	3.5352	3.6452
Ω_4	4.1264	4.1854	4.2527	4.3276	4.4093	4.4973

Source: After [23].

3.1.8 Synthesis Methods Based on Taylor Patterns: Elliott's Modified Taylor Patterns and the Iterative Method of Elliott

A variety of methods can be used to synthesize generalized patterns. Particularly well-documented and convenient procedures have been developed by Elliott for the synthesis of patterns that can be generated beginning with Taylor and Bayliss patterns. Since the details are given in an available text, the following lists only the final formulas and definitions. Moreover, although this book makes reference only to patterns generated from Taylor and Bayliss line source starting patterns, Elliott has also derived analogous methods for circular aperture arrays. These are well documented in [23]. A particular advantage of this technique is that it only changes the locations of the innermost set of zeros, not the zeros far from the main beam, which are left at locations $z_n = \pm n$. This ensures that the far sidelobes are well behaved and follow the sin $\pi z/(\pi z)$ dependence of the uniform array for $z = uL/\lambda$.

In the first instance, to produce a generalized sum pattern, Elliott [24] derived a more general pattern function than the Taylor pattern, one that behaves like

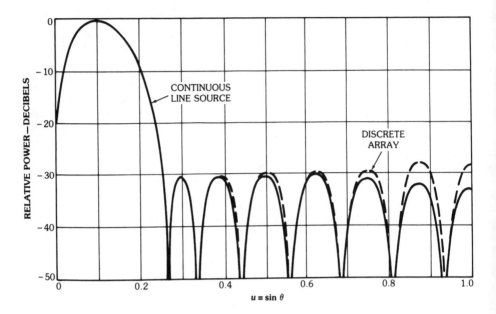

Figure 3.8 Bayliss $\bar{n} = 4$ line source and array ($n = 16$) difference patterns for -40-dB sidelobe level (after [10]).

Taylor patterns with different characteristics at either side of the main beam. For example, the pattern could resemble a Taylor 20-dB pattern with $\bar{n} = 2$ to the left of the main beam and a Taylor 40-dB pattern with $\bar{n} = 5$ to the right. To represent the new pattern, Elliott writes a form that is equivalent to (3.31) if the null locations are kept at the location of the Taylor patterns, but which is modified by removal of some of the Taylor pattern nulls and multiplication by factors that produce new nulls at desired locations. Elliott then expresses the sum pattern $S(z)$, which he terms a modified Taylor pattern:

$$S(z) = Cf(z) \prod_{-(\bar{n}_L - 1)}^{\bar{n}_R - 1} (1 - z/z_n) \tag{3.49}$$

where

$$f(z) = \frac{(\sin \pi z)/(\pi z)}{\displaystyle\prod_{\substack{-(\bar{n}_L - 1) \\ (n \neq 0)}}^{\bar{n}_R - 1} (1 - z/n)} \tag{3.50}$$

The constant C is a normalization. The pattern of the starting distribution $S_0(z)$ has zeros at locations z_m for all $-(\bar{n}_L - 1) \leq m \leq (\bar{n}_R - 1)$. The subscripts R and L refer to the assumption of different numbers of zeros controlled to the right and left of the pattern.

The zero locations for the modified Taylor distribution are given by

$$
\begin{aligned}
z_n &= -\bar{n}_L \frac{[A_L^2 + (n + 1/2)^2]^{1/2}}{[A_L^2 + (\bar{n}_L - 1/2)^2]} & n &= -[1, 2, \ldots, (\bar{n}_L - 1)] \\
&= \bar{n}_R \frac{[A_R^2 + (n - 1/2)^2]^{1/2}}{[A_R^2 + (\bar{n}_R - 1/2)^2]} & n &= 1, 2, \ldots, (\bar{n}_R - 1)
\end{aligned}
\tag{3.51}
$$

and

$$
\begin{aligned}
A_L &= \frac{1}{\pi} \cosh^{-1} r_L \\
A_R &= \frac{1}{\pi} \cosh^{-1} r_R
\end{aligned}
\tag{3.52}
$$

for (voltage) sidelobe levels r_L and r_R on the left and right sides of the main beam. The values $\bar{n}_L$ and $\bar{n}_R$ must both be one greater than the number of controlled sidelobes of the left and right of the main beam.

The aperture illumination required to produce this pattern is again readily found from a Fourier series approximation, and is given by Elliott as

$$
g(x) = \frac{1}{L} \sum_{-(\bar{n}_L - 1)}^{\bar{n}_R - 1} S(m) e^{-j2m\pi x/L}
\tag{3.53}
$$

where here the sum includes $m = 0$.

The $S(m)$ is obtained from (3.49) evaluated at $z = m$, and truncates at $\bar{n}$ on either side. Since evaluating the function f involves a limiting process, the resulting equation for the function f is

$$
f(m) = \frac{-(-1)^m}{\displaystyle\prod_{\substack{-(\bar{n}_L - 1) \\ m \neq n}}^{\bar{n}_R - 1} (1 - m/n)}
\tag{3.54}
$$

for $m \neq 0$, and

$$
f(0) = 1
$$

This modified Taylor pattern is itself a convenient illumination, since it can produce patterns with different sidelobe levels at either side of the main beam. Figure 3.9 shows typical patterns produced using the above expression (here applied to a 32-element array) and compared with the pattern computed by the function (3.49). This pattern is for an array with $\bar{n}_L = 5$ and $\bar{n}_R = 7$, with four sidelobes at the left of the main beam set at -20 dB and six at the right set to -40 dB. The patterns evaluated from the sampled illumination (3.53) are indeed an excellent representation of the exact line source pattern, but it is clear that the line source distribution itself has difficulty reproducing sidelobe levels different by a factor of 100, with the result that the sidelobes to the left are not constant, those to the left are lower than required, and those to the right are too high. Although this modified Taylor distribution of Elliott achieves a useful degree of pattern control, the iterative procedure used to derive generalized patterns can give a far greater degree of accurate pattern control.

The modified Taylor pattern is used in Elliott's iterative procedure [25] as a starting pattern, and in this context he uses the notation z_m^0 to index the zeros of the starting pattern or the pattern from the previous iteration.

For a consistent notation, Elliott defines the position of the peaks in the starting pattern as z_m^p. With these definitions, one can show that if the perturbations are small, the values of the new pattern $S(z)$ can be written at the location of the peaks of the starting pattern z_m^p as

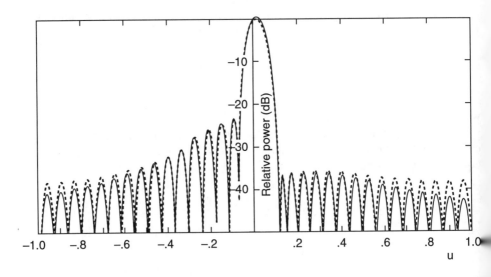

Figure 3.9 Modified Taylor pattern of Elliott. Pattern design for $\bar{n}_L = 5$, $\bar{n}_R = 7$ with -20-dB left sidelobes and -40-dB right (line source pattern solid, array pattern dashed).

$$\frac{S(z_m^p)}{S_0(z_m^p)} - 1 = \frac{\Delta C}{C_0} + \sum_{\substack{n=-(\bar{n}_L-1) \\ n\neq 0}}^{(\bar{n}_R-1)} \frac{z_m^p/(z_n^0)^2}{1 - z_m^p/z_n^0} \Delta z_n \tag{3.55}$$

for $-(\bar{n}_L - 1) \leq m \leq (\bar{n}_R - 1)$, where the constant C has been written to account for the perturbation in the pattern normalization amplitude

$$C = C_0 + \Delta C \tag{3.56}$$

and the Δz_n are the changes in null locations, so that the nulls of the new pattern are at

$$z_n = z_n^0 + \Delta z_n \tag{3.57}$$

The $S(z_m^p)$ are the known heights of the peaks in the desired pattern. These peaks must obviously be included between the nulls at $z_{\bar{n}_L} \leq z \leq z_{\bar{n}_R}$, and so the technique is not intended for control of far sidelobes, but only those within the controlled nulls of the original modified Taylor patterns.

The starting pattern null locations z_n^0 are known and the peak locations z_m^p can be accurately found by a numerical search. Except for the main beam peak location, I have found it entirely adequate to set each z_m^p by choosing the location halfway between the adjacent zeros. However, in order to properly normalize the pattern sidelobes to the pattern peak, one must accurately determine the main beam location, and here a numerical search is often necessary.

Given these coefficients, the only remaining unknowns are the ΔC and the values of the null shifts Δz_n, so the equation above is written at each pattern peak to form a matrix equation with $(\bar{n}_L + \bar{n}_R - 1)$ rows and the same number of unknowns. Once the new nulls z_n are found, the pattern $S(z)$ for the continuous aperture can be computed from (3.49) and used as a new starting pattern for another iteration if necessary.

The amplitude illumination is obtained from (3.52) and is exact for the continuous (line source) case.

Figure 3.10 shows an iterated pattern of a line source 16λ long, in which the three sidelobes on the left are set to -20 dB and the first three on the right set to -40 dB, the next two set to -30 dB, and the next two to -40 dB. Beyond these, the sidelobes are allowed to revert to whatever level is dictated by the $\sin(\pi z)/(\pi z)$ pattern function, so these actually increase at the right, but decrease at left. The procedure is begun using the modified Taylor pattern of Elliott with $\bar{n}_L = 4$ and $\bar{n}_R = 7$. Only three iterations were necessary to obtain ± 0.1-dB accuracy.

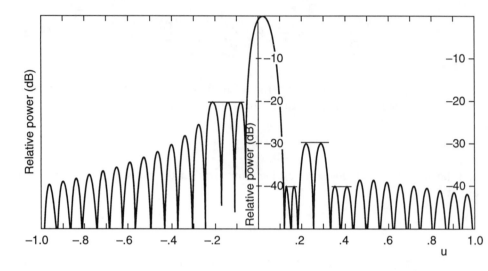

Figure 3.10 Iterated sum pattern of Elliott. Pattern design for $\bar{n}_L = 4, \bar{n}_R = 7$ (solid horizontal lines indicate constrained sidelobe levels).

Generalized Patterns Synthesized From Bayliss Difference Patterns

Following the same procedure as that used for the sum patterns, Elliott [26] has obtained an equally convenient iterative procedure to facilitate the synthesis of difference patterns with arbitrary sidelobe levels. Only final results are given below, and, again, the starting pattern is the Bayliss pattern written to express the factors to the left and to the right of the origin separately, so that $\bar{n}_L$ need not be equal to $\bar{n}_R$.

Written in this fashion, (3.44) becomes

$$D_0(z) = C_0 z f(z) \prod_{\substack{-(\bar{n}_L-1) \\ n\neq 0}}^{\bar{n}_R-1} (1 - z/z_n^0) \qquad (3.58)$$

where z_n^0 is the location of the original nth Bayliss root σz_n from (3.47) and

$$f(z) = \frac{\pi \cos(\pi z)}{\displaystyle\prod_{n=-(\bar{n}_L-1)}^{\bar{n}_R-1} [1 - z/(n + \tfrac{1}{2})]} \qquad (3.59)$$

The zero locations of the modified version of the Bayliss pattern and the constants A_R and A_L are given by the Bayliss equations (3.47) and (3.48), with the sidelobe levels as appropriate to the two sides of the pattern.

The desired pattern is expressed in terms of parameters Δz_n and ΔC to give the result

$$\frac{D(z)}{D_0(z)} - 1 = \frac{\Delta C}{C_0} - \frac{\Delta z_0}{z} + \sum_{\substack{n=-(\bar{n}_L-1) \\ n\neq 0}}^{\bar{n}_R-1} \frac{[z/(\sigma z_n^0)^2]}{1 - z/(\sigma z_n^0)} \Delta z_n \qquad (3.60)$$

Using the peak positions z_m^p and nulls of the lobes in the starting pattern produces a set of $\bar{n}_L + \bar{n}_R$ simultaneous linear equations, which can be solved to produce the desired perturbed solution, and used, if necessary, as the starting point for further iterations.

The required aperture distribution for this synthesis is

$$g(x) = \sum_{-(\bar{n}_L-1)}^{\bar{n}_R-1} F(n + 1/2)e^{-j(n+1/2)2\pi x/L} \qquad (3.61)$$

with the end result (for $0 \leq m \leq \bar{n}_R - 1$) being

$$F(m + 1/2) = (-1)^m(m + 1/2)\pi C$$
$$\times \frac{\displaystyle\prod_{-(\bar{n}_L-1)}^{\bar{n}_R-1} [\sigma z_n - (m + 1/2)]}{\displaystyle\prod_{\substack{n=0 \\ n\neq m}}^{\bar{n}_R-1} \left(1 - \frac{m + 1/2}{n + 1/2}\right) \prod_{n=0}^{\bar{n}_L-1} \left(1 + \frac{m + 1/2}{n + 1/2}\right)} \qquad (3.62)$$

and in the range $-\bar{n}_L \leq m \leq -1$, the result is

$$F(m + 1/2) = (-1)^m(m + 1/2)\pi C$$
$$\times \frac{\displaystyle\prod_{-(\bar{n}_L-1)}^{\bar{n}_R-1} [\sigma z_n - (m - 1/2)]}{\displaystyle\prod_{n=0}^{\bar{n}_R-1} \left(1 - \frac{m + 1/2}{n + 1/2}\right) \prod_{\substack{n=0 \\ n\neq -(m+1)}}^{\bar{n}_L-1} \left(1 + \frac{m + 1/2}{n + 1/2}\right)} \qquad (3.63)$$

Figure 3.11 shows an iterated pattern, which was constructed from a Bayliss 30-dB, $\bar{n} = 10$ pattern, but with the four innermost sidelobes suppressed to -40

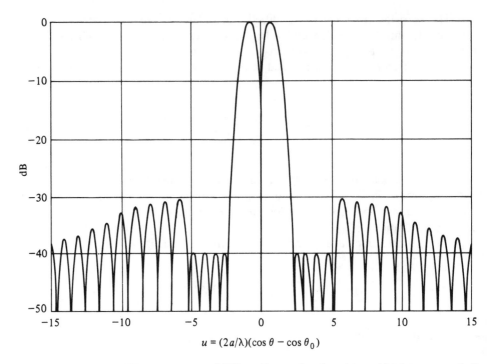

Figure 3.11 Iterated difference pattern of Elliott. Pattern has four inner sidelobes symmetrically depressed (after [26]).

dB. The horizontal axis, denoted by u in the figure, corresponds to our $z - z_0$ for a beam scanned to the direction cosine u_0. The pattern is accurate for the continuous line source. A discrete array would necessarily suffer the approximation that results from discretizing this continuous distribution.

3.1.9 Discretization of Continuous Aperture Illuminations by Root Matching and Iteration

The Taylor and Bayliss patterns and the patterns derived from them for arbitrary sidelobe distributions are exactly reproduced by applying the continuous aperture illuminations given in previous sections, but the sampling process required to discretize these continuous illuminations results in some errors. In the case of the Taylor pattern, an equivalent discrete array formulation has been derived by Villeneuve [27]. This result is not reproduced here because the several discretizing procedures [28,29] are sufficiently accurate for most purposes.

The simplest way to approximate the pattern of a continuous illumination with a discrete array is to periodically sample the continuous illumination. In the previous sections, this sampling led to excellent approximations, but this procedure may not be adequate for small arrays with very low sidelobes, or for relatively widely spaced elements, or for synthesized patterns that have severe changes in sidelobe levels (e.g., Figure 3.11).

A procedure that yields improved results for patterns with real roots, especially for small arrays, is to choose the roots of the discrete array pattern so that they match those of the pattern of the continuous aperture. The continuous aperture root locations are given by (3.31) and (3.47) for the Taylor and Bayliss patterns and are also readily derived for other continuous aperture distributions, such as those derived from iterative solutions.

Referring to the series form (3.6), an array's pattern is given by the product of its zeros, which can be represented by a power series in the exponential form

$$F(u) = \sum a_n z^n \tag{3.64}$$

with coefficients a_n representing the array (complex) amplitude distribution for

$$z = \exp(jkd_x u) \tag{3.65}$$

$$F(u) = a_{N-1}(z - z_1)(z - z_2) \cdots (z - z_{N-1}) \tag{3.66}$$

Since this polynomial is also given by the product of the roots (3.6), it is only necessary to multiply the terms in (3.66) and then to identify them as the coefficients a_n in (3.64). This procedure is tedious for large arrays, and in some cases it may be easier to match roots using the adaptive procedure to be described in Section 3.4 or the set of simultaneous equations to find the array illumination corresponding to the desired real roots. Since the N-element array has $N - 1$ independent roots within the region $-\lambda/(2d_x) \le u \le \lambda/(2d_x)$, one can write the homogeneous equation below at the required location of the nulls corresponding to the desired pattern.

$$F(u_n) = 0 \tag{3.67}$$

In addition, one must satisfy a normalizing condition to fix the value of the main beam peak at u_0:

$$F(u_0) = \sum_{p=1}^{N_T} a_p \tag{3.68}$$

The solution of this set of simultaneous equations yields the excitations a_n.

Elliott [29] gives several examples of the utility of the real root matching technique, which can result in an excellent approximation to a given pattern if the sidelobe topography is not too severe. In such cases where it is not adequate, Elliott presents an iterative procedure to individually reset sidelobe levels to account for discretization.

3.1.10 Synthesis of Patterns With Complex Roots and Power Pattern Synthesis

The specialized distributions that produce pencil beam patterns while optimizing sidelobe characteristics have proven very useful for antenna design. The Chebyshev, Taylor, Bayliss, and the various iterative schemes are excellent, efficient solutions and offer enough pattern selection options to satisfy most needs. These patterns, as described in the previous section, have real zeros and well-defined main beams and nulls. However, there is also a need to develop patterns that do not have zeros, but have shaped beams. This is often done using the Fourier series method or the Woodward synthesis method with $(\sin \pi z)/(\pi z)$ type patterns (Section 3.1.3), which have the added advantage of being implemented with the lossless networks of Chapter 7. But it can also be done by moving the roots of the array polynomial off the unit circle. A convenient procedure for accomplishing this is given in the work of Elliott and Stern [30], which will not be specifically described here.

Power pattern synthesis [1] offers real advantages in the synthesis of shaped antenna patterns, where a wide area of the pattern needs to be approximated. The advantage results from the fact that the array factor is a complex function, with both magnitude and phase, while most pattern control procedures need to synthesize only the pattern amplitude. Most field pattern synthesis procedures assume real pattern functions, and this reduces the number of degrees of freedom available to approximate the desired pattern. The notation used in the following description follows Steyskal [31].

Steyskal's Synthesis Procedure

An array with an odd number $(N + 1)$ of elements has the field pattern

$$F = \sum_{-N/2}^{N/2} I_m \exp(jm\pi u) \tag{3.69}$$

which is a set of $N + 1$ harmonics. The corresponding power pattern q is

$$q = FF^* = \sum_{-N}^{N} q_n \exp(jn\pi u) \tag{3.70}$$

The coefficients q_n are related to the element currents by the expression

$$q_n = \begin{cases} \sum_{m=n}^{N} I_m I_{m-n}^* & n \geq 0 \\ \sum_{m=0}^{N+n} I_m I_{m-n}^* & n < 0 \end{cases} \tag{3.71}$$

Since the power pattern has $(2N - 1)$ terms for an $N + 1$ element array, there are many more degrees of freedom in the power pattern expression than in the field pattern expression. Power pattern synthesis, by allowing amplitude and phase of every element to be determined by the synthesis, recaptures all of the available degrees of freedom.

Steyskal [31, 32] shows that if some desired shape is synthesized within a given region using conventional field pattern synthesis, the resulting power pattern includes the added zeros outside of the synthesized region. The added degrees of freedom present in the power pattern are therefore not used to better match the desired pattern shape. Steyskal considered minimization of the Gaussian or weighted mean square error ϵ between the desired power pattern $p_d(u)$ and the realizable actual pattern $q(u)$ subject to weighting criteria. The chosen error is the weighted mean square

$$\epsilon = \int_{-1}^{1} [p_d(u) - q(u)]^2 w(u) du \tag{3.72}$$

where $u = \sin \theta$ and $w(u)$ is a weighting function chosen according to the relative accuracy of the approximation over the interval in u. This error is the mean square error when $w(u) = 1$.

Steyskal formulates the optimization in Hilbert space, in which a function is interpreted as a vector. The inner product $(\mathbf{x} \cdot \mathbf{y})$ of two vectors $\mathbf{x}$ and $\mathbf{y}$ is defined as the weighted integral

$$(\mathbf{x}, \mathbf{y}) = \int_{-1}^{1} \mathbf{x} \mathbf{y}^* w du \tag{3.73}$$

He further defines the length or norm $\|\mathbf{x}\|$ of a vector $\mathbf{x}$ as

$$\|\mathbf{x}\| = (\mathbf{x}, \mathbf{x})^{1/2} \tag{3.74}$$

and the square of the "distance" of some point p_D to some point q as

$$\|p_D - q\|^2 = \int_{-1}^{1} |p_D - q|^2 \, wdu \qquad (3.75)$$

which corresponds with the Gaussian measure of the error in approximating point p_D by q.

The problem solved by Steyskal is to find the best approximation to the desired, but possibly nonphysical, pattern p_D, with a realizable pattern q_D such that

$$\|p_D - q_D\| \text{ is a minimum} \qquad (3.76)$$

This is done in two steps. First the desired pattern p_D is expanded (projected onto) a set of orthonormal functions, resulting in an approximation p, which is the best sum of $2N + 1$ harmonics to approximate p_D. This best approximation satisfies the condition

$$\|p_D - p\| = \text{minimum} \qquad (3.77)$$

The approximation p is not achievable physically because it may lead to negative power. What is required is the solution q_D, which is the best nonnegative approximation to p_D. Therefore, the second step is to now find a realizable q_D that is a best nonnegative approximation to the harmonic approximation p, and therefore to minimize

$$\|p - q_D\| \qquad (3.78)$$

This process leads to the best realizable approximation to the desired minimization.

From (3.70), the power pattern is composed of the finite set of the harmonics

$$\{\exp(j2\pi nu)\}_{-N}^{N} \qquad (3.79)$$

One way of expanding the power pattern would be to construct an orthogonal set of basis functions from the harmonic functions using the Gram-Schmidt process, but Steyskal introduced the following technique, which is computationally simpler. Since the best harmonic pattern p minimizes the error $\epsilon_1 = \|p_D - p\|^2$, that minimization problem is solved for the basis vectors.

Using the notation $e_{n+N} = \exp(jn\pi u)$, the harmonic pattern p is written

$$p = \sum_{n=1}^{2N+1} p_n e_n \qquad (3.80)$$

then the error is written

$$\epsilon_1 = \|p_D - p\|^2 = \int_{-1}^{1} \left| p_D - \sum_{n=1}^{2N+1} p_n e_n \right|^2 w \, du \tag{3.81}$$

The solution to the minimization problem is the result

$$(p_D, e_n) = (p, e_n) \qquad n = 1, 2, 3, \ldots, 2N + 1$$

$$= \left(\sum_{m=1}^{2N+1} p_m e_m, e_n \right) \tag{3.82}$$

$$= \sum_{m=1}^{2N+1} p_m (e_m, e_n)$$

Since the left side (p_D, e_n) is known for all n, this is a set of $2N + 1$ equations that can be solved for the unknown p_m. The solution gives the best harmonic approximation to p_D, but is not necessarily realizable. Next is to find a nonnegative expression for the pattern (call it q) and evaluate the currents such that the error term

$$\epsilon_2 = \|p - q\|^2 \tag{3.83}$$

is minimized.

In this regard, it is significant that the set of patterns Q is convex, and therefore any local minimization of ϵ_2 is also the global minimum. This property allows q to converge uniformly to p without fear of finding an incorrect minimum.

Writing q in terms of the element currents from (3.70), the minimization of ϵ_2 is performed by a gradient descent method. Setting the currents to some initial value I_n and writing a set of increments Δ_n so that the new currents are

$$I_n = \Delta_n + I_n \tag{3.84}$$

defines the new approximation to the desired pattern. The incremental change to the currents is given by

$$\Delta I = -\frac{\text{grad } \epsilon_2}{|\text{grad } \epsilon_2|} \cdot s \tag{3.85}$$

where the scalar constant s is a measure of the step size and is progressively decreased as the minimum is approached.

Steyskal gives several examples of patterns synthesized using this weighted or Gaussian error minimization. Figure 3.12 shows two examples of the use of different weighting functions to provide significantly different approximations of the desired pulse pattern (dashed). The curves at left use a 30-dB weighting function, while those at right use a 60-dB weighting function. The severe weighting of the curves at right reduces all sidelobes to about -37 dB as opposed to -20 dB for the less severe weighting. Within the shaped region, it is clear that the price paid for this sidelobe suppression is a higher ripple level and a narrowed pulse region. The patterns shown at the bottom of the set of figures are the best harmonic approximation to the desired pattern, and the negative portions are shown dashed.

The Procedure of Orchard et al.

A new technique for power pattern synthesis was recently proposed by Orchard et al. [33]. The procedure differs from that of Steyskal and others in that it combines the intuition presented by Schelkunov's unit circle with the added degrees of freedom accorded by power pattern synthesis.

Figure 3.13(a) indicates that the pattern range is divided into two parts, a shaped beam region (I) and a region of controlled sidelobes (region II). The goal of this synthesis, as indicated in Figure 3.13(a), is to best approximate the shaped pattern (denoted by $S(\Phi)$) in region I, while maintaining control of all the sidelobes in region II.

The procedure begins with the antenna array factor, expressed

$$F = \sum_{n=0}^{N} I_n \exp[jknd \cos \theta] \tag{3.86}$$

in which the elements each support excitation currents I_n and are spaced d apart. The angle θ is here measured from endfire to agree with the definitions and figures from the reference.

Using $\Phi = kd \cos \theta$ and $w = \exp(j\Phi)$, one can write F in the form of a product of its zero locations

$$F = \sum_{n=0}^{N} I_n w^n = I_N \prod_{n=1}^{N} (w - w_n) \tag{3.87}$$

In general, the zero locations are not assumed real, so Orchard writes

$$w_n = \exp(a_n + jb_n) \tag{3.88}$$

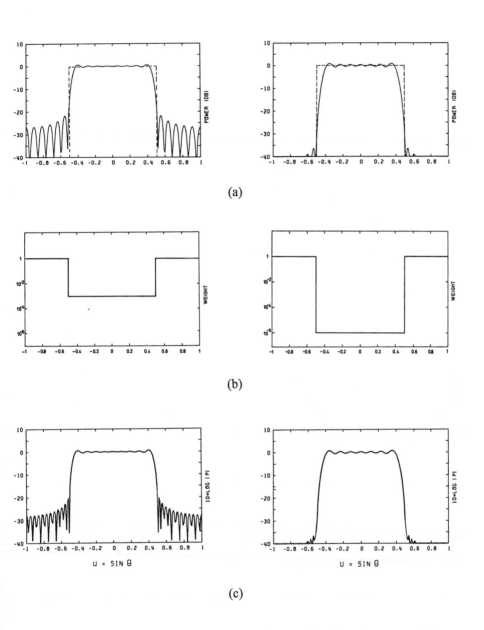

Figure 3.12 Gaussian power pattern synthesis (after [31]). Curves at left use 30-dB weighting function. Curves at right use 60-dB weighting function. Upper curves (a) show desired pattern (dashed) and optimum realizable power pattern (solid). Center curves (b) show weighting functions. Lower curves (c) show best harmonic approximation (magnitude only).

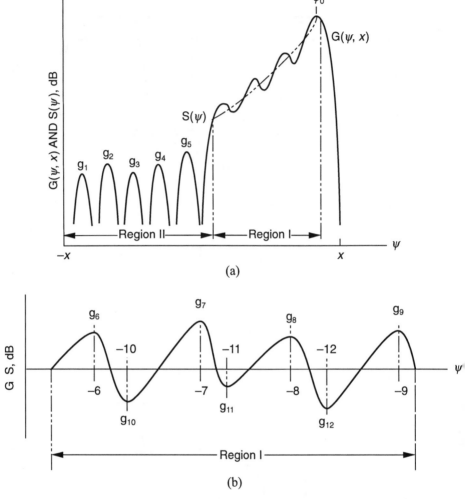

Figure 3.13 Synthesis of shaped power pattern (after [33]): (a) complete pattern showing shaped region, sidelobe region, desired contour (dashed), and sidelobe topography; (b) ripple peaks and troughs in region I relative to the desired contour.

where a_n and b_n are both real. This facilitates visualization on the Schelkunov unit circle.

The power pattern is given by

$$|F|^2 = |I_N|^2 \prod_{n=1}^{N} [1 - 2e^{a_n} \cos(\Phi - b_n) + e^{2a_n}] \tag{3.89}$$

As a convention, Orchard sets the Nth root at $w_N = 1$ (so $a_N = 0$ and $b_N = \pi$). The power pattern expressed in decibels is

$$G = \sum_{n=1}^{N-1} 10 \log_{10}[1 - 2e^{a_n} \cos(\Phi - b_n) + e^{2a_n}]$$
$$+ 10 \log_{10}[2(1 + \cos \Phi)] + C_1 \tag{3.90}$$

The added constant C_1 allows the value of G at the main beam peak to be set at a given value, typically 0 dB.

For the purpose of ordering the pattern zeros, and with no loss in generality, the shaped beam edge (region I) is arranged to end at $\Phi = \pi$ (as in the figure). Of the total $N - 1$ roots, choose N_1 roots to lie in region I and N_2 roots in region II. In the shaped beam region (I), the N_1 roots are arranged to be outside of the unit circle so that $a_n > 0$ and both a_n and b_n are adjustable, while in region II the N_2 zeros are constrained to lie on the unit circle so that the b_n are adjustable while the a_n are zero. There are thus a total of $2N_1 + N_2 + 1 = N_3$ constants that need to be evaluated, and these are grouped together in a column vector $\mathbf{x}$. The vector $\mathbf{x}$ consists of N_1 values of a_n and b_n in region I, N_2 values of b_2 in region II, and the constant C_1.

The pattern G is a function of the angular parameter Φ and the vector $\mathbf{x}$. The performance of G must be specified by means of the desired values of G at the N_2 maxima of G in region II and the $N_1 + 1$ maxima and N_1 minima of $(G - S)$ in region I (see Figure 3.13(b)). These N_3 known values, which are denoted by g_i $(i = 1, 2, 3, \ldots, N_3)$, are grouped as the components of the column vector $\mathbf{g}$.

The solution proceeds using a matrix form of the Taylor series as an iterative scheme to arrive at an estimate of the a_n and b_n coefficients.

$$A\Delta\mathbf{x} = \mathbf{g} - \hat{\mathbf{g}} \tag{3.91}$$

where $\mathbf{g}$ is the column vector of the desired values (sidelobe peaks) of G in region II and of the ripple peaks and troughs of $(G - S)$ in region I. The $\hat{\mathbf{g}}$ is the present approximation to $\mathbf{g}$, and so is known once the procedure has started. The matrix A is the matrix of derivatives, so the coefficients a_{ij} are

$$a_{ij} = \frac{dG(\Phi_i, \mathbf{x})}{dx_j} \quad (i, j = 1, 2, 3, \ldots, N_3) \tag{3.92}$$

Once solved, the updated vector $\mathbf{x} + \Delta\mathbf{x}$ can be taken as a better approximation vector.

The required derivative of G is readily obtained from (3.90) as

$$\frac{dG}{da_n} = Me^{a_n} \frac{[e^{a_n} - \cos(\Phi - b_n)]}{D_n} \tag{3.93}$$

and

$$\frac{dG}{db_n} = -Me^{a_n} \frac{\sin(\Phi - b_n)}{D_n}$$

$$= \frac{-M \sin(\Phi - b_n)}{2(1 - \cos(\Phi - b_n))} \quad \text{if} \quad a_n = 0 \tag{3.94}$$

where

$$M = 20/\ln 10 = 8.686 \tag{3.95}$$

and

$$D_n = 1 - 2e^{a_n} \cos(\Phi - b_n) + e^{2a_n} \tag{3.96}$$

In addition, since this procedure does not require that the main beam peak be scaled to 0 dB, an extra constant C_2 is added to the $G(\Phi, x)$ and before every new iteration C_2 is decreased by the value $G(\Phi_0, x)$ at the beam peak. This scaling assures that each new iteration begins with the peak at 0 dB.

The procedure begins by selecting values of a_n and b_n to lead to the proper number of maxima and minima for $(G - S)$ in region I, and G in region II. A good first choice is

$$b_n = [2n/(N + 1) - 1]\pi \quad n = 1, 2, 3, \ldots, N - 1 \tag{3.97}$$

which makes N_2 roots lie within region II and on the unit circle, so in region II,

$$a_n = 0 \quad \text{for } n = 1, 2, 3, \ldots, N_2 \tag{3.98}$$

The next N_1 roots are required to be in region I and are chosen slightly outside the unit circle. It is usually sufficient to choose

$$a_n = 0.01 \quad n = N_2 + 1, N_2 + 2, \ldots, N - 1 \tag{3.99}$$

The initial value of C_1 can be taken as zero.

It is now necessary to find the values of Φ_i that are the locations of the maxima or minima. These are obtained numerically by using the Newton-Raphson technique to find the location of each zero of the derivative of g in region II and of the derivative of $(G - S)$ in region I. This requires both first and second derivatives of G and S with respect to the angular variable Φ. The derivative of G is readily obtained from (3.90):

$$\frac{dG}{d\Phi} = M \sum_{n=1}^{N-1} e^{a_n} \frac{\sin(\Phi - b_n)}{D_n} - M \frac{\sin \Phi}{2(1 + \cos \Phi)} \tag{3.100}$$

$$\frac{d^2G}{d\Phi^2} = M \sum_{n=1}^{N-1} e^{a_n} \frac{[(1 + e^{2a_n}) \cos(\Phi - b_n) - 2e^{a_n}]}{D_n^2} - M \frac{1}{2(1 + \cos \Phi)} \tag{3.101}$$

Derivatives of S depend on the specific mathematical description chosen for S. Orchard et al. choose a polynomial form for S and so derive useful expressions for the derivatives in the reference. These are not repeated here, since use of the polynomial approximation is not fundamental to the method, but is a useful mathematical convenience. It is assumed here that the function S is well defined within the region I.

Figure 3.14 shows four patterns that demonstrate control of ripple level to various degrees which approximate a $(\text{cosec } \theta)(\cos \theta)$ pattern over a region. The first four sidelobes on one side of the main beam are at -30 dB, while all other sidelobes are set to -20 dB. The four patterns demonstrate that it is possible to restrict the ripple level amplitude within a controlled region. A detailed analysis of the patterns also reveals, however, that requiring the extremely tight ripple level of Figure 3.14(d) (± 0.1 dB) within the controlled region produced two undesirable results in comparison with a less restricted (± 1.5 dB) approximation of the desired pattern. One is that the main beam width is reduced from 41 deg for the ± 1.5-dB ripple to 34 deg for the lesser ripple. The second disadvantage of the result of Figure 3.14(d) is that the relative amplitude of the currents is only 4.34 dB for the ± 1.5-dB case, but is about 9.3 dB for the ± 0.1-dB case.

These disadvantages can be minimized by placing more pattern zeros in the shaped region I at the cost of sidelobe levels in region II.

Unlike the method developed by Steyskal, the procedure outlined above does not produce a mathematically optimum result. It does, however, introduce substantial flexibility and individual control of each ripple level or sidelobe level, and, most importantly, controls the entire radiation pattern.

3.2 CIRCULAR PLANAR ARRAYS

3.2.1 Taylor Circular Array Synthesis

A technique analogous to the Taylor line source method was also developed by Taylor [34] in 1960. The synthesized pattern is derived as a modification of the

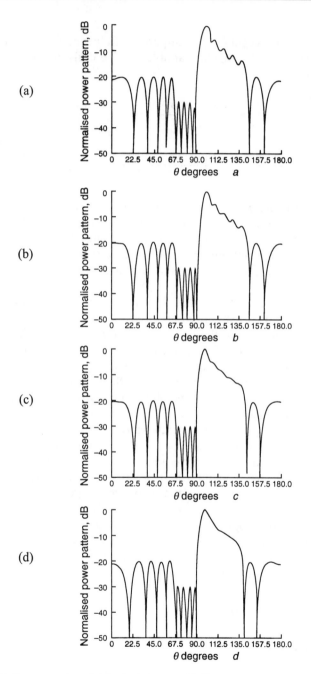

Figure 3.14 Representation of shaped beam with cosec² θ cos θ pattern with a 16-element array; $d =$ $\lambda/2$. Ripple requirement: (a) ± 1.5 dB; (b) ± 1.0 dB; (c) ± 0.5 dB; (d) ± 0.1 dB (after [33]).

pattern of a uniformly illuminated circular aperture, which is $J_1(\pi z)/(\pi z)$. Taylor's expansion removes the first zeros (to $\bar{n} - 1$) and substitutes the new zeros. This synthesized pattern is

$$F(z, A, \bar{n}) = 2 \frac{J_1(\pi z)}{\pi z} \prod_{n=1}^{\bar{n}-1} \frac{1 - z^2/z_n^2}{1 - z^2/\mu_n^2} \qquad (3.102)$$

for $z = (D/\lambda)\sin \theta$, and $J_1(w)$ is the Bessel function of order one.
The zeros of the function $F(z, A, \bar{n})$ are given by

$$z_n = \pm\sigma[A^2 + (n - 1/2)^2]^{1/2} \qquad (3.103)$$

for $1 \le n \le \bar{n}$, the zeros μ_n are the natural zeros of the $J_1(\pi z)$ function and are listed in Table 3.7. The parameter σ is defined as

$$\sigma = \frac{\mu_{\bar{n}}}{[A^2 + (\bar{n} - 1/2)^2]^{1/2}} \qquad (3.104)$$

The parameter A is defined as in the Taylor line source method:

$$A = \frac{1}{\pi} \cosh^{-1}(r) \qquad (3.105)$$

for voltage sidelobe level r, as in the earlier sections.

Table 3.7
Zero Locations μ_m for $J_0(\pi\mu_m)$

m	μ_m	m	μ_m
1	1.2196699	11	11.2466228
2	2.2331306	12	12.2468985
3	3.2383155	13	13.2471325
4	4.2410629	14	14.2473337
5	5.2439216	15	15.2475086
6	6.2439216	16	16.2476619
7	7.2447598	17	17.2477974
8	8.2453948	18	18.2479181
9	9.2458927	19	19.2480262
10	10.2462933	20	20.2481237

The beamwidth is given as with the Taylor line source method as

$$\theta_3 = \sigma\beta_0\lambda/D \tag{3.106}$$

with β_0 defined in (3.30).

The aperture distribution is given by

$$g(x) = \frac{2}{\pi^2} \sum_{m=0}^{\bar{n}-1} \frac{F_m J_0(x\mu_m)}{[J_0(\pi\mu_m)]^2} \tag{3.107}$$

for $x \leq 2\pi\rho/D$, with ρ the radial measure within the circular aperture. Here J_0 is the Bessel function of order zero.

It is necessary to compute $\bar{n} - 1$ values of the coefficients F_m given by

$$F_0 = 1 \qquad F_m = -J_0(\pi\mu_m) \frac{\displaystyle\prod_{n=1}^{\bar{n}-1} 1 - \mu_m^2/z_n^2}{\displaystyle\prod_{\substack{n=1 \\ n\neq m}}^{\bar{n}-1} 1 - \mu_m^2/\mu_n^2} \tag{3.108}$$

The required zero locations of the $J_0(\pi\mu_m)$ function are given in Table 3.7. Once again, the useful values of $\bar{n}$ are limited to maintain balance between efficiency, sidelobe levels, and realizability of the amplitude distribution.

The aperture efficiency for the circular Taylor pattern is

$$\epsilon_a = \frac{1}{1 + \displaystyle\sum_{n=1}^{\bar{n}-1} \frac{F_n^2}{J_0^2(\pi\mu_n)}} \tag{3.109}$$

Ruddick [35] presents a table showing aperture efficiency for various sidelobe levels and $\bar{n}$ values.

3.2.2 Bayliss Difference Patterns for Circular Arrays

In his classic paper [7], E. T. Bayliss presented the development of a two-parameter difference pattern for circular aperture antennas. The pattern is expressed in a Fourier-Bessel series of $\bar{n}$ terms similar to Taylor's treatment of the sum pattern. The development of the line source pattern (presented here in Section 3.1.7) is given in the appendix to the Bayliss paper. The synthesized patterns are again described in terms of the two parameters A and $\bar{n}$, which control the sidelobe level and decay behavior.

The synthesized pattern is given by:

$$F(z, \phi) = C(\cos \phi)2\pi z J_1'(\pi z) \frac{\prod\limits_{n=1}^{\bar{n}-1} [1 - (z/\sigma z_n)^2]}{\prod\limits_{n=0}^{\bar{n}-1} [1 - (z/\mu_n)^2]} \qquad (3.110)$$

for

$$z = \frac{2a}{\lambda} \sin \theta \qquad \sigma = \frac{\mu_{\bar{n}}}{z_{\bar{n}}} = \frac{\mu_{\bar{n}}}{[A^2 + \bar{n}^2]^{1/2}} \qquad (3.111)$$

The μ_n are the zeros of the Bessel function derivatives $J_1'(\pi \mu_m) = 0$. The first twenty roots are given in the Bayliss paper. Table 3.8 lists these zero locations. The zero locations of the synthesized function are at σz_n, with σ defined above, and z_n is defined as in the line source description (3.47).

The circular aperture excitation is given by [8]:

$$g(p, \phi) = \cos \phi \sum_{m=0}^{\bar{n}-1} B_m J_1(\mu_m, p) \qquad p < \pi \qquad (3.112)$$

for $p = \pi r/a$ and r is the radial variable.

After evaluating an indeterminate form, the coefficients B_m are given as [8, p. 635]:

$$B_m = \frac{-2jC\mu_m^2}{J_1(\mu_m \pi)} \frac{\prod\limits_{n=1}^{\bar{n}-1} [1 - (\mu_m/\sigma z_n)^2]}{\prod\limits_{\substack{L=0 \\ L \neq m}}^{\bar{n}-1} [1 - (\mu_m/\mu_L)^2]} \qquad m = 0, 1, \ldots, \bar{n} - 1$$

$$= 0 \qquad m \geq \bar{n} \qquad (3.113)$$

As described in the earlier sections of this chapter, the selected value of $\bar{n}$ has a primary effect on the aperture efficiency and the level of specific sidelobes, although the maximum sidelobe level is primarily determined by the parameter A.

Table 3.8
Zero Locations of $\pi\mu_m$ for Bessel Function Derivatives

m	μ_m	m	μ_m
0	0.5860670	10	10.7417435
1	1.6970509	11	11.7424475
2	2.7171939	12	12.7430408
3	3.7261370	13	13.7435477
4	4.7312271	14	14.7439856
5	5.7345205	15	15.7443679
6	6.7368281	16	16.7447044
7	7.7385356	17	17.7450030
8	8.7398505	18	18.7452697
9	9.7408945	19	19.7455093

Bayliss gives a relative directivity expression ϵ, defined relative to the maximum directivity of a circular aperture (uniform illumination).

$$\epsilon = \frac{8}{\pi^4}\left\{\sum_{L=0}^{\bar{n}-1}|B_L|^2 J_1(\mu_L\pi)[1 - (\mu_L\pi)^{-2}]\right\}^{-1} \tag{3.114}$$

Figure 3.15 shows the relative directivity in decibels (10 log ϵ) as a function of sidelobe level. The selection of larger values of $\bar{n}$ does not lead to increased efficiency for any given sidelobe level, but it is clear that to achieve progressively lower sidelobe levels, one must increase the selected values of $\bar{n}$ in order to maintain good efficiency. The maximum relative directivity that can be achieved in any difference pattern is -2.47 dB [36].

3.3 METHODS OF PATTERN OPTIMIZATION/ADAPTIVE ARRAYS

3.3.1 Pattern Optimization

Patterns synthesized by the various procedures described above are commonly used with passive feed networks because of their overall good qualities. However, it is sometimes desirable to select optimized distributions subject to special circumstances which might include external interference or receiver noise. The mathematics of pattern optimization were developed beginning in the 1960s. Among the earlier papers in this area is one by C. T. Tai [37] on gain optimization of linear arrays. The results of this work, noted in Chapter 2, showed that the uniformly illuminated array has the highest gain except in the "superdirective" limit. A formal

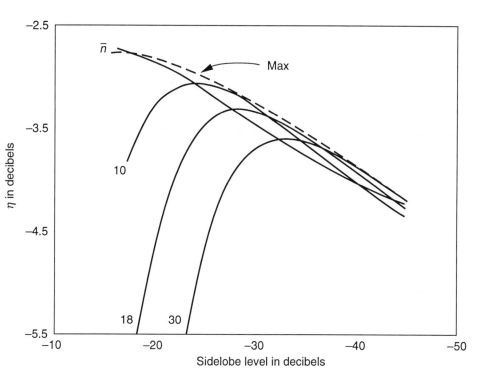

Figure 3.15 Relative directivity of Bayliss patterns (after [7]).

treatment leading to the same result is given by Uzkov [38]. The optimization of various array power measures like directivity, gain, efficiency, or signal to noise ratio is obtained by casting these parameters into an expression that is the ratio of Hermitian quadratic forms. The theorem for gain optimization is stated concisely in a paper by Cheng and Tseng [39]. Their method of presentation is followed here.

The generalized gain of an antenna array with signal amplitudes a_n at each nth array element is given by the ratio

$$G = \frac{\sum\limits_{n=1}\sum\limits_{m=1} a_m^* \alpha_{mn} a_n}{\sum\limits_{n=1}\sum\limits_{m=1} a_m^* \beta_{mn} a_n} \tag{3.115}$$

where the numerator represents the power density radiated at a point in space, and the denominator represents the input power. In general,

$$\alpha_{mn} = \exp\{jku_0(\hat{\mathbf{r}} \cdot [\mathbf{r}'_\mathbf{m} - \mathbf{r}'_\mathbf{n}])\} \tag{3.116}$$

where the scalar product

$$\hat{\mathbf{r}} \cdot \mathbf{r}'_\mathbf{m} = \hat{x}u(x - x'_m) + \hat{y}v(y - y'_m) + \hat{z}(z - z'_m)\cos\theta \tag{3.117}$$

and

$$\beta_{mn} = \frac{1}{4\pi}\int_0^{2\pi} d\phi \int_0^{\pi} d\theta \, \exp\{jku(\mathbf{r} \cdot [\mathbf{r}'_\mathbf{m} - \mathbf{r}'_\mathbf{n}])\}g(\theta, \phi)$$

Here, $g(\theta, \phi)$ is the element pattern, which is normalized to unity at peak $u = u_0$.

In matrix form, G can be written in terms of a matrix vector (column matrix **a**) as

$$G = \frac{\mathbf{a}^\dagger \mathbf{A} \mathbf{a}}{\mathbf{a}^\dagger \mathbf{B} \mathbf{a}} \tag{3.118}$$

where $\mathbf{a} = [a_1 \, a_2 \, a_3 \, . . .]^T$, $\mathbf{a}^\dagger$ is the conjugate transpose of **a**, and the matrices **A** and **B**,

$$\mathbf{A} = [\alpha_{mn}] \qquad \mathbf{B} = [\beta_{mn}]$$

are both Hermitian $N \times N$ square matrices. (The dagger symbol means conjugate transpose.) In addition, **B** is positive definite, and so the roots of the characteristic equation (eigenvalues of the "regular pencil") are defined $(\mathbf{A} - \lambda\mathbf{B})$:

$$\det(\mathbf{A} - \lambda\mathbf{B}) = 0 \tag{3.119}$$

where $\lambda_1 \geq \lambda_2 \geq \cdots \lambda_N$, are real.

With the eigenvalues ordered as shown above, λ_1 and λ_N represent the upper and lower bounds of the value of G, with the upper and lower bounds of gain determined from the equations

$$\mathbf{A}\mathbf{a} = \lambda_1 \mathbf{B}\mathbf{a} \qquad \mathbf{A}\mathbf{a} = \lambda_N \mathbf{B}\mathbf{a} \tag{3.120}$$

This theorem is used for maximizing gain, directivity, or other array parameters and requires only the evaluation of the maximum eigenvalues and the associated eigenvector, which becomes the complex excitation vector **a**.

Cheng and Tseng further show that for a linear array with uniformly spaced elements and using the coordinate system of Figure 2.9 (Chapter 2) to facilitate gain computation, the matrices **A** and **B** have terms

$$\alpha_{mn} = e^{+jku(m-n)d_x} \tag{3.121}$$

and

$$\beta_{mn} = \frac{1}{4\pi} \int_0^{2\pi} d\phi \int_{-1}^1 du \, g(\theta, \phi) e^{+jku(m-n)d_x} \tag{3.122}$$

where here u is defined as $u = \cos\theta$.

The matrix **A** can be written as an outer product using the column vector notation

$$\mathbf{e} = [e_1 \quad e_2 \quad \cdots \quad e_N]^T \tag{3.123}$$

where

$$e_m = e^{jkmd_x u_0} \tag{3.124}$$

then

$$\mathbf{A} = \mathbf{ee}^\dagger \tag{3.125}$$

All of the roots of the characteristic equation (3.120) are zero except

$$\lambda_1 = G_{\max} = \mathbf{e}^\dagger \mathbf{B}^{-1} \mathbf{e} > 0 \tag{3.126}$$

and the optimum array excitation is

$$\mathbf{a} = \mathbf{B}^{-1} \mathbf{e} \tag{3.127}$$

The method applies to arbitrarily oriented arrays and can include nonisotropic element patterns. It is also applicable to arrays in which the total power is evaluated in terms of circuit parameters using array element impedances. This formulation is presented in a paper by Harrington [40].

The method has been applied to produce gain optimization in the presence of random errors in the design parameters [41], and by Lo et al. [42] to optimize directivity and signal-to-noise ratio.

McIlvenna and Drane [43,44] have used similar matrix methods to achieve maximization of gain while constraining the antenna pattern to have specific null locations. These techniques were later extended to include the use of measured array element scattering matrices [45]. Other summaries of developments in this area are presented in [21] and [46].

3.3.2 Adaptive Arrays

Often pattern optimization is done by real-time active weighting of the received signal and can adapt to changes in the outside environment. Although in principle it is possible to adapt transmit patterns to optimize the transmission subject to some received signal or noise distribution, this is seldom done except for the formation of so-called *retrodirective beams*, which automatically transmit in the direction of a received signal or pilot tone.

Adaptive array theory has undergone extensive development, and only the most skeletal descriptions of the theory are included here in order to facilitate the calculation of radiation patterns that would result from adaptation to steady-state interference. The mathematics of pattern optimization is based on matrix theory and specifically on the optimization of quadratic forms [45,46,47]. An excellent treatment of optimization is given by Harrington [48], while detailed descriptions of the real-time response of adaptive arrays in a transient environment are given in the texts [49,50], as well as in many journal publications and tutorial papers [51–53].

Among the algorithms chosen for adaptive optimization, the most commonly selected are derived from the so-called *Howells-Applebaum* [54] method and a procedure due to Widrow et al. [55,56] that minimizes the least mean square (LMS) difference between the array output signal and some known reference signal. The technique of Howells and Applebaum performs signal-to-noise optimization subject to the constraint of a specified *quiescent array pattern* formed by the array weighting network in the absence of interference. In principle, both types of optimization could be used for either radar or communication, but in practice the Howells-Applebaum algorithm is often used for radar systems because the direction of the desired return signal is known, while the LMS algorithm can be used for communication systems, where the direction of the desired incident signal may not be known. In this case, the reference signal can be some replica of the format of the received signal, a pilot tone or code sequence. The LMS algorithm is also used in radars, with the transmit waveform as a reference.

The weighting of the received signals can either be done using analog circuits or by digital operations on the output signals, and by closed-loop feedback methods or open-loop procedures that seek to optimize the returns based on measurements of the signal environment.

Figure 3.16 shows several adaptive array configurations and serves to point out the distinction between *fully* and *partially* adaptive arrays. Fully adaptive arrays, whether organized with element level controls, as in Figure 3.16(a), or as multiple beam arrays with a network to excite the orthogonal sin(*x*)/*x* "Woodward" beams, as in Figure 3.16(b), have every element port controlled adaptively; $N_T - 1$ available degrees of freedom are used for pattern control, and the remaining degree of freedom points the main beam.

In order to reduce the cost of adaptive systems, one can use partially adaptive arrays, defined as those in which only some of the elements are controlled adaptively, as shown in Figures 3.16(c,d). Such arrays include the so-called *sidelobe cancelers*, which use one or several elements weighted to cancel interference at the level of the array sidelobes. In practice, the elements used for cancellation can either be within (see Figure 3.16(c)) or outside (see Figure 3.16(d)) of the array, and can be grouped into subarrays (rows, columns, areas) or randomly oriented elements. In each case, the analysis is unchanged, but the practical results of such choices are of major importance in determining antenna performance. Often, the use of individual elements for adaptivity is called *element space adaptation*, but if a large number of elements are grouped together passively and then used adaptively for cancellation, then this is called *subarray level adaptation*. In any case, an array with N_T elements, and fewer than $N_T - 1$ available for the adaptive process, is a partially adaptive array and usually suffers some limitations that are the price to be paid for the desired simplifications.

3.3.3 Generalized S/N Optimization for Sidelobe Cancelers, Phased and Multiple Beam Arrays

Assume that the *n*th port of the receiving system receives a signal,

$$E_n = e_n(u_j, v_j) \tag{3.128}$$

corresponding to an incident or interfering signal of unity amplitude at an angle given by the direction cosines u_j and v_j. The angular function e_n contains the amplitude and phase of the signal received by the *n*th port. The port can either be an element port, as in Figures 3.16(a,b), or a multiple beam port, as in Figure 3.16(d).

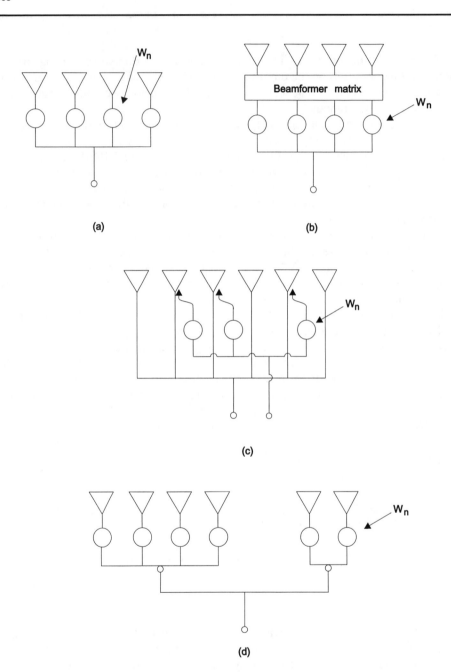

Figure 3.16 Adaptive array configurations: (a) fully adaptive array (element weighting); (b) fully adaptive array with multiple-beam feed (beam space adaptivity); (c) partially adaptive array; (d) array with multiple canceler elements.

For a fully adaptive array (Figure 3.16(a)), in which the output ports are the N_T element ports, or a linear array used with canceler elements (Figure 3.16(d)), the e_n include the complex element patterns f_n sampled at the interfering angles. For a linear row array of elements, as in Figure 3.16(a),

$$e_n = f_n(u_j, v_j)z_j^n \tag{3.129}$$

where

$$z_j = \exp(jk_0 d_x u_j/\lambda) \tag{3.130}$$

In the more general situation of a two-dimensional array or an array with arbitrary element locations,

$$e_n = f_n(u_j, v_j) \exp(jk_0 \mathbf{r_n} \cdot \hat{\mathbf{r}}_{0j}) \tag{3.131}$$

using the notation of Chapter 1.

For a multiple-beam matrix forming specific beams, the e_n are the complex received patterns at the nth beam ports. As a simple example, for a linear array forming the orthogonal beams (as will be described in Chapter 8), this term is (for assumed isotropic element patterns):

$$e_n(u) = F_n(u) \tag{3.132}$$

where

$$F_n(u) = \frac{\sin[N_T \pi(u - u_n)d_x/\lambda]}{N_T \sin[\pi(u - u_n)d_x/\lambda]}$$

In either case of multiple-beam or element-level adaptation, upon receiving an interfering signal from a source S_j at angle (u_j, v_j), the adaptive network applies a set of weights w_n to obtain the resulting received interfering signal.

$$E = \sum w_n e_n(u_j, v_j) = \mathbf{w}^T \mathbf{e} \tag{3.133}$$

Adaptive Weights—Howells-Applebaum Method

For an array of N elements with adaptive control at the element level, the beam former selects the set of weights w_n to receive some desired signal at angle θ_0. For the linear row array, if the interfering source were not present, the set would be

$$w_n^0 = \left| w_n^0 \right| \exp(-jknd_x \sin \theta_0) = \left| w_n^0 \right| z_0^n \tag{3.134}$$

or, in vector form,

$$\mathbf{W_0} = (w_1^0, w_2^0, w_3^0, w_4^0, \ldots, w_N^0)^T \tag{3.135}$$

where the superscript T indicates transpose, so that $\mathbf{W_0}$ is a column vector.

The amplitude of the weights $\left| w_n^0 \right|$ are chosen to produce some desired quiescent pattern. This excitation (3.135) is referred to as the *quiescent steering vector*. The quiescent steering vector for a multiple beam array depends on whether the beam ports are used separately or combined to form shaped patterns as in the Woodward-Lawson synthesis procedure. For example, the quiescent steering vector

$$\mathbf{W_0} = (0, 0, 0, \ldots, 1, 0, 0, \ldots)^T \tag{3.136}$$

excites a single beam in the direction of the one beam switched on.

Neglecting receiver noise, and if the interference and desired signals are monochromatic, then for a single interfering signal, the ratio of signal to interference is maximized by choosing array weights to move one of the pattern zeros to the interfering angle θ_j. This argument holds when there are a number of interfering sources, and so there has been much work on antenna pattern synthesis and optimization subject to the constraints of setting pattern nulls at arbitrary locations. An N-element array can have up to $(N - 1)$ nulls, and in principle can cancel up to $(N - 1)$ interfering signals. In practice, one cannot place too many of the nulls close together without incurring severe pattern distortion.

In the more general case, there may be a number of wideband interfering sources and a noisy receiver in each channel. The treatment of wideband signal response is beyond the scope of this text, and so for the purposes of illustration, it has been assumed that the signals are all represented as narrowband modulation about some carrier at frequency f. It will be assumed that all interfering signals are uncorrelated with each other and with the desired signal and the channel noise. In terms of a total signal (noise plus interference plus desired signal) e_n at the nth channel of the receive array, the weighted and combined signal at the output port is given by

$$E = \sum w_n e_n \tag{3.137}$$

or, in vector form,

$$E = \mathbf{W^T e} \tag{3.138}$$

for column vector $\mathbf{e} = (e_1, e_2, \ldots, e_N)^T$, and the sum is taken over all N output ports. Note that in this form e_n may include signals from input ports of an array with arbitrary element locations or be the output of a multiple-beam matrix.

The average power received at the combined output port is $\overline{E^*E}$, where the overbar indicates a time average over the correlation interval. It is this average power that determines the signal-to-noise-plus-interference ratio for the system. Excluding the signal power, the average noise-plus-interference power is given by

$$\overline{E^*E} = \overline{\{\mathbf{W^Te}\}^*\{\mathbf{W^Te}\}} = \Sigma_m\Sigma_n \, \overline{w_m^*e_m^*w_ne_n} \tag{3.139}$$
$$= \mathbf{W^\dagger MW}$$

where the matrix M is the noise *covariance matrix* and is given by the outer product

$$\mathbf{M} = \overline{\mathbf{e^*e^T}} = \begin{bmatrix} \overline{e_1^*e_1} & \overline{e_1^*e_2} & \cdots & \overline{e_1^*e_N} \\ \overline{e_2^*e_1} & \overline{e_2^*e_2} & \cdots & \overline{e_2^*e_N} \\ \vdots & \vdots & & \vdots \\ \overline{e_N^*e_1} & \overline{e_N^*e_2} & \cdots & \overline{e_N^*e_N} \end{bmatrix} \tag{3.140}$$

where the e_n terms include only noise and interference, with the desired signal excluded. Once again, the symbol † means conjugate transpose. Note that some texts use the matrix form $\mathbf{ee^\dagger}$ instead of the above $\mathbf{e^*e^T}$. In that case, the solution vector is $\mathbf{W^*}$ instead of the $\mathbf{W}$ obtained in the following results.

The received power of the desired signal after passing through the same weighting network is similarly given by

$$\overline{E_s^*E_s} = \mathbf{W^\dagger M_s W} \tag{3.141}$$

where $\mathbf{M_s}$ is the signal covariance matrix and has the same form as the noise covariance matrix above, but only includes the signal terms and is evaluated at the beam peak.

The above expressions $\overline{E^*E}$ and $\overline{E_s^*E_s}$ are quadratic forms, and the ratio of signal to noise plus interference is the ratio of these two quadratic forms.

$$S/N = \frac{\mathbf{W^\dagger M_s W}}{\mathbf{W^\dagger MW}} \tag{3.142}$$

The procedure for maximizing this ratio is well known and was outlined in the previous discussion of gain optimization. Subject to these conditions, the optimum weight vector $\mathbf{W}$ is given as

$$\mathbf{W} = \mathbf{M}^{-1}\mathbf{W_0} \tag{3.143}$$

For a linear array of N_T elements receiving uncorrelated noise, narrow-band interference, and a monochromatic desired receive signal, the total undesired signal at the nth port is made up of the sum of noise (n_n) and interference signals as

$$e_n = n_n + \Sigma_j A_j \exp[j2\pi(d_x/\lambda)u_j n] \tag{3.144}$$

The covariance matrix is made up of terms

$$M_{nm} = \overline{e_n^* e_m} = N_n\delta(n, m) + \Sigma_j P_j \exp[j2\pi(dx/\lambda)u_j(m - n)] \tag{3.145}$$

where $N_n = \overline{n_n^* n_n}$ and $P_j = |A_j|^2$, and $\delta(N, m) = 1$ for $n = m$ and zero otherwise.

The expression for a two-dimensional array with arbitrarily located elements is given by

$$M_{nm} = \overline{e_n^* e_m} = N_n\delta(n, m) + \Sigma_j P_j \exp[j2\pi\hat{\mathbf{r}}_j \cdot (\overline{\mathbf{r}}_m - \overline{\mathbf{r}}_n)/\lambda_j] \tag{3.146}$$

for $\mathbf{r_n}$, the position vector of the nth element in the two-dimensional array, and $\hat{\mathbf{r}}_j$, the unit vector denoting the interfering source of wavelength λ_j.

Adaptive optimization of a multiple-beam array is also controlled by (3.143), but in this case the terms of the noise covariance matrix are

$$M_{nm} = \overline{e_n^* e_m} = N_n\delta(n, m) + \Sigma_j P_j F_n^*(u_j)F_m(u_j) \tag{3.147}$$

It is assumed that the interfering signals are narrow-band and uncorrelated. The F_n could be of the form of the orthogonal beams of (3.132) or any more general form.

3.3.4 Operation as Sidelobe Canceler

If one considers the simplest case of a single sidelobe canceler, in which the covariance matrix has but four terms, then the inverse matrix is written

$$M^{-1} = \frac{1}{M_{11}M_{22} - M_{12}M_{21}} \begin{bmatrix} M_{22} & -M_{12} \\ -M_{21} & M_{11} \end{bmatrix} \tag{3.148}$$

since the quiescent weight vector is just the input to the main antenna

$$\mathbf{W_0} = [1, 0]^T \tag{3.149}$$

and the weights of an adaptively optimized two-element system are

$$\mathbf{W} = M^{-1}\mathbf{W_0} = \frac{1}{M_{11}M_{22} - M_{12}M_{21}} \begin{bmatrix} M_{22} \\ -M_{21} \end{bmatrix} \tag{3.150}$$

The analysis of a two-antenna canceler system applies to a two-element array (Figure 3.16(a)) or to an array with scanned beam and a single sidelobe canceler, or a two-beam multiple-beam system. In either of these cases, the analysis shows that the system weights are adjusted so that the signal at terminal 1 is

$$e_1 w_1 = \frac{e_1}{M_{11}M_{22} - M_{12}M_{21}} \overline{e_2^* e_2} \tag{3.151}$$

while that from the canceler output port 2 is

$$e_2 w_2 = \frac{-e_2}{M_{11}M_{22} - M_{12}M_{21}} \overline{e_2^* e_1} \tag{3.152}$$

For a monotonic interfering signal, these two signals are equal and opposite, and their sum cancels. The resulting pattern from the antenna plus canceler has a zero at the angular location of the interference. In some cases, the covariance matrix can become singular. This would occur in the above case if the uncorrelated noise were not included in the receiver channels. In the expressions above, the denominator would be zero except for the uncorrelated noise terms in the covariance matrix coefficients.

For *multiple sidelobe canceler*, locating elements at positions x_n (either within or outside of the array) and labeling the main antenna port #1, then the signal at any port (canceler or main antenna) is given as

$$e_n = n_n + \sum_J A_J F_n(J) \tag{3.153}$$

where $F_n(J) = \exp\{j2\pi(x_n/\lambda)u_J\}$ for $n \neq 1$, and where $F_n(J) =$ the main antenna pattern at $u = u_J$ for $n = 1$. With this substitution, the power in each port is given by the previous expression for the multiple-beam case.

The steering vector is simply the weighting of the antenna port alone (in the absence of interfering signals) as

$$\mathbf{W_0} = [1, 0, 0, 0, \ldots]^T \tag{3.154}$$

Although the mathematics of the canceler circuit is the same for any of the configurations, the pattern performance of this basic canceler circuit is very different for the several configurations. Figure 3.17(a) shows pattern nulling with a low-gain canceler element pattern. As the figure shows, the low-gain element with its broad pattern can cancel interference that enters at the level of the main antenna side-lobes. Since the canceler pattern is much broader than the main antenna pattern, sidelobe cancellation produces an effective total pattern with some sidelobe distortion, but with a null at the location of the interference. Low-gain sidelobe cancelers are not usually used to cancel interference entering through high sidelobes of the pattern main beam because of the extreme sidelobe distortion that results. For the case shown, in which an omnidirectional canceler is used to produce a null at an interfering signal entering the first sidelobe (-13 dB), the resulting pattern distortion is significant and the gain obviously lowered because of the resulting sidelobe at the -16-dB level. For this reason, sidelobe cancelers are primarily useful for cancellation in regions where sidelobes are small relative to the rest of the pattern structure to be left undistorted.

If several sources of interference are present, the single sidelobe canceler still optimizes S/N, but the system does not have sufficient degrees of freedom to satisfactorily complete the task. However, as long as the number of interfering signals is not too large, there is still the possibility of using a multiplicity of sidelobe antenna cancelers to produce closely spaced nulls for interference cancellation. Figure 3.17(b) shows an example of two cancelers used effectively to suppress sources of interference in a relatively low-sidelobe pattern region.

Effective pattern nulling with low distortion can be obtained using one of a set of multiple beams as a sidelobe canceler. In this case, the pattern distortion is reduced, as compared with the low-gain canceler example, because the beam pattern acting as canceler has high gain and a narrow beam, and so produces only localized distortion of the total pattern. The difficulty in using a single multiple-beam or any kind of narrow-beam antenna in the canceler mode is that one must choose the appropriate beam or scan the canceler to perform the cancellation.

Although the abbreviated treatment in this text does not consider wide-band cancellation of interfering signals, it is worth mentioning that another limitation of the use of cancelers, especially those located outside of the main antenna aperture, is that there may be substantial distances between the phase centers of the main antenna and the canceler. In this case, wide-band cancellation is only achieved using tapped delay lines to equalize the electrical line length between phase centers for all interference angles.

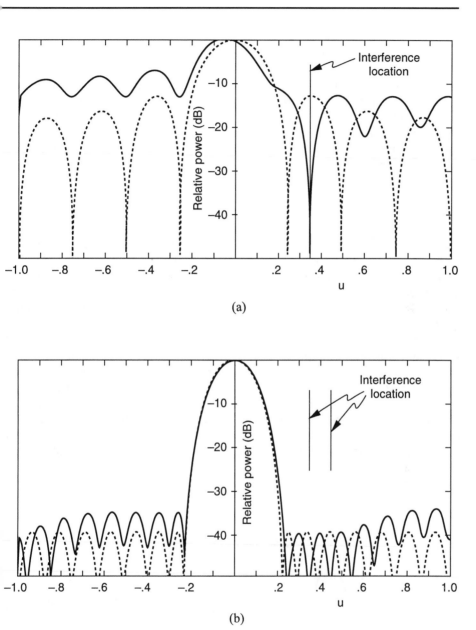

(a)

(b)

Figure 3.17 Pattern nulling with sidelobe cancelers: (a) single canceler at location 2λ from center of 8-element array (interference at peak of first sidelobe, initial pattern from uniform array (dashed)); (b) two cancelers for 16-element low-sidelobe array (cancelers at 5, 5.5λ from array center, -40-dB Chebyshev pattern dashed).

3.3.5 Fully Adaptive Phased or Multiple-Beam Arrays

Applebaum has shown that the cancellation process that takes place in a fully adaptive array with a single source of interference is equivalent to forming a uniformly illuminated canceler pattern with the full array and weighting that beam to exactly cancel the interfering signal. Figure 3.18 illustrates the quiescent pattern and a cancellation pattern chosen by the optimization process to suppress a single source of interference.

Phased Array

Figure 3.19 shows two patterns of a 16-element array with a quiescent steering vector chosen to form a 40-dB Chebyshev pattern. If interfering sources are located very near the natural nulls of the quiescent pattern, the resulting pattern, although not shown, is nearly unchanged. The solid curve of Figure 3.19(a) results from placing 10 interfering sources (see solid vertical lines) separated by about one-quarter beamwidth; a placement that results in a wide trough in the pattern. Still, there are many degrees of freedom not being used, and the pattern is not yet significantly distorted. Placing up to 15 sources of interference between one-eighth and one-quarter wavelength apart (at locations shown by the solid and the dashed vertical lines) produces very little change in the pattern if the interfering power is maintained at 10 times the noise (curve not shown). However, if the interfering sources are all 100 times as large as the noise, one additional null is moved into the region (see dashed curve) and the average level is reduced over the trough region. The pattern has not been adapted to place nulls at each of the interferers, but to optimize the signal to a noise-plus-jammer power ratio.

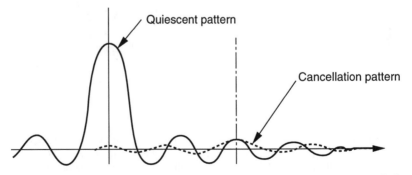

Figure 3.18 Quiescent pattern and canceler beam for single source of interference (after [54]).

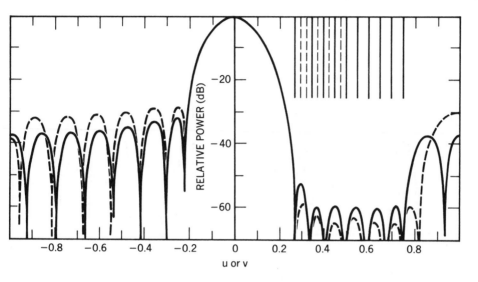

Figure 3.19 Adapted pattern of a 16-element array with quiescent − 40-dB Chebyshev pattern. Response to 10 (solid line) and 15 (dashed line) interfering sources (after [57]).

Increasing pattern distortion results when there are not enough degrees of freedom to place zeros very close together in the array factor. In principle, there can be $N - 1$ zeros for an N-element array, but one cannot move all the zeros to a small area of the pattern without radically changing the rest of the pattern. The limitation in available degrees of freedom places ultimate limits on the width and depth of nulled sectors and upon the bandwidth of adaptive pattern control.

Multiple-Beam Array

Adaptive cancellation with multiple-beam systems has advantages for certain applications, especially when the beams are used to cover only a restricted section of space (as in a satellite antenna). Most often this "beam space" nulling is used when the antenna configuration is a multiple-beam lens or reflector, but the beam forming could also be done digitally. Mayhan [58] has outlined the advantages of beam space nulling and has shown that cancellation with a multiple-beam antenna produces a lower sidelobe level outside of the angular extent of the multiple-beam set. Figure 3.20 shows two patterns of a 16-beam array with four interfering sources at $u = 0.3, 0.35, 0.40, 0.45$. The quiescent pattern is a single orthogonal beam with peak at $u_0 = 0.625$. The two patterns are for interfering power levels of approximately 100 (solid) and 1,000 (dashed) times the quiescent level. The higher

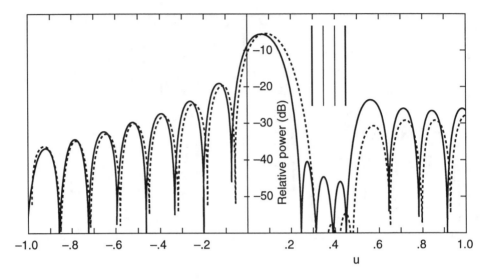

Figure 3.20 Adapted patterns of 8-element multiple-beam array. Quiescent beam at $u = 0.0625$ interference levels (relative to quiescent pattern) 99 (solid) and 1,000 (dashed).

level of interference drives the sidelobes lower throughout the trough region formed by the four sources and produces a slight lowering of sidelobes near the trough, but does not alter the pattern or main beam significantly.

3.3.6 Wide-Band Adaptive Control

Although this discussion has necessarily been limited to narrow-band signals and interfering sources, adaptive arrays provide wide-band cancellation. This is done naturally by using extra degrees of freedom or through the use of special techniques to process the various frequency components.

If the array has a sufficient number of degrees of freedom, the pattern of an array subject to wide-band interference will adapt by placing additional nulls in the vicinity of the interfering signal.

Figure 3.21 illustrates why this relates to broadband cancellation. If the adaptive array weights are fixed as a function of frequency, then the array pattern is unchanged with frequency, except for a compression in scale as frequency is increased. If the pattern of Figure 3.21 has a trough of width Δ_u centered at u_n

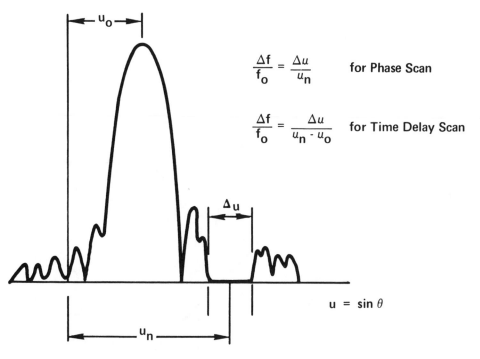

Figure 3.21(a) Broadband interference: relations between bandwidth and angular null width (after [59]).

then one can readily show that the bandwidth over which good suppression of the interference at the angle u_n is given as

$$\frac{\Delta f}{f_0} = \frac{\Delta u}{u_n} \tag{3.155}$$

If the array is time-delay steered to some angle u_0, then the bandwidth represented by the trough region is

$$\frac{\Delta f}{f_0} = \frac{\Delta u}{u_n - u_0} \tag{3.156}$$

because the pattern is stationary about the u_0 position.

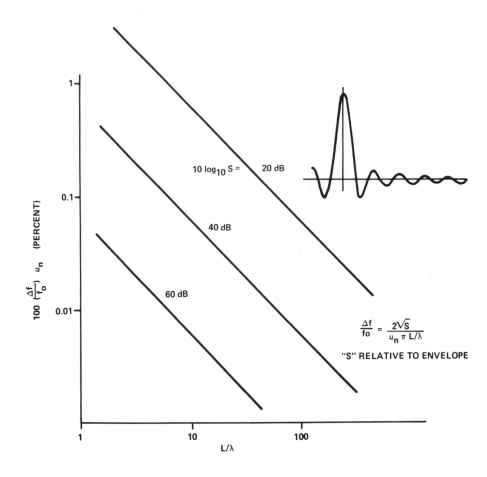

Figure 3.21(b) Broadband interference: bandwidth of single null (after [59]).

To produce such a trough in an antenna pattern requires one to relocate a number of the pattern nulls into the region of the trough. One can show, for example [59], that the bandwidth for a simple pattern null measured to some depth S below the local pattern amplitude is given by

$$\frac{\Delta f}{f_0} = \frac{2(S)^{1/2}}{(\pi L/\lambda_0)u_n} \qquad (3.157)$$

which is plotted in Figure 3.21(b). In this expression, L is the array length.

If we were to place many nulls together to form a trough, then clearly one must move the nulls close together to form a deep trough, and further apart if the trough need not be so deep. Hence, the number of required nulls must increase if either the required trough width (frequency bandwidth) is increased or if the required trough depth is increased. Stated another way, this is seen to lead to a very fundamental question: for a given array, what is the minimum number of degrees of freedom required to suppress pattern interference to some given level over a specified bandwidth? Steyskal [60] has investigated the relationship between the number of pattern nulls and the width and depth of a pattern trough. Figure 3.22 shows the cancellation relative to the local sidelobe level for several equivalent alternative abscissas, and parametrically with the number of equispaced pattern nulls M. The alternative choices of abscissa are the desired number of canceled sidelobes ν, the spatial angle normalized to the beamwidth $(\ell/\lambda)\Delta u$, and the normalized bandwidth $(\ell/\lambda)u_j\Delta f/f_0$, where $\Delta f/f_0$ and u_j denote the bandwidth and direction of the interference. Using this set of curves, one can estimate the number of

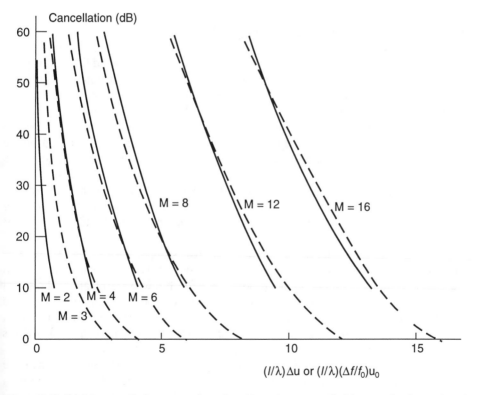

Figure 3.22 Sidelobe cancellation vs. number of equispaced pattern nulls M, normalized array length ℓ/λ, and nulling sector Δu (after [60,61]).

degrees of freedom (nulls) required to place a trough of given depth and width in the pattern of a linear array.

Steyskal's results were obtained through a numerical search using an optimum mean square approximation of the pattern. A simple formula that approximates Steyskal's results was obtained by Franchi [61] using a polynomial representation based on the Schelkunov method. Since Franchi's procedure is too complicated to repeat here, only the result is stated below. Given an original pattern with sidelobe level SLL_0 throughout some pattern region Δu which is M fundamental beamwidths (λ/L) wide, the new pattern is made to have a lower sidelobe region SLL_1 by forcing a larger number of nulls N into the same region. The relationship between the two sidelobe levels and the number of nulls N and M is given by Franchi [61] as

$$\frac{SLL_1}{SLL_0} = \left\{ (e/2)^{[1-(M/N)^2]} \frac{M}{N} \right\}^{2N} \tag{3.158}$$

This equation is plotted (dashed) in Figure 3.22.

Fully adaptive arrays are expensive in terms of hardware costs and processing costs, and even with partially adaptive arrays, the cost of adding additional degrees of freedom can be significant. Therefore, it is sometimes convenient to provide wide-band cancellation by using several other approaches. The first is to use a filter bank behind each array channel and then optimally process each band separately. If p such subbands are used, the processing is essentially the same as for p separate arrays. The optimized outputs of these virtual arrays are then combined to produce the wide-band output of the full array. This approach can be implemented in the analog version described, or digitally by first taking a Fourier transform of the array element outputs and then processing the spectral components separately. Frequency domain processing, whether analog or digital, has the effect of producing an array pattern that changes with frequency to keep essentially a single null at the source of interference.

An alternative to the frequency domain processing of a wide-band signal, but one that achieves similar results, is to use a programmable tapped delay line in each array channel. This alternative produces a pattern that is stationary at the signal frequency and the frequency of the interfering source. The required number of taps at each channel increases with the array size and with the instantaneous bandwidth. The delay-line function could also be implemented digitally. Mayhan et al. [62] presents a discussion of the tapped delay-line matching of adaptive cancelers.

The details of these and other adaptive optimization procedures constitute an exciting field of antenna research. Some of these are described in recent antenna texts [63,64], as well as in three new books [49,50,65] dedicated to this exciting new subject.

3.4 GENERALIZED PATTERNS USING COVARIANCE MATRIX INVERSION

The adaptive procedure provides a generalized method for the adaptive control of patterns. Although it was introduced in connection with pattern optimization in the presence of system noise and interfering signals, if the interfering signals are much larger than the system noise, then the method of covariance matrix inversion becomes a pattern nulling scheme. The technique can be used as a generalized root matching procedure and so is directly applicable for discretizing the continuous aperture distributions of the Taylor and Bayliss patterns or the more generalized procedures described earlier. This is done by fully constraining all the pattern zeros and so can lead to relatively large matrices to invert.

A procedure first suggested by Sureau and Keeping [66] involves the use of an interference spectrum for the synthesis of sum and difference patterns from cylindrical arrays. This innovative use of the very general adaptive array process allows the direct incorporation of array element patterns and arbitrary (conformal) element location in the synthesis process. Sureau and Keeping distributed inter-fering sources around the entire cylindrical array, except for a window arc that defined the width of the main beam. The number of interfering sources was chosen to be several times the number of array elements. Although there was no explicit way of directly synthesizing desired sidelobe structures, the authors found that by inversely tapering the "noise" interference amplitude, it was possible to produce a variety of radiated patterns, including some with nearly constant sidelobe levels over specified regions. Dufort [67] also investigated the use of adaptive methods for pattern control, but for periodic arrays of isotropic elements. By making the angular interference spectrum equal to the reciprocal of the desired pattern, he obtained improved pattern sidelobe control.

A recent paper by Olen and Compton [68] extended these numerical approaches and iteratively tailored the interference spectrum. In this procedure, an initial quiescent pattern is computed and the sidelobes compared with the desired sidelobe level. The number of interfering sources is again taken to be several times the number of elements in the array. Figure 3.23 shows a sample case of 10 elements with isotropic element patterns. The desired sidelobe level is -30 dB; but with the interference power set uniform for the initial (zeroeth) iteration, the optimum array pattern is that of a uniformly illuminated array, shown in Figure 3.23 (upper left). Figure 3.23 (upper right) shows the first iteration of the interference spectrum which results from choosing an interference-to-noise ratio spectrum that has no interference within the main beam region, and outside of the main beam has an interference level proportional to the difference between the desired and existing pattern level. Further iterations proceed in like manner, choosing the interference at each location to be varied with each iteration in proportion to the difference between calculated and desired sidelobe structure for the previous iteration. The

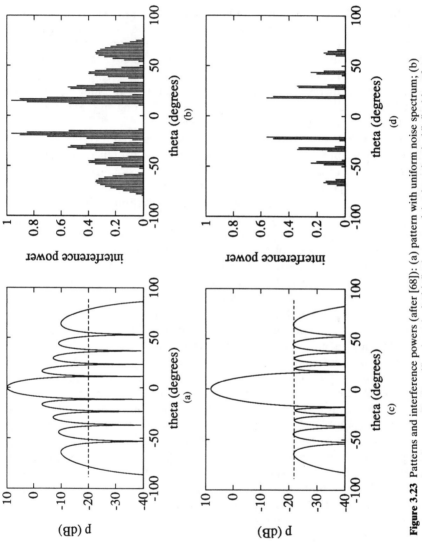

Figure 3.23 Patterns and interference powers (after [68]): (a) pattern with uniform noise spectrum; (b) interference spectrum (first iteration); (c) final pattern (nine iterations); (d) final interference spectrum (ten iterations). *Dashed line is desired −30-dB level.

cycle is repeated until the pattern is judged satisfactory. Figure 3.23 (lower left) shows the converged pattern after nine iterations, and 3.23 (lower right) shows the interference spectrum that would have been used at the next iteration, had it been necessary.

To illustrate the flexibility of the scheme, Figure 3.24(b) shows the pattern of an array of 17 elements with λ/2 spacing, designed to have a sidelobe envelope

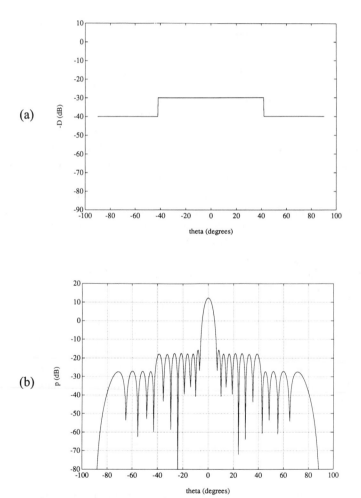

Figure 3.24 Synthesized pattern with −30- and −40-dB sidelobe levels (after [68]): (a) desired level; (b) resulting pattern.

like that of 3.24(a), with -30- and -40-dB levels. The resulting pattern is an excellent representation of this two-level sidelobe structure and illustrates the power of this relatively simple technique. This degree of control, coupled with the ability to insert known or measured element patterns and to readily address conformal array structures, is a major advantage of this new technique.

3.5 PATTERN SYNTHESIS USING MEASURED ELEMENT PATTERNS

Most of the synthesis procedures outlined in this chapter require that the array pattern can be written as a product of an element pattern and an array factor. The pattern of an element in an array is not, however, the same as the pattern of that same element when used alone. As indicated in Chapter 6, this is because exciting one element in the array produces radiation from that element and additional radiation from all other elements, because of the currents induced on them by the excited radiator. Nevertheless, the synthesis procedures are still valid if the radiating currents or aperture field has the same distribution on each element. In this case, one can speak of there being only a single "mode" of excitation on each element, with higher order terms negligible. The assumption of a single mode is approximately true for most small elements when higher order current distributions are relatively small compared with that part of the current that is common to all elements. Assuming that the element current distributions are the same, in an actual array, the interelement coupling results in currents that are not proportional to the applied sources, and this in turn produces the unequal element patterns. One can, however, synthesize a required array factor and then invert the coupling matrix, as indicated in Section 2.2 (Chapter 2), to solve for the required excitation.

Experimentally, one could use measured reflection coefficient or impedance data to determine the array coupling and then readjust excitations. Alternatively, one can obtain measured array element patterns and use these measured data to remove the deleterious effects of mutual coupling. Using measured element patterns, one can follow the procedure of Steyskal and Herd [69] and expand the element pattern of the nth element of a linear array in the form of a set of radiating signals from each of the N-array elements. Corresponding to an incident signal A_n at the nth element is the radiation

$$g_n(u) = e_0(u) \sum_{m=1}^{N} C_{nm} e^{jkmdu} \qquad (3.159)$$

where $e_0(u)$ is the isolated element pattern and C_{mn} is an unknown coupling coefficient relating the signals incident at the nth element to the radiating signal at the mth element.

The radiated signal from the whole array is

$$F(u) = \sum_n F_n(u)$$
$$= \sum_n A_n \sum_m C_{nm} e^{jkmdu}$$

(3.160)

The above element pattern relationship can be inverted to solve for the coefficient C_{nm} based on a measured element pattern $F_n(\theta)$. This result is given below and is the same as (3.3).

$$C_{nm} = \frac{1}{2\pi} \int_{-\pi/kd}^{\pi/kd} \frac{q_n(u)}{e_0(u)} e^{-jkmdu} du$$

(3.161)

When the spacing is less than $\lambda/2$, the convenient orthogonality is lost.

Once the coefficients C_{nm} are known, the desired array excitation A_n is obtained by equating the radiated pattern to that which would be radiated from some desired source excitation D_n (which might be any of the illuminations studied in this chapter).

$$F(u) = \sum_m D_m e^{jkmdu}$$

(3.162)

Comparing this with (3.160) leads to the relationship between the desired excitation coefficients D_m and the required incident signals A_n:

$$D_m = \sum_n A_n C_{nm}$$

(3.163)

and so the A_n are evidently given by the matrix inversion

$$A = [C^{-1}]D$$

(3.164)

Steyskal and Herd [69] give an example of this method applied to correct the excitation of an 8-element waveguide array. Typical element patterns for a central and edge element are given in Figure 3.25(a). The element is an open-ended waveguide measured in the E-plane. In Figure 3.25(b,c), the array is excited with a 30-dB Chebyshev taper and scanned to -30 deg from broadside. In Figure 3.25(b), the Chebyshev excitation is applied to the array input ports without correction, but such a small array is dominated by edge effects and the actual sidelobe level is well above the desired -30-dB level. In Figure 3.25(c), the correction

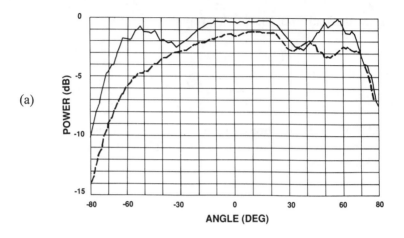

(a)

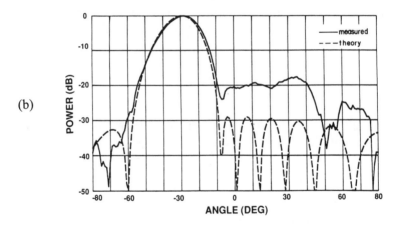

(b)

Figure 3.25 Pattern control using measured element patterns (after [69]): (a) the measured pattern magnitudes for center (solid line) and edge (dashed line) elements in an 8-element array; (b) 30-dB Chebyshev patterns without coupling compensation (solid line = measured; dashed line = theory; dashed line is ideal).

above has been computed from the measured array element patterns, and the resulting sidelobes are improved to the point of approaching the design sidelobes. The ability to perform such correction requires precise control of array amplitude and phase and is one of the major advantages of digital beam forming technology.

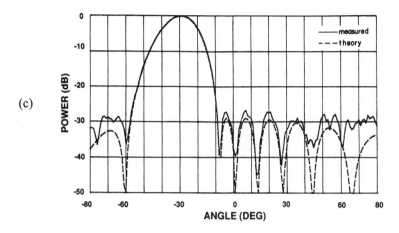

Figure 3.25 (cont.) Pattern control using measured element patterns (after [69]): (c) 30-dB Chebyshev patterns with coupling compensation. *Dashed pattern is ideal.

REFERENCES

[1] Schell, A. C., and A. Ishimaru, "Antenna Pattern Synthesis," *Antenna Theory*, Part 1, R. Collin and F. J. Zucker, eds., New York: McGraw-Hill, 1969.

[2] Ma, M. T., *Theory and Application of Antenna Arrays*, New York: Wiley Interscience, 1979.

[3] Rhodes, D. R., *Synthesis of Planar Antenna Sources*, London: Clarendon Press, Oxford, 1974.

[4] Silver, S., *Microwave Antenna Theory and Design*, MIT, Rad. Lab., Series, Vol. 12, New York: McGraw-Hill, 1979.

[5] Schelkunov, S. A., "A Mathematical Theory of Linear Arrays," *Bell System Tech. J.*, 1943, pp. 80–107.

[6] Taylor, T. T., "Design of Line Source Antennas for Narrow Beamwidth and Low Sidelobes," *IEEE Trans.*, Vol. AP-3, Jan. 1955, pp. 16–28.

[7] Bayliss, E. T., "Design of Monopulse Antenna Difference Patterns With Low Sidelobes," *Bell System Tech. J.*, Vol. 47, 1968, pp. 623–640.

[8] Woodward, P. M., "A Method of Calculating the Field Over a Plane Aperture Required to Produce a Given Polar Diagram," *Proc. IEE*, Part IIIA, Vol. 93, 1947, pp. 1554–1555.

[9] Woodward, P. M., and J. P. Lawson, "The Theoretical Precision With Which an Arbitrary Radiation Pattern May Be Obtained From a Source of Finite Size," *Proc. IEEE*, Vol. 95, P1, Sept. 1948, pp. 362–370.

[10] Mailloux, R. J., "Periodic Arrays," Ch. 13 in *Antenna Handbook*, Y. T. Lo and S. W. Lee, eds., Van Nostrand Reinhold, 1988.

[11] Stutzman, W. L., "Synthesis of Shaped-Beam Radiation Patterns Using the Iterative Sampling Method," *IEEE Trans.*, Vol. AP-19, No. 1, Jan. 1971, pp. 36–41.

[12] Dolph, C. L., "A Current Distribution for Broadside Arrays Which Optimizes the Relationship Between Beamwidth and Sidelobe Level," *Proc. IRE*, Vol. 34, June 1946, pp. 335–345.

[13] Stegen, R. J., "Excitation Coefficients and Beamwidths of Tschebyscheff Arrays," *Proc. IRE*, Vol. 41, Nov. 1953, pp. 1671–1674.

[14] Brown, L. B., and G. A. Scharp, "Tschebyscheff Antenna Distribution, Beamwidth, and Gain Tables," Naval Ordnance Lab., Corona, CA, NAVORD Rept. 4629 (NOLC Rept. 383), Feb. 1958.

[15] Hansen, R. C., "Linear Arrays," Ch. 9 in *The Handbook of Antenna Design*, A. Rudge, ed., London: Peter Peregrinus, 1983, p. 22.

[16] Stegen, R. J., "Gain of Tschebyscheff Arrays," *IEEE Trans.*, Vol. AP-8, 1960, pp. 629–631.

[17] Drane, C. J., Jr., "Useful Approximations for the Directivity and Beamwidth of Large Scanning Dolph-Chebyshev Arrays," *Proc. IEEE*, Vol. 56, Nov. 1968, pp. 1779–1787.

[18] Elliott, R. E., "The Theory of Antenna Arrays," Ch. 1 in *Microwave Scanning Antennas*, Vol. II, R. C. Hansen, ed., New York: Academic Press, 1966, pp. 29, 32.

[19] Van der Mass, C. J., "A Simplified Calculation for Dolph-Chebyshev Arrays," *J. Appl. Phys.*, Vol. 25, No. 1, pp. 121–124.

[20] Tang, R., and R. W. Burns, "Phased Arrays," Chap. 20 in *Antenna Engineering Handbook*, R. C. Johnson and H. Jasik, eds., McGraw-Hill, 1984.

[21] Hansen, R. C., "Linear Arrays," Ch. 9 in *The Handbook of Antenna Design*, Vol. 2, A. Rudge, ed., London: Peter Peregrinus, 1983, p. 309.

[22] Taylor, T. T., "One Parameter Family of Line Sources Producing Modified Symmetry Patterns," Rept. No. TM 324, Hughes Aircraft Co., Culver City, CA, 1953.

[23] Elliott, R. S., *Antenna Theory and Design*, Englewood Cliffs, NJ: Prentice-Hall, 1981.

[24] Elliott, R. S., "Design of Line Source Antennas for Narrow Beamswidth and Asymmetric Low Sidelobes," *IEEE Trans.*, Vol. AP-23, pp. 100–107, 1975.

[25] Elliott, R. S., "Design of Line-Source Antennas for Sum Patterns With Sidelobes of Individually Arbitrary Heights," *IEEE Trans.*, Vol. AP-24, 1976, pp. 76–83.

[26] Elliott, R. S., "Design of Line Source Antennas for Difference Patterns With Sidelobes of Individual Arbitrary Heights," *IEEE Trans.*, Vol. AP-24, 1976, pp. 310–316.

[27] Villeneuve, A. T., "Taylor Patterns for Discrete Arrays," *IEEE Trans.*, Vol. AP-32, 1984, pp. 1089–1093.

[28] Winter, C. F., "Using Continuous Aperture Illuminations Discretely," *IEEE Trans.*, Vol. AP-25, Sept. 1977, pp. 695–700.

[29] Elliott, R. S., "On Discretizing Continuous Aperture Distributions," *IEEE Trans.*, Vol. AP-25, Sept. 1977, pp. 617–621.

[30] Elliott, R. S., and G. J. Stern, "A New Technology for Shaped Beam Synthesis of Equispaced Arrays," *IEEE Trans.*, Vol. AP-32, 1984, pp. 1129–1133.

[31] Steyskal, H., "On Antenna Power Pattern Synthesis," *IEEE Trans.*, Vol. AP-18, No. 1, Jan. 1970, pp. 123–124.

[32] Steyskal, H., "On the Problem of Antenna Power Pattern Synthesis for Linear Arrays," FDA Reports, Vol. 5, No. 3, Research Institute of National Defence, Stockholm 80, Sweden, May 1971, pp. 1–16.

[33] Orchard, H. J., R. S. Elliott, and G. J. Stern, "Optimizing the Synthesis of Shaped Antenna Patterns," IEEE Proc. (London), Pt. H, No. 1, 1984, pp. 63–68.

[34] Taylor, T. T., "Design of Circular Apertures for Narrow Beamwidth and Low Sidelobe," *IRE Trans.*, Vol AP-8, 1960, pp. 17–22.

[35] Rudduck, R. C., et al., "Directive Gain of Circular Taylor Patterns," *Radio Science*, Vol. 6, 1971, pp. 1117–1121.

[36] Kinsey, R. R., "Monopulse Difference Slope and Gain Standards," *IRE Trans. Antennas and Propagation*, Vol. AP-10, May 1962, pp. 343–344.

[37] Tai, C. T., "The Optimum Directivity of Uniformly Spaced Broadside Arrays of Dipoles, *IEEE Trans.*, Vol. AP-12, 1964, pp. 447–454.

[38] Uzkov, A. I., "An Approach to the Problem of Optimum Directive de L'Academie de Sciences des l'RSS, Vol. 3, 1946, p. 35.

[39] Cheng, D. K., and F. I. Tseng, "Gain Optimization for Arbitrary Antenna Arrays, *IEEE Trans.*, Vol. AP-13, Nov. 1965, pp. 973–974.

[40] Harrington, R. F., "Antenna Excitation for Maximum Gain," *IEEE Trans.*, Vol. AP-13, No. 6, 1965, pp. 896–903.

[41] Tseng, F. I., and D. K. Cheng, "Gain Optimization for Antenna Arrays With Random Errors in Design Parameters," *Proc. IEEE*, Vol. 54, 1966, pp. 1455–1456.

[42] Lo, Y. T., S. W. Lee, and Q. H. Lee, "Optimization of Directivity and Signal-to-Noise Ratio of an Arbitrary Antenna Array," *Proc. IEEE*, Vol. 54, 1966, pp. 1033–1045.

[43] McIlvenna, J. F., and C. J. Drane Jr., "Maximum Gain, Mutual Coupling and Pattern Control in Array Antennas," *The Radio and Electronic Engineer*, Vol. 41, No. 12, 1971, pp. 569–572.

[44] Drane, C. J., and J. F. McIlvenna, "Gain Maximization and Controlled Null Placement Simultaneously Achieved in Aerial Array Patterns," *The Radio and Electronic Engineer*, Vol. 39, No. 1, 1970, pp. 49–57.

[45] McIlvenna, J. F., J. Schindler, and R. J. Mailloux, "The Effects of Excitation Errors in Null Steering Antenna Arrays," RADC-TR-76-183, Rome Air Development Center.

[46] Lo, Y. T., "Array Theory," Ch. 11 in *Antenna Handbook*, Y. T. Lo and S. W. Lee, eds., Van Nostrand Reinhold, 1988.

[47] Gantmacher, F. R., *The Theory of Matrices*, translated by K. A. Hirsch, Vol. 1, New York: Chelsea Publishing, 1959.

[48] Harrington, R. F., *Field Computation by Moment Methods*, New York: Macmillan, 1968.

[49] Monzingo, R. A., T. W. Miller, *Introduction to Adaptive Arrays*, New York: John Wiley and Sons, 1980.

[50] Hudson, J. E., *Adaptive Array Principles*, London: Peter Peregrinus, 1981.

[51] Gabriel, W. F., "Adaptive Arrays—An Introduction," *Proc. IEEE*, Vol. 64, No. 2, Feb. 1976, pp. 239–272.

[52] Barton, P., "Adaptive Antennas," AGARD Lecture Series 151, "Microwave Antennas for Avionics," AGARD-LS-151, IBSN 92-835-1547-1.

[53] Griffiths, J. W. R., "Adaptive Array Processing: A Tutorial," *IEEE Proc.*, Vol. 130, Pts. F. and H, No. 1, Feb. 1983.

[54] Applebaum, S. P., "Adaptive Arrays," *IEEE Trans.*, Vol. AP-24, Sept. 1976, pp. 585–598.

[55] Widrow, B., and J. M. McCool, "Comparison of Adaptive Algorithms Based on the Methods of Steepest Descent and Random Search," *IEEE Trans.*, Vol. AP-24, No. 5, Sept. 1976, pp. 615–637.

[56] Widrow, B., P. E. Mantay, L. J. Griffiths, and B. B. Goode, "Adaptive Antenna Systems," *Proc. IEEE*, Vol. 55, Dec., 1967, pp. 2143, 2159.

[57] Mailloux, R. J., "Array Antennas," Sec. 1, Ch. 12, *Handbook of Microwave and Optical Components*, Vol. 1, K. Chang, ed., New York: John Wiley and Sons, 1989.

[58] Mayhan, J. T., "Adaptive Nulling With Multiple Beam Antennas," *IEEE Trans.*, Vol. AP-26, No. 2, March 1978, p. 267.

[59] Mailloux, R. J., "Phased Array Theory and Technology," *IEEE Proc.*, Vol. 70, No. 3, March 1972, pp. 246–291.

[60] Steyskal, H., "Wide-Band Nulling Performance Versus Number of Pattern Constraints for an Array Antenna," *IEEE Trans.*, Vol. AP-31, No. 1, Jan. 1983, pp. 159–163.

[61] Franchi, P. R., "Degree of Freedom Requirements for Angular Sector Nulling," *Proc. 1992 Antenna Applications Symp.*, Sept. 1992.

[62] Mayhan, J. T., A. J. Simmons, and W. C. Cummings, "Wide-Band Adaptive Antenna Nulling Using Tapped Delay Lines," *IEEE Trans.*, Vol. AP-29, No. 6, Nov. 1981, pp. 923–935.

[63] Davies, D. E. N., K. G. Corless, D. A. Hicks, and K. Milne, "Array Signal Processing," Ch. 13 in *The Handbook of Antenna Design*, A. Rudge et al., eds., London: Peter Peregrinus, 1983, pp. 408–417.

[64] Ricardi, L., "Adaptive Antennas," Ch. 22 in *Antenna Engineering Handbook*, R. C. Johnson and H. Jasik, eds., New York: McGraw-Hill, 1961, 1984.

[65] Compton, R. J., Jr., *Adaptive Arrays—Concepts and Performance*, Englewood Cliffs, NJ: Prentice-Hall, 1988.

[66] Sureau, J. C., and K. J. Keeping, "Sidelobe Control in Cylindrical Arrays," *IEEE Trans.*, Vol. AP-30, No. 5, Sept. 1982, pp. 1027–1031.

[67] Dufort, E. C., "Pattern Synthesis Based on Adaptive Array Theory," *IEEE Trans.*, Vol. AP-37, 1989, pp. 1017–1018.

[68] Olen, C. A., and R. T. Compton, Jr., "A Numerical Pattern Synthesis Algorithm for Arrays," *IEEE Trans.*, Vol. AP-38, No. 10, Oct. 1990, pp. 1666–1676.

[69] Steyskal, H., and J. S. Herd, "Mutual Coupling Compensation in Small Array Antennas," *IEEE Trans.*, Vol. AP-38, No. 12, Dec. 1990, pp. 1971–1975.

Chapter 4
Patterns of Nonplanar Arrays

4.1 INTRODUCTION

An important class of applications for arrays requires them to conform to some shaped surface, often the surface of an aircraft, missile, or some other mobile platform. The conformality may be required for aerodynamic reasons or to reduce the antenna's radar cross section. Sometimes arrays are conformal to a stationary shaped surface in order to increase the angular sector served by a single array. Arrays required to provide 180 deg azimuth coverage may be conformal to a cylinder, depending on the elevation coverage required, while a spherical surface may be required for full hemispherical coverage.

Arrays on nonplanar surfaces can be categorized according to the two sketches shown in Figure 4.1. If the array dimensions are small compared to the radius of curvature as in Figure 4.1(a), the array is treated as locally planar, with planar array element patterns summed in accordance with the geometry of the curved surface. Such nearly planar arrays also have coverage limited by the field of view of the planar array. Arrays that are large with respect to the radius of curvature (Figure 4.1(b)) conform to the surface and may be used to scan over a far larger sector if the illuminations are somehow commutated around on the surface. This commutation is accomplished by several means, which are discussed briefly in this chapter. For such large arrays, the analysis and synthesis is significantly more complex than for a nearly planar or a conventional planar array.

The analysis and synthesis of nonplanar arrays differ from planar arrays in several aspects. Pattern synthesis is complicated because the element positions are in one plane and the element spacings are not always equal. For these arrays, the array factor and element patterns are not separable, and the array factor is not generally a simple polynomial. This situation alone is not a major detriment, and procedures are developed for properly handling the synthesis to almost any degree of accuracy. Further, to produce a low-sidelobe pattern with an array that is large with respect to the radius of curvature, one must commutate the illumination around the radiating surface in order to utilize the elements that radiate efficiently in the

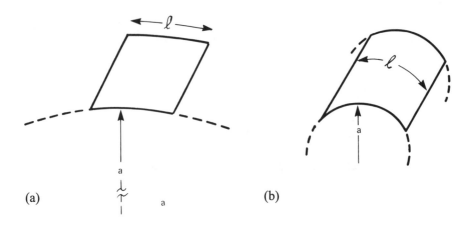

(a)   (b)

Figure 4.1 Conformal arrays: (a) aperture dimensions much less than local radius of curvature; (b) aperture dimensions comparable with local radius of curvature.

direction of desired radiation. A third aspect is that the polarization radiated by elements on surfaces that are not parallel to one another will not generally be aligned. This can cause high cross polarization. Finally, the element patterns on shaped surfaces may all be different and can also be very distorted. This can lead to high sidelobes and poor scanning performance.

To these practical aspects, one must add a fundamental analytical difficulty. Except for relatively simple nonplanar surfaces like circular cylinders, it is generally not possible to obtain convergent, accurate Green's functions for the sources on or above these surfaces. In such cases, it is customary to use the Geometrical Theory of Diffraction to obtain approximate element patterns for synthesis or analysis. The detailed calculations required for computing the radiation and mutual coupling of elements conformal to nonplanar surfaces are beyond the scope of this text. These evaluations are implicit in the use of element patterns throughout this chapter, and all of the electromagnetics are included in the evaluation of active element patterns. A detailed discussion of these more complex mathematical issues is included in the work of Borgiotti [1].

4.2 PATTERNS OF CIRCULAR AND CYLINDRICAL ARRAYS

Circular and cylindrical arrays possess the advantage of symmetry in azimuth, which makes them ideally suited for full 360 deg coverage. This advantage has been exploited for the development of broadcast antennas and direction-finding antennas. A recent book chapter by D. E. N. Davies [2] summarizes practical developments in circular arrays, and the *Conformal Array Antenna Array Design Hand-book* [3], edited by R. C. Hansen, presents an extensive literature search and practical pattern results for both circular and circular arc arrays, as well as other conformal array geometries.

Figure 4.2(a) shows a group of elements disposed around a circle. The array pattern for the circular (or ring) array of radius a with N elements at locations $\phi' = n\Delta\phi$ is given by the usual array expression (1.47), with

$$r_n = R_0 - a \sin \theta \cos(\phi - n\Delta\phi) \tag{4.1}$$

The resulting pattern is

$$F(\theta, \phi) = \sum_{n=0}^{N-1} I_n f_n(\theta, \phi) e^{+jka \sin \theta \cos(\phi - n\Delta\phi)} \tag{4.2}$$

In this expression, the element patterns are shown as scalar, although in the general case they would be vector. Further, because of symmetry, the element patterns are dependent on the element location and have the form:

$$f_n(\theta, \phi) = f(\theta, \phi - n\Delta\phi) \tag{4.3}$$

and generally include element interaction and the effects of ground plane curvature. The element patterns are typically not hemispherical, and often their phase center is not well known, so this must be accounted for in determining the excitation current.

The excitation I_n contains the amplitude and phase required for array taper and collimation. To produce an inphase collimated beam at the angle (θ_0, ϕ_0), one selects

$$I_n f_n(\theta_0, \phi_0) = |I_n f_n(\theta_0, \phi_0)| e^{-jka \sin(\theta_0) \cos(\phi_0 - n\Delta\phi)} \tag{4.4}$$

while for a near constant radiation as a function of ϕ, one selects constant I_n. Notice that the ring array can be focused at an elevation angle θ_0 that is not necessarily in the plane of the array ($\theta = \pi/2$).

The circular array is of particular importance because it is also the basic element of cylindrical arrays and even conical and spherical arrays, or arrays on generalized bodies of revolution, as shown in Figure 4.2(c). In the generalized system that is depicted Figure 4.2(c), one can write the far-field pattern of the kth circu-

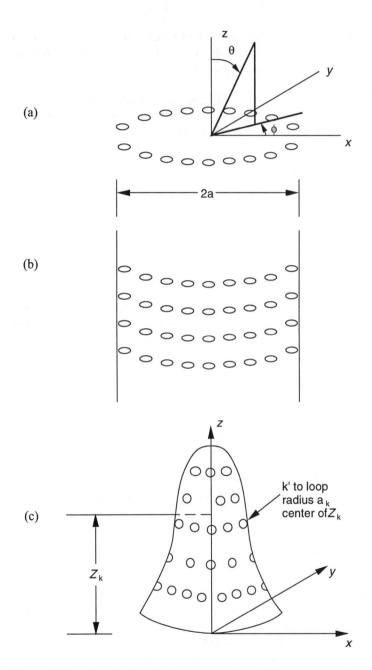

Figure 4.2 Circular and cylindrical array geometries: (a) circular array; (b) cylindrical array; (c) generalized array conformed to a body of revolution.

lar array by using the local radius a_k of the array and the position vector $\mathbf{r}'$ that is measured to the nth element of the kth circular array (x_{nk}, y_{nk}, z_k) on the array. Using

$$\mathbf{r}'_{nk} = \hat{\mathbf{x}} x_{nk} + \hat{\mathbf{y}} y_{nk} + \hat{\mathbf{z}} z_k \qquad (4.5a)$$

where

$$x_{nk} = a_k \cos \phi_{nk} \qquad y_{nk} = a_k \sin \phi_{nk} \qquad (4.5b)$$

and the position vector in space at the angle (θ, ϕ)

$$\hat{\boldsymbol{\rho}} = \hat{\mathbf{x}} u + \hat{\mathbf{y}} v + \hat{\mathbf{z}} \cos \theta \qquad (4.6)$$

one obtains

$$\begin{aligned} \mathbf{r}' \cdot \hat{\boldsymbol{\rho}} &= a_k \cos \phi_{nk} \sin \theta \cos \phi + a_k \sin \phi_{nk} \sin \theta \sin \phi + z_k \cos \theta \qquad (4.7) \\ &= a_k \sin \theta \cos(\phi - \phi_{nk}) + z_k \cos(\theta) \end{aligned}$$

and the resulting equation for the field of the kth loop with N_k elements located at equally spaced angles $\phi_{nk} = n\Delta\phi_k$:

$$F_k(\theta, \phi) = \sum_{n=0}^{N_k-1} I_{nk} f_{nk}(\theta, \phi) e^{+jk[a_k \sin \theta \cos(\phi - n\Delta\phi_k) + z_k \cos \theta]} \qquad (4.8)$$

The element patterns f_{nk} of any kth circle are assumed identical except for displacement in the angle ϕ.

Specific characteristics of the patterns of circular and cylindrical arrays are discussed in the following sections.

4.2.1 Fourier Series Expansion of the Fields of a Single Ring Array ($\theta = \pi/2$)

Most circular (and cylindrical) arrays are made up of directional elements. This section, however, will introduce the circular array in its most elementary form, with omnidirectional elements. For further simplicity, the array pattern will be written in the principal plane ($\theta = \pi/2$). The pattern at completely general locations can be recovered from these results by replacing the radius a with $a \sin \theta$, as in the previous equation, and $a \sin \theta_0$ in the current distribution. For a single ring array of N omnidirectional elements, this principal plane pattern is

$$F(\phi) = \sum_{n=0}^{N-1} I_n e^{+jka \cos(\phi - n\Delta\phi)} \tag{4.9}$$

Since the pattern is evidently periodic in space, one can write the far field in the form of an exponential Fourier series:

$$F(\phi) = \sum_{q=-\infty}^{\infty} A_q e^{jq\phi} \tag{4.10}$$

where the coefficients A_q are

$$A_q = \frac{1}{2\pi} \int_{-\pi}^{\pi} F(\phi) e^{-jq\phi} d\phi \tag{4.11}$$

In this Fourier series form, each term is called a *phase mode* of the radiation pattern. The qth phase mode of $F(\phi)$ is a harmonic term that has a $2\pi q$ variation in phase as ϕ varies from 0 to 2π.

The pattern function can be rewritten using the Fourier-Bessel series to expand the exponent:

$$e^{jka \cos(\phi - n\Delta\phi)} = J_0(ka) + \sum_{m=1}^{\infty} 2j^m J_m(ka) \cos[m(\phi - n\Delta\phi)] \tag{4.12}$$

In this form, the series can be integrated to yield the coefficients A_q of the far-field Fourier series

$$A_q = \sum_{n=0}^{N-1} I_n \{\delta(q)J_0(ka) + \sum_{m=1}^{\infty} j^m J_m(ka)[e^{-jmn\Delta\phi}\delta(q - m) + e^{+jmn\Delta\phi}\delta(q + m)]\} \tag{4.13}$$

where δ is the Kronecker delta function

$$\begin{aligned} \delta(k) &= 1 \quad \text{for } k = 0 \\ &= 0 \quad \text{for all other } k \end{aligned} \tag{4.14}$$

The three terms in this expression have exactly the same form. This is so since $j^{-q} = (-1)^q j^q$ and $J_{-q} = (-1)^q J_q$, therefore, $j^{-q} J_{-q} = j^q J_q$. Thus, for any q, one obtains

$$A_q = \sum_{n=0}^{N-1} I_n \{ j^q e^{-jqn\Delta\phi} J_q(ka) \} \tag{4.15}$$

Array Excitation With Phase Modes of Current

Although the expression above is not generally summable for all I_n, there are special choices of the set of currents that correspond to symmetries of the array, and it is convenient to write the current in terms of these special symmetrical sets, which are the phase mode excitation of the array. Without loss in generality, the currents I_n are written as the finite sum

$$I_n = \sum_{p=-P}^{P} I_n^p = \sum_{p=-P}^{P} C_p e^{jpn\Delta\phi} \tag{4.16}$$

where

$$P = (N - 1)/2$$

The phase mode currents are thus

$$I_n^p = C_p \exp[jp(n\Delta\phi)] \tag{4.17}$$

This expansion is more than just a mathematical artifice, since the phase mode currents have precisely the range of periodic phases obtainable from an $N \times N$ Butler matrix (see Chapter 8). This application of Butler matrices is described in [4].

One can find the far-field pattern for each phase mode using the expression (4.15) and writing the Fourier coefficients of the far field A_q as a series of contributions due to each phase mode current, or

$$A_q = \sum_{p=-P}^{P} A_q^p \tag{4.18}$$

where the A_q^p is now the contribution of the pth phase mode to the qth far-field Fourier series term, and

$$A_q^p = j^q J_q(ka) C_p \sum_{n=0}^{N-1} e^{j(p-q)n\Delta\phi} = j^q J_q(ka) C_p N \qquad \text{for } p - q = NI$$

$$= 0 \qquad \text{for other } q \tag{4.19}$$

where I is any integer and N is the number of elements in the ring.

The far-field pattern can be written as the sum of patterns due to each of the phase modes. The pattern $F^p(\phi)$ corresponding to the pth phase mode is written

$$F^p(\phi) = \sum_{q=-\infty}^{\infty} A_q^p e^{jq\phi} \tag{4.20a}$$

and the total pattern is

$$F(\phi) = \sum_p F^p(\phi) \tag{4.20b}$$

Converting this into a summation over index I eliminates most of the terms of the summation

$$F^p(\phi) = C_p N \sum_{q=-\infty}^{\infty} j^q J_q(ka) e^{jq\phi} = C_p N \sum_{I=-\infty}^{\infty} j^{(p-NI)} J_{p-NI}(ka) e^{j(p-NI)\phi}$$

$$= C_p N [j^p J_p(ka) e^{jp\phi} + \sum_{I=1}^{\infty} j^{(p-NI)} J_{p-NI}(ka) e^{j(p-NI)\phi} \tag{4.21}$$

$$+ \sum_{I=1}^{\infty} j^{(p+NI)} J_{p+NI}(ka) e^{j(p+NI)\phi}]$$

Equations (4.20) and (4.21) reveal a great deal about the behavior of circular arrays. The first term of the summation has the same angular dependence $[\exp(jp\phi)]$ as the phase mode excitation. The additional terms have angular dependence $[\exp(j(p \pm NI)\phi)]$, and so have either much slower or much faster angular variation than the first term if the number of elements N is large. The order of the Bessel function $J_p(ka)$ is critical for determining the amplitude of the radiated signal. Figure 4.3 shows the values of several functions $J_p(ka)$ and indicates that the amplitudes corresponding to large values of p (higher order phase modes) are small unless the array radius a is large accordingly. The array will not radiate phase modes higher than about ka, and so modes with faster angular variation correspond

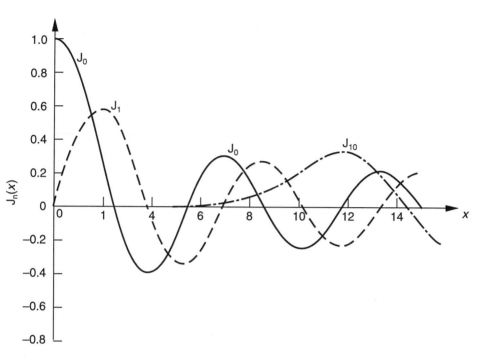

Figure 4.3 Bessel functions $J_p(ka)$ versus radial parameter ka for $p = 1, 2, 10$.

to superdirective excitation. The application of only one phase mode thus results in a far field with one term having the same angular dependence as the applied phase mode and a set of other radiating modes which can be considered as distortions to the far-field pattern.

Synthesis and Scanning Using Phase Modes of Continuous Current Sheets

The first term of the above is the radiation pattern of a continuous current sheet with phase mode currents

$$I_n^p = C_p e^{jp\alpha} \tag{4.22}$$

where here the discrete location $n\Delta\phi$ is replaced by the continuous variable α (for infinitesimal separation of current elements). It is this first term that is often used in pattern syntheses by recognizing the mathematical similarity between the phase mode radiation of the continuous current sheet and the corresponding far-field pattern of a finite linear array.

A finite linear array of $N = 2Q + 1$ isotropic element has the pattern

$$F_L(\phi) = \sum_{p=-Q}^{Q} B_p e^{jp(kd_x \sin \phi)} \tag{4.23}$$

Comparing this with the circular-array far-field pattern of all phase modes in (4.20), but using only the first term of (4.21), one has

$$F(\phi) = N \sum_{p=-P}^{P} C_p j^p J_p(ka) e^{jp\phi} \tag{4.24}$$

The similarity of the two expressions (4.23) and (4.24) is apparent, and so by exciting phase modes with the coefficients

$$C_p = \frac{B_p}{j^p J_p(ka)} \tag{4.25}$$

and identifying ϕ of the circular array with $kd_x \sin \phi$ of the linear array, one can select a group of phase mode excitations C_p for a given cylindrical array to produce approximately the same pattern in ϕ space as the linear array produces in $\sin \phi$ space. Notice that for the linear array the summation is over array elements, while for the circular array the summation is over phase modes. The synthesis in the $\theta = \pi/2$ plane does not produce the same pattern for other θ because the elevation pattern of each phase mode is different. This constitutes another difference from linear array pattern synthesis, where the linear array element currents all have identical elevation element patterns except for second-order mutual coupling effects.

Examples of this synthesis are given in the literature [4,5]. The synthesis is exact for the circular current sheet loop, but the presence of higher order terms in (4.21) distorts the actual radiated pattern of the discrete array.

The expression above also indicates the choice of excitation to scan the beam to a particular direction, since the excitation coefficients that scan the linear array to a given angle ϕ_0 are obtained by multiplying the current mode excitation by $\exp[-jp\phi_0]$ to obtain

$$C_p = \frac{e^{-jp\phi_0}|B_p|}{j^p J_p(ka)} \tag{4.26}$$

The particular case of a uniform element illumination of a linear array $|B_p| = 1$ leads to a radiated pattern of the form

$$\frac{\sin[(kNd_x/2)(\sin \phi - \sin \phi_0)]}{[kNd_x/2] \sin(\sin \phi - \sin \phi_0)} \tag{4.27}$$

while choosing equal current modes in a circular array leads to the radiated pattern

$$F(\phi) = \frac{\sin[N/2(\phi - \phi_0)]}{N \sin[(\phi - \phi_0)/2]} \tag{4.28}$$

in the plane $\theta = \pi/2$.

The synthesis achieved by selecting the mode coefficients C_p (4.26) is not always ideal because, for larger arrays, the $J_p(ka)$ might be zero for a given frequency and circle radius. Phase modes corresponding to these ka cannot be excited.

Array Bandwidth

The argument of the Bessel function also restricts the array bandwidth and elevation patterns. The bandwidth criterion given by Davies [2] is that the argument (ka) of the Bessel functions not change more than about $\pi/8$ to avoid excessive changes in the coefficients. This leads to the bandwidth criterion

$$\Delta f/f_0 \approx \lambda/8a \tag{4.29}$$

This corresponds to an extremely narrow bandwidth, for even a moderately large array, and is one reason why circular arrays of omnidirectional elements do not suit many applications. The severe limitation is due to cancellation effects between those elements at opposite sides of the circle.

4.2.2 Patterns and Elevation Scanning

The expressions of the previous section are readily generalized to arbitrary elevation angles. The pth phase mode radiation becomes

$$F_p(\phi) = C_p N[j^p J_p(ka)e^{jp\phi} + \sum_{l=1}^{\infty} j^{(p-Nl)} J_{p-Nl}(ka \sin \theta)e^{j(p-Nl)\phi}$$

$$+ \sum_{l=1}^{\infty} j^{(p+Nl)} J_{p+Nl}(ka \sin \theta)e^{j(p+Nl)\phi}] \tag{4.30}$$

Again using only the first term,

$$F(\theta, \phi) \approx N \sum_{p=-P}^{P} C_p j^p J_p(ka \sin \theta) e^{jp\phi} \qquad (4.31)$$

The elevation pattern of each pth mode is very narrow, with a peak at the maximum of $J_p(ka \sin \theta)$. The argument $(ka \sin \theta)$ limits the pattern bandwidth, as indicated earlier, but since it is the only expression in which the elevation angle θ enters, it contains the elevation pattern shape, which is shown as severely narrowed compared to the elevation pattern of a linear array. Again using the criterion of a $\pm \pi/8$ change in the Bessel function argument $(ka \sin \theta)$, Davies obtains an expression for the phase mode vertical beamwidth as

$$\theta_3 \approx (\lambda/2a)^{1/2} \qquad (4.32)$$

which is the same as that of a linear endfire array of length equal to the diameter. This severe pattern narrowing also makes the elevation pattern very frequency-dependent, and, moreover, introduces significant complications in the synthesis of azimuth patterns at elevations other than $\pi/2$.

The pattern is scanned by choosing modes that add at θ_0, ϕ_0.

$$C_p = \frac{e^{-jp\phi_0} |B_p|}{j^p J_p(ka \sin \theta_0)} \qquad (4.33)$$

Figure 4.4 shows the elevation and azimuth patterns of an array of 30 omnidirectional elements with $\lambda/2$ spacing, arrayed in a ring or loop array and scanned to various elevation angles. The excitation chosen to scan the array is that of (4.4), with an equal amplitude distribution $|I_n| = 1$. The array radius is 4.775λ. Figure 4.4(a) shows ϕ-plane patterns at the scan planes (θ) for chosen scan angles $\theta_0 = 90$ deg (horizon) and $\theta_0 = 45$ deg. Aside from slight broadening, the patterns are similar. Figures 4.4(b) and 4.4(c) show elevation patterns for a progression of elevation scan angles from $\theta_0 = 90$, 60 deg (Figure 4.4(b)) to $\theta_0 = 45$, 30 deg (Figure 4.4(c)). These patterns show the bidirectional scanned beam with symmetry about the plane of the circle, since there is no ground screen. The elevation pattern is broadest at the horizon $\theta_0 = 90$ deg, where the vertical projection of the array is minimal. Scanning up from the horizon narrows the pattern and forms two distinct beams that narrow with increasing angle from the horizon (decreasing θ_0).

4.2.3 Circular and Cylindrical Arrays of Directional Elements

Since the pattern characteristics of circular and cylindrical arrays cannot be represented in terms of the product of an element pattern and an array factor, it is

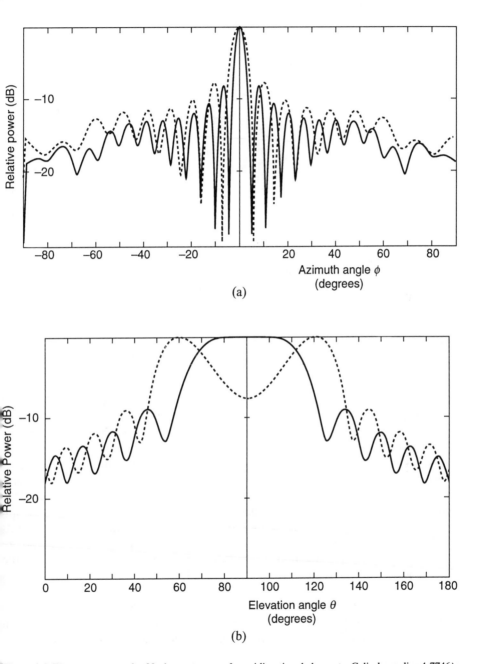

Figure 4.4 Element patterns for 30-element array of omnidirectional elements. Cylinder radius 4.7746λ: (a) plane patterns $(\theta, \theta_0) = (90, 90)$ solid curve, $(\theta, \theta_0) = (45, 45)$ dashed curve; (b) elevation patterns for $\phi = \phi_0 = 0$ dashed curve and for array scanned to $\theta_0 = 90$ deg (solid), 60 deg (dashed).

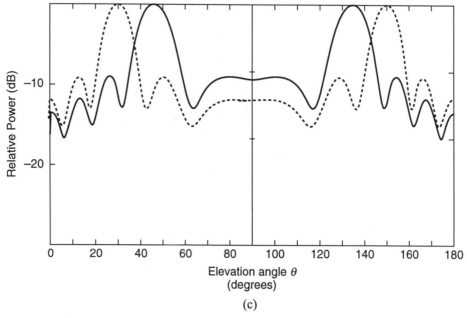

Figure 4.4 (cont.) Element patterns for 30-element array of omnidirectional elements. Cylinder radius 4.7746λ: (c) elevation patterns for $\theta_0 = 45$ deg (solid) and 30 deg (dashed).

especially important to consider the array patterns with directional elements. Beyond this general statement, however, there are two special reasons why the directive properties of elements are particularly important in such conformal arrays. First, the mutual coupling between elements narrows the active element pattern, and so in general one cannot design omnidirectional elements. Although this is true for planar arrays as well, it is much more important in conformal arrays because all elements "point" in different directions. Second, the very limited bandwidth and narrowed elevation pattern of the circular arrays discussed in the last section are due to the interaction between widely separated omnidirectional elements located at opposite sides of the array. If the array is built using elements that radiate primarily in the radial direction, or at least into some forward sector, then the circular array characteristics are substantially different and the bandwidth significantly improved.

For generalized elements, the far field is given below, with $f(\phi - n\Delta\phi)$ the element pattern of the nth element located at $\phi' = n\Delta\phi$. By symmetry, all element patterns are identical except for angular displacement due to element location.

$$F(\phi) = \sum_{n=0}^{N-1} I_n f(\phi - n\Delta\phi)e^{+jka\,\cos(\phi - n\Delta\phi)} \qquad (4.34)$$

Inclusion of an element pattern in the phase mode representation is a relatively simple operation if the element pattern is written as a Fourier series.

$$f(\phi) = \sum_{K=-M}^{M} g_K e^{jK\phi} \tag{4.35}$$

where M is the maximum spatial harmonic of the element pattern.

The derivation of the far field for these cases follows directly from the one included for omnidirectional elements and will not be included here. Davies [2] gives the following expression for the far field based on the above Fourier series element pattern.

For the pth phase mode,

$$F^p(\phi) = C_p |A_p| e^{j(p\phi + \Phi_p)} \tag{4.36}$$
$$+ C_p \sum_{l=1}^{\infty} \{|A_{p+Nl}| e^{j(p+Nl)\phi + \Phi_{p+Nl}} + |A_{p-Nl}| e^{j(p-Nl)\phi + \Phi_{p-Nl}}\}$$

and where the constant $|A_p| \exp(j\Phi_p)$ is defined

$$|A_p| \exp(j\Phi_p) = \sum_{K=-M}^{M} g_K j^{p-K} J_{p-K}(ka)$$

A particular case studied by Rahim and Davies [6] using an element pattern of the form $(1 + \cos \phi)$ yields an especially simple form for the phase mode pattern:

$$F^p(\phi) = C_p j^p e^{jp\phi} [J_p(ka) - jJ'_p(ka)] \tag{4.37}$$

where $J'_p(x)$ is the derivative of $J_p(x)$.

Bandwidth and Elevation Patterns

Davies [2] provides a useful interpretation of 4.37 and points out that the sum $J_p(ka) - jJ'_p(ka)$ is not strongly dependent on (ka), and so the bandwidth of an array of directional elements can be much wider than that of the same array with omnidirectional elements. A striking demonstration of this effect is given by Figure 4.5 [6], which shows the amplitude of a phase mode ($p = 1$) versus frequency for omnidirectional elements (dashed) and directional $(1 + \cos \phi)$ elements as a function of frequency. The directional elements remove all of the zeros that were present in the omnidirectional element array wide-band excitation. Similarly, the elevation beamwidth of each phase mode is no longer limited by interactions

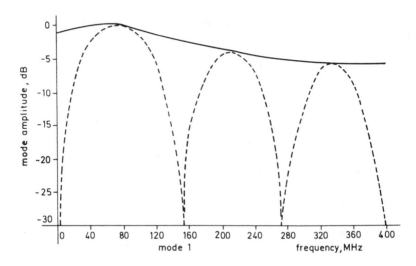

Figure 4.5 Theoretical results showing the stability of mode 1 versus frequency due to the use of directional elements of the form $(1 + \cos \phi)$ (solid curve) and omnidirectional (dashed curve) elements for 2λ radius array at 300 MHz (after [2]).

between elements separated on the order of the array diameter and so can be substantially broadened. Both of these effects (increased bandwidth and broadened elevation patterns) imply that the array is behaving more like a line source array.

Like the array of omnidirectional elements, however, it is difficult in practice to synthesize low-sidelobe patterns with the full array excited, because of the radiation from elements with contributions in the sidelobe region. This topic is addressed in Section 4.2.4.

4.2.4 Sector Arrays on Conducting Cylinders

Practical Means for Commutation

Cylindrical arrays require commutation [7,8] of an illuminated region around the array. Practical surveillance and communication systems with azimuth scan requirements of 360 deg use cylindrical array geometries, but with only a restricted sector of the cylinder illuminated. Typical illumination regions span between 90 and 120 deg of the cylinder. The illumination is commutated around the cylinder by means of a switching network. Phase mode excitation is not often used for large circular or cylindrical arrays because of the complexity of large Butler matrices (see Chapter 8) and the difficulty in obtaining sufficient accuracy to cancel all radiation in the back direction. Figure 4.6 shows several networks for commutating a given illu-

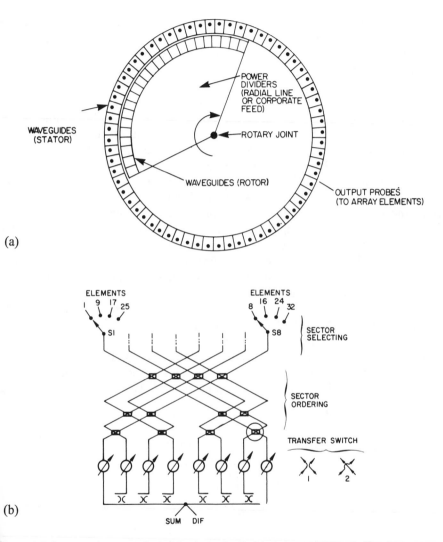

Figure 4.6 Commutating networks for circular and cylindrical arrays: (a) waveguide commutator; (b) switch network (after [9]).

mination around a cylinder. Four basic approaches are used: mechanical rotation of a fixed illumination, diode or ferrite switch networks, lens scanning, and matrix beam formers to excite phase modes. Not shown in the figure, but extremely important to note, is that conformal arrays can also be controlled by digital beam formers, and in many ways this is the most flexible mode of control.

The simplest kind of commutator is typified by the mechanically scanned waveguide structure of Figure 4.6(a) using a power distribution network that rotates in contact (or proximity) with a fixed stator. In this design it is important that the stator and rotor have different element spacings to avoid excessive modulation of the scanned radiation.

Switching networks using diode or ferrite switches and microwave hybrid power dividers have also been used for commutating an amplitude and phase distribution around a circular array. Giannini [9] describes a technique that uses a band of switches to bring a given illumination taper to one sector of the array (usually a 90- or 120-deg arc), and a set of switches to provide beam steering between those characteristic positions determined by the sector switching network. For a 32-element array, the circuit shown in Figure 4.6(b) requires eight phase shifters and twelve transfer switches (double-pole, double-throw), and achieves sector selection using eight single-pole, four-throw switches. This network excites an eight-element quadrant of the array that can be moved in increments of one element to provide coarse beam steering. Fine steering (selecting angles with separation less than the angular separation between radiating elements on the cylinder) is provided by the phase controls.

Several lens-fed circular arrays have been constructed using R-2R, Luneberg, and geodesic lenses. The R-2R lens of Figure 4.6(c) described by Boyns et al. [10] forms as many beams as there are elements in the array, but does not provide fine steering unless additional phase controls are added to each element.

Holley et al. [11] show that lens systems can provide fine steering by using an amplitude illumination with a movable phase center. This is accomplished using

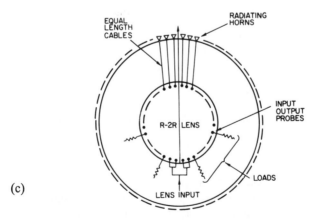

(c)

Figure 4.6 (cont.) Commutating networks for circular and cylindrical arrays: (c) R-2R Lens (after [10]).

a set of phase shifters at the input to a Butler matrix. With all input elements excited with zero relative phase, the amplitudes of signals into the Butler matrix are chosen to produce the required array excitation. By inserting a progressive phase distribution at the Butler matrix input, the amplitude distribution at the output ports can be moved with very little change in the shape of the distribution. This phase center shift is adequate for fine steering between the normal increments of one element. An alternative point of view to explain this operation is to consider the relative weighting of the multiple beams available from the lens feed. From this perspective, the net resulting phase tilt at the input of the Butler matrix can synthesize intermediate beams from a composite of the available lens beams and so provide high-quality fine steering of the lens-radiated pattern.

Holley et al. [11] applied this principle to a geodesic lens, as shown in Figure 4.7(a), which offered the advantage that the desired array amplitude distribution could be formed with very few probes. With the geodesic lens feed, only 8 elements need to be switched to move an illumination spanning about 100 elements of the cylindrical array (of 256 elements). Figure 4.7(b) shows the broadside amplitude distribution for the illuminated lens (solid) and the displaced illumination due to inserting a progressive phase at the input terminals. This economy, added to the phase center motion produced by the Butler matrix, resulted in good-quality scanned beams with a near-minimum number of controls.

There have been a number of developments in the area of multimode electronic commutators for circular arrays. These systems derive from techniques similar to that first used by Honey and Jones [12] for a direction-finding antenna application, where several modes of biconical antennas were combined to produce a directional pattern with full 360-deg azimuthal rotation. Recent efforts by Bogner [13] and Irzinski [14] specifically address the use of such a commutator combined with phase shifters and switches at each element. The phase shifters provide collimating and fine steering, and the switches are used to truncate the illumination so that only a finite sector of the array is used at any time, a procedure that is required for sidelobe control.

Butler matrices (see Chapter 8) have been used to excite the phase modes of circular arrays directly. As originally proposed by Shelton [15] and developed by Sheleg [4], a matrix-fed circular array with fixed phase shifters can excite current modes around the array, and variable phase shifters can then be used to provide continuous scanning of the radiated beam over 360 deg. The geometry is shown in Figure 4.8. A more recent extension of this technique proposed by Skahil and White [16] excites only that part of the circular array that contributes to the formation of the desired radiation pattern. The array is divided into a given number of equal sectors, and each sector is excited by a Butler matrix and phase shifters. With either of these circuits, sidelobe levels can be lowered by weighting the input excitations to the Butler matrix. The technique by Skahil and White was demon-

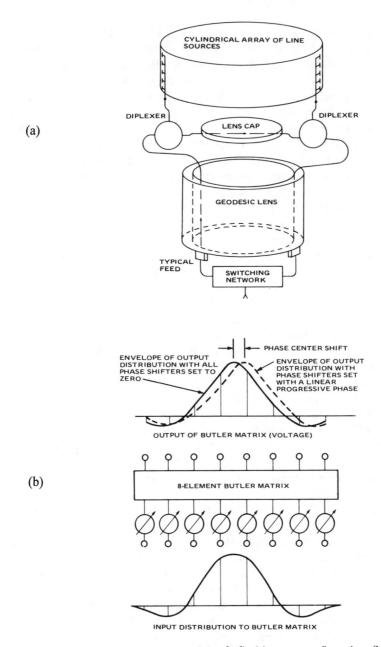

Figure 4.7 Electronically scanned array (after [11]): (a) system configuration; (b) fine beam steering with diode phase shifters and Butler matrix.

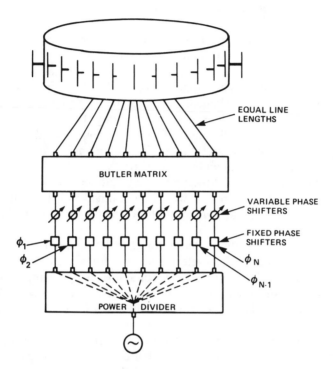

Figure 4.8 Matrix scanning system (after [4]).

strated by using an 8 × 8 Butler matrix, eight phase shifters, and eight single-pole, four-throw switches to feed four 8-element sectors of a 32-element array. The design sidelobes were −24 dB and measured data showed sidelobes below −22 dB.

Cylindrical sector arrays are excited by currents to focus the far-field distribution for each ring $F_k(\theta, \phi)$ (see (4.8)) to some point (θ_0, ϕ_0). Assuming element patterns with constant far-field phase, one uses

$$I_{nk} = \left| I_{nk} \right| e^{-jk[a_k \sin \theta_0 \cos(\phi_0 - n\Delta\phi_k) + z_k \cos \theta_0]} \tag{4.38}$$

This excitation is applied only to the desired illuminated sector, while the other elements of the array are, ideally, terminated in matched loads. In this manner, all element patterns in any particular ring array are equal except for angular displacement.

Patterns of Elements and Arrays on Cylinders

Since the elements of a cylindrical array point in different directions, the active element pattern is far more important than in planar arrays. This is illustrated in Figure 4.9(a), which shows a sector array that occupies 120 deg of the cylinder. If the array is collimated to radiate broadside ($\phi_0 = 0$), then the elements near the top of the cylinder have their element pattern peaks at the desired scan direction, while those at ± 60 deg have their peaks at ± 60 deg. However, to form a beam at $\phi_0 = 0$, the elements near the ends of the array have a local scan angle of 60 deg, and so these elements are operating as if in a wide-angle scanned array. Notice from the dashed lines that if, in addition, the array were scanned to 60 deg (Figure 4.9(b)), then the end elements at the right side would be locally scanned to broadside, while the ones at the left end would be scanned well beyond endfire and shadowed by the cylinder, and thus would have essentially no contribution to the radiation.

The above description and sketches should make two points clear. First, even if the array is not scanned, the pattern synthesis for a sector array is critically dependent on the array element pattern. Second, it is generally impractical to build a cylindrical array with only a few stationary sectors and then scan the array to gain angular coverage. These two facts are the basis for separate discussions to follow.

Active element patterns in cylindrical sector arrays behave similar to those in finite planar arrays, but exhibit additional effects due to array curvature. These effects from mutual coupling are discussed in the following paragraphs, but, in addition, there are certain bounds imposed even on isolated elements because of the cylindrical surface.

Isolated Element Patterns

If the circular sector array is small compared to the radius of the cylinder, and if the cylinder itself is large compared to wavelength, then the element patterns will be similar to those in a planar array, but modified by the presence of the cylinder. Figure 4.10 [17] shows the upper hemisphere element power pattern for a single slot with axial or circumferential polarization at the top of a large cylinder. The figure indicates that for a large cylinder neither polarization radiates substantially into the lower hemisphere, while in the upper hemisphere the pattern is nearly unchanged from that over an infinite ground plane (denoted by $f(\phi)$), except in a small transition region near the horizon of width approximately $(ka)^{-1/3}$ on either side of the horizon. In this transition region, the circumferential polarized radiation is reduced from unity for the infinite ground plane case to about -3.2 dB for the

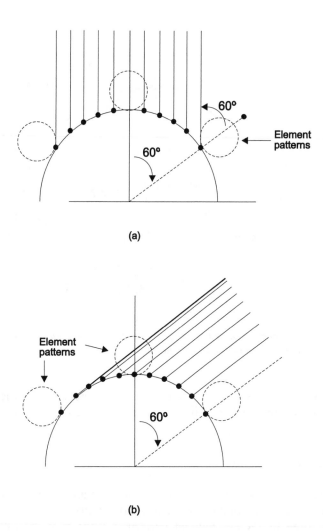

(a)

(b)

Figure 4.9 Conformal cylindrical sector array: (a) array filling 120-deg sector of cylinder; (b) conformal array scanned to 60-deg from broadside.

cylinder, and the axial polarization, which is zero for the infinite ground screen, is only reduced to about

$$0.4(2/ka)^{1/3} \qquad (4.39)$$

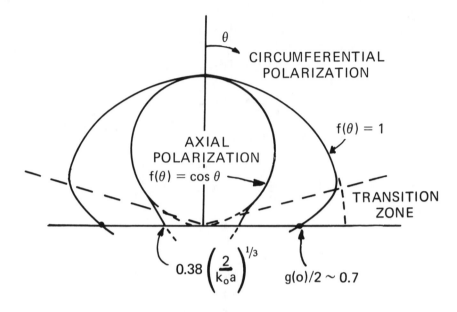

Figure 4.10 The approximate pattern of slot on cylinder of radius a for circumferential polarization $[f(\phi) = 1]$ and axial polarization $[f(\phi) = \cos \phi]$.

for the cylinder. This result is obtained from the geometrical theory of diffraction and is valid as long as the cylinder radius is large compared to wavelength.

Active Element Patterns

If the sector array is large compared to the cylinder radius, or if the element is in an illuminated region of an array fully wrapped around the whole cylinder (Figure 4.11(a)), then the active element patterns are those of the full cylindrical array with a ground plane. Active element patterns in a cylindrical array can be significantly different from those in a planar array and have been the subject of careful research. Figures 4.11(b) through 4.11(d) show data of Herper et al. [18] that describe the element pattern behavior of axial dipoles in the cylindrical phased array shown in Figure 4.11(a). The dipoles are mounted a distance s from the conducting cylinder ground screen and separated by the circumferential distance b and axial distance d. Figure 4.11(b) compares the H-plane ($\theta = 90$ deg) voltage element gain of a cylindrical array ($b/\lambda = 0.6$; $d/\lambda = 0.7$; $ka = 120$) with that of a planar array with the same lattice dimensions. The essential similarity of the patterns is obvious, since both exhibit a substantial dropoff near 42 deg due to an

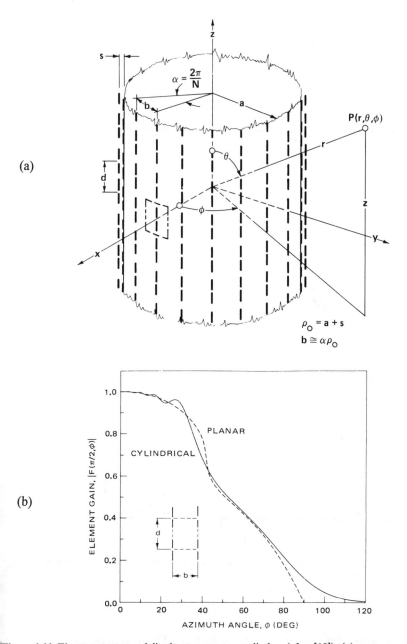

Figure 4.11 Element patterns of dipoles on arrays on cylinders (after [18]): (a) geometry of the circular array of dipoles in a rectangular lattice; (b) H-plane voltage element gain patterns for cylindrical and reference planar arrays ($b/\lambda = 0.6$, $d/\lambda = 0.7$, $ka = 120$).

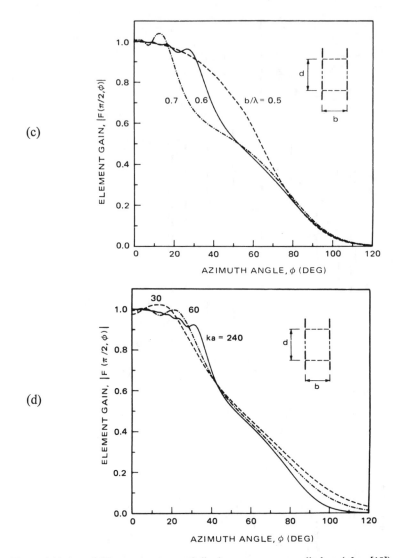

Figure 4.11 (cont.) Element patterns of dipoles on arrays on cylinders (after [18]). H-plane voltage element gain pattern: (c) parameter: ($d/\lambda = 0.7$, $ka = 120$); azimuth spacing $b/\lambda = 0.5$, 0.6, 0.7; (d) parameter: ($b/\lambda = 0.6$, $d/\lambda = 0.7$); $ka = 30$, 60, 240.

endfire grating lobe of the planar structure. In the cylindrical array case, the slope is less steep because of the array curvature that shadows distant elements and so reduces the number of elements that play a role in the endfire grating lobe effect.

The cylindrical array element pattern also has a periodic ripple that is not due to edge effects, but to interference of the single element with the grating lobes of other localized sections of the array excited by creeping waves. This is discussed further in Chapter 6.

Figure 4.11(c) exhibits the dependence of element circumferential spacing (b) on the pattern dropoff and ripple in the $\theta = 90$ deg plane for a fixed-cylinder radius ($ka = 120$). As expected, the pattern broadens with decreasing b/λ and becomes smooth, exhibiting neither the ripple or grating lobe falloff when the spacing is made $\lambda/2$. The broadening is similar to planar array behavior, but is an even more important phenomenon in cylindrical arrays because it is not possible to synthesize low-sidelobe azimuth patterns using element patterns with angle- and frequency-dependent ripples. These results are extremely important, because they first demonstrated that it was possible to obtain well-behaved element patterns by reducing the spacing to $\lambda/2$, and this revealed the potential for forming low-sidelobe radiation patterns with cylindrical arrays.

Figure 4.11(d) shows the dependence of circumferential element patterns on the cylinder radius for fixed element spacing. Several effects are observed. For larger radii, the period of the element pattern ripple gets shorter, its amplitude is reduced, and the endfire grating lobe pattern dropoff moves out to wider angles. The ripple period is apparently reduced because for a given change in observation angle ϕ, a larger number of elements are traversed at the array surface. The ripple amplitude is reduced because of the increased creeping wave loss with increasing ka. In the planar limit, the ripple disappears. Herper et al. give other convincing data that show the pattern slope being constant within the shadow region and proportional to $ka(\sin \theta)$.

A very significant issue in the design of microwave lenses and reflectors is the phase center data shown in Figure 4.12. This figure, due to Tomasic and Hessel [19], shows element pattern data for two arrays of monopoles fed within a parallel plane, as shown in Figure 4.12(a). In this analysis, the arrays were considered infinitely long and were alike except for their element spacings, which were taken as $d/\lambda_c = 0.4$ and 0.6 at center frequency f_c. The other dimensions were $l/\lambda_c = 0.233$ and 0.25, $s/\lambda_c = 0.163$ and 0.245, and $h/\lambda_c = 0.369$, and were kept constant. Figure 4.12(b,c) give the results of the array with $0.6\lambda_c$ separation at frequencies within a 20% frequency band centered at f_c, and depicts significant element pattern distortion at angles ϕ_{EGL} that correspond to an endfire grating lobe condition.

$$\phi_{EGL} = \sin^{-1}(\lambda/d - 1) \tag{4.40}$$

For angles less than ϕ_{EGL}, the far-field amplitude is smoothly varying and phase nearly constant, signifying a well-defined phase center. At approximately ϕ_{EGL}, the amplitude and phase of the element pattern undergo significant changes, and for larger angles there is no phase center and a significantly distorted element

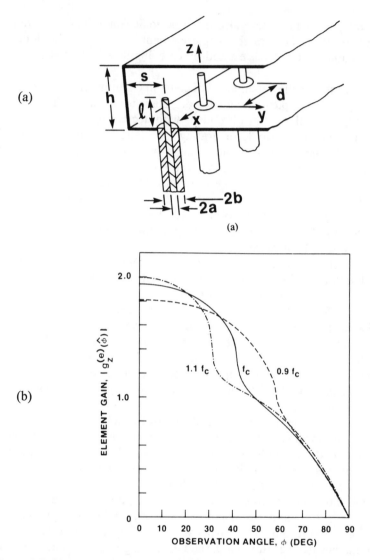

(a)

(a)

(b)

Figure 4.12 (a) Linear array of coaxial monopole elements in semi-infinite parallel plate waveguide (after Tomasic and Hessel [19]); (b) voltage element pattern amplitude for array with $d/\lambda_0 = 0.6$. (Note: backplane spacing optimized for each d/λ_0.)

pattern. The figure also shows the severe frequency dependence of the behavior of such elements, where performance is based primarily on the location of the endfire grating lobe.

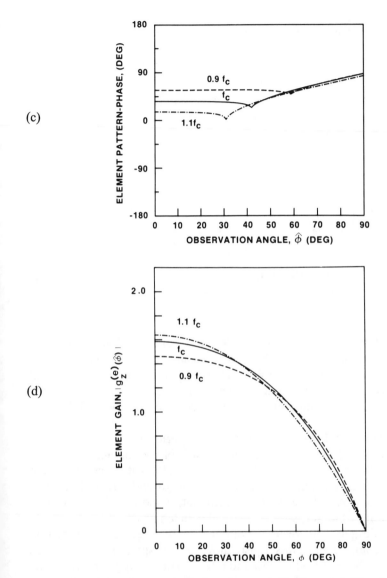

Figure 4.12 (cont.) (c) element pattern phase for array with $d/\lambda_0 = 0.6$; (d) voltage element pattern amplitude for array with $d/\lambda_0 = 0.4$. (Note: backplane spacing optimized for each d/λ_0.)

Figures 4.12(d,e) show the primary result of this important work. For spacings less than 0.5λ, the element pattern amplitude and phase are well behaved out to wide scan angles. These figures show the array pattern of elements spaced nominally

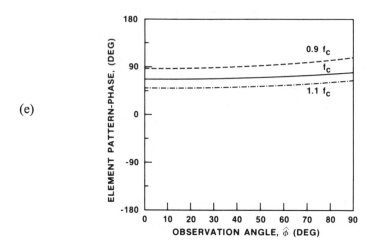

Figure 4.12 (cont.) (e) element pattern phase for array with $d/\lambda_0 = 0.4$. (Note: backplane spacing optimized for each d/λ_0.)

0.4λ at frequencies $0.9f_c$ to $1.1f_c$. The dramatic phase center and amplitude distortion seen in the previous figures for the larger spacing is now eliminated. These results were the first to emphasize the role of element phase centers in conformal arrays and showed conclusively that reduced spacing on the order of 0.4λ to 0.5λ, as is normally required for scanning arrays, is also required for low-sidelobe conformal arrays.

Herper et al. [18] and Tomasic and Hessel [19] performed a detailed study of phase center location and showed that for arrays of monopoles or dipoles mounted in front of the cylindrical ground screen or parallel plate back plane, the phase center of each array element pattern is not at the ground screen, but much closer to the element itself. These data are extremely important for conformal arrays used in the design of lenses, where electrical path lengths must be accurately known for optimum designs. This result is not due to curvature, but to interaction in the large array. Tomasic and Hessel also show that, if spacing is maintained sufficiently small, then both the cylindrical and planar arrays have nearly the same patterns.

Taken together, these two basic references present significant conclusions about conformal array design. Among the most important are:

- One can reliably predict impedance and grating lobe effects using the planar equivalent array.
- The dipole phase center is located near the element and is not on the cylinder surface.

- Mutual coupling effects are less severe for a cylindrical array than for a planar array, provided the element spacing is kept small.
- The active element pattern ripple near broadside and the pattern slope in the shadow region are both determined by creeping wave phenomena and are prime determinants of array minimum sidelobe levels. These two effects require element spacings to be reduced to or below half wavelength for low-sidelobe arrays.

Array Pattern Comparison: Cylindrical Versus Planar Arrays

Figure 4.13(a) shows the pattern of uniformly illuminated sector cylindrical arrays with various cylinder radii. The chosen element pattern has the $\cos(\phi - \phi_n)$ dependence in azimuth out to $|\phi - \phi_n| = 90$ deg and is zero in the shadow region ($|\phi - \phi_n| > 90$ deg). The element spacing is 0.5λ and the array has 36 elements. The two chosen radii are 11.46λ (solid) and 200λ (dashed). On the smaller cylinder, the array occupies a 90-deg sector of the cylinder. These patterns are to be compared with Figure 4.13(b), a 30-element array with 0.6λ spacing between elements, so that the broadside beamwidths are approximately the same when the array is mounted on a very large cylinder (and so is nearly planar). These figures reveal several important characteristics of arrays on cylinders.

The beamwidth is narrowest for the nearly planar arrays, which have the greatest projected length. As the cylinder radius is decreased, the array is wrapped around the cylinder and the beamwidth broadens. Sidelobes rise because many of the element patterns now have their peaks at angles other than $\phi = 0$. As the array curvature increases, the array with 0.6λ (Figure 4.13) spacing develops regions of substantial radiation at wide angles. These broad peaks are grating lobes and result from the fact that near the ends of the array there is a rapid phase variation in the field applied to each element. Figure 4.9 shows that for elements at some angle Φ_0 from broadside ($\phi = 0$), the local phase progression required to form a broadside beam has to be such that a beam coheres at the angle $-\Phi_0$ from the local array normal. Assuming that the array is nearly planar (locally), then this local array section would form a grating lobe at the angle

$$\Phi_{GL} = \Phi_0 + \sin^{-1}[\lambda/d_x - \sin \Phi_0] \qquad (4.41)$$

Since the array is curved, and the local Φ_0 a variable, the grating lobe angle varies, and instead of a replicated main beam at a well-defined angle, there is a broad range of increased radiation. For the array with 0.6λ element spacing, the contribution begins at about 90 deg and ends at about 130 deg. This grating lobe effect becomes more pronounced for larger spacing and greater curvature.

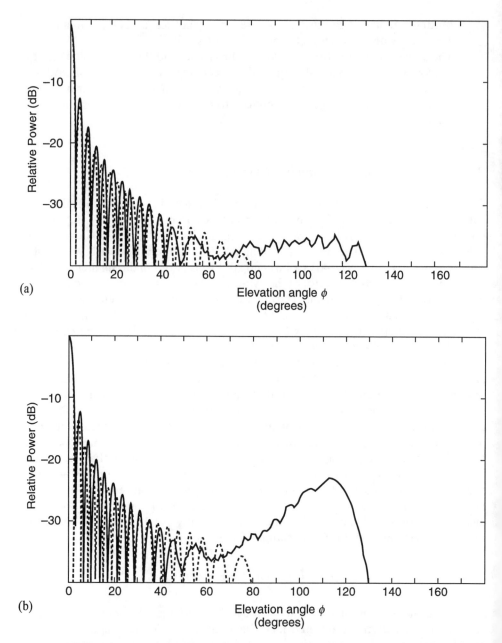

Figure 4.13 Patterns of uniformly illuminated sector arrays on cylinders with radius 11.46λ (90-deg sector—solid curve) and 200λ (dashed curve): (a) 36 element array with 0.5λ spacing; (b) 30 element array with 0.6λ spacing.

Normalized Gain of Cylindrical Sector Arrays

In addition to the obvious differences in array patterns indicated above, there are significant differences in array gain. For a linear array, the antenna gain increases with array length, but for an array wrapped around a given size cylinder there is little advantage to increasing the sector size much beyond 90 deg. For a large array with element patterns $f(\phi - \phi_n)$, all elements matched, and uniform broadside illumination, the array gain at $\phi = 0$ is roughly proportional to the integral of the normalized element pattern

$$\text{Gain} \approx 2a \int_0^{\phi_{\max}} f(\phi')d\phi' \qquad (4.42)$$

Figure 4.14 shows this normalized gain for three element patterns $\cos^n(\phi - \phi')$ for $n = 1, 2, 3$, with all curves normalized to the circle diameter. Notice that for $n = 1$ the integral is just $2a \sin \phi_{\max}$, which is the projection of the array arc onto the plane perpendicular to $\phi = 0$, so the uppermost curve is also the normalized broadside gain of the equivalent planar array and represents the maximum achievable gain. Thus, the projected planar array has the same gain as the cylindrical array, even though in the cylindrical array many of the elements have their maximum gain pointed away from the beam peak, while the planar array has all element peaks at the broadside direction. However, with the cylindrical sector array, there is a higher density of elements near the edges of the projected aperture that compensates for the cosine element pattern. This is discussed again in the next section.

The figure shows that for included angles less than 60 deg, all element patterns yield approximately the same relative gain (since all elements are within ± 30 deg of broadside). At $2\phi_{\max} = 60$ deg, the array with cosine element patterns has 3 dB less gain than the array that spans the half-cylinder $2\phi_{\max} = 180$ deg. Between $2\phi_{\max} = 90$ deg and 120 deg, only about 1 dB is gained using the cosine element pattern.

An example, shown in Figure 4.15 is the data by Hessel [20] that indicate that increasing the array size from a sector with included angle 60 deg to one with an included angle of 120 deg and doubling the number of array elements only increases the array gain by 1 dB, even though the array size is doubled. This result is similar to what would be predicted by Figure 4.12, even though (4.42) is an approximation. An additional disadvantage of the large sector is the large sidelobe near $\phi = 100$ deg. These results are not equally limiting, since the chosen array spacing is 0.65λ and closer element spacing can relieve the grating lobe problem and broaden the element pattern as described earlier; but Figure 4.14 shows that the gain restrictions are fundamental limitations. Unless the cylinder diameter is

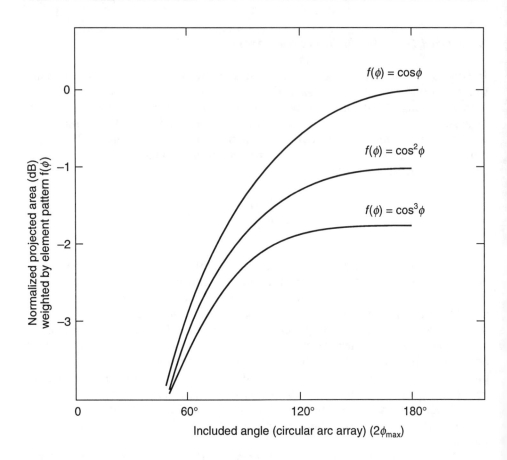

Figure 4.14 Normalized gain for various element patterns.

restricted by mobility or some other constraint, it is clearly advantageous to build a larger array on a larger cylinder rather than cover a larger angular sector of a smaller cylinder.

In the [3], directivity data are given for ring arrays with various sidelobe levels and constant projection ($a \sin \phi_{max}$ constant) for different sector angles ($2\phi_{max} = $ 90, 118 deg) for assumed $\cos(\phi - \phi_n)$ element patterns. These results show that, for a given taper, the directivity is nearly a constant, independent of the cylinder radius a and dependent only on the projection length. This fact again is supported by Figure 4.14, because for these large angular sectors the tapered region near the array edges is compacted by the cylinder curvature.

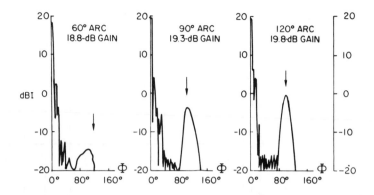

Figure 4.15 Radiation characteristics of sector arrays (after [20]). Maximum gain E-plane patterns for sector array of $ka = 86$, $b/\lambda = 0.65$. Arrays occupy sectors of 60, 90, and 120 deg (assumed $\cos(\phi - \phi_n)$ element pattern).

Pattern Synthesis for Sector Arrays

One of the major problems with circular sector array synthesis is that all the elements of the array have different element patterns according to their location. This situation is depicted in Figure 4.16 and precludes use of all of the standard synthesis methods. One can, however, control near sidelobes by projecting the array element locations and element patterns onto a plane tangent to the cylinder, as shown in Figure 4.16.

For example, Figure 4.16 depicts a circular sector array with elements located equally spaced around the circumferential sector with angular separation $\Delta\phi$. Projected onto the array tangential plane, the element locations of an array of NT elements are

$$y_n = a \sin \phi_n$$
$$= a \sin[n\Delta\phi] \tag{4.43}$$

Defining the length of the projected array as

$$L = 2a \sin[(NT/2)\Delta\phi] \tag{4.44}$$

and sampling the aperture distribution at points y_n automatically accounts for the extra one-half element on each side of the array when sampling the Taylor or other traditional distribution, as indicated in Chapter 3.

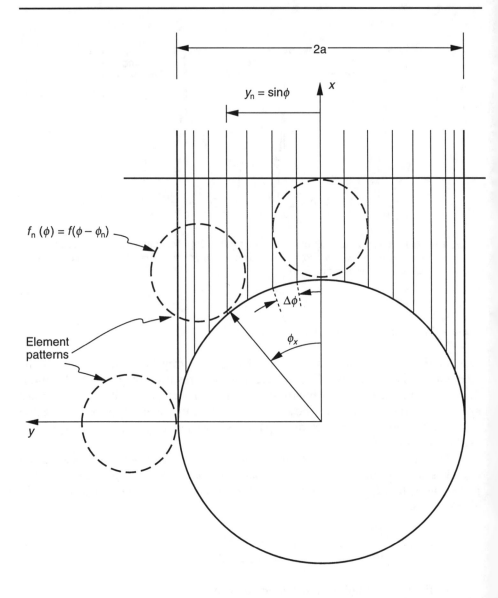

Figure 4.16 Array element patterns and projection to synthesize low-sidelobe pattern.

The projection tends to make the points y_n closer together near the ends of the array, but the element pattern tends to make up for that effect. For example, for relatively large ϕ_n, the projected spacing between elements varies approximately like $\cos \phi_n$, and so the density of elements has a $1/\cos \phi_n$ dependence. If the array element patterns vary like cos ϕ_n, then near broadside the projected array weighting will be correct without altering the weights from those of a linear array with omnidirectional element patterns. If the element pattern is very different from a cosine, that fact must be included in choosing the weights.

Figure 4.17 shows a sequence of patterns that are synthesized by projecting a -40-dB Taylor pattern with $\bar{n} = 8$ onto cylinders with various radii and 36 elements. The element patterns relative to the local cylinder normal at ϕ_n has the form

$$f(\theta, \phi) = \sin \theta \cos(\phi - \phi_n) \tag{4.45}$$

which corresponds to cosine element patterns in both planes. The sequence of curves shows a nearly perfect Taylor pattern at radius $a = 100\lambda$, which degrades little for $a = 20\lambda$. The patterns at $a = 11.46\lambda$ and 8.59λ correspond to arrays that occupy 90- and 120-deg sectors of the cylinder, and these differ significantly from the Taylor pattern on the large cylinder (100λ). Significant changes in these patterns are beam broadening that results from shortening the array length and increased sidelobe levels near the main beam, with the first sidelobe starting to merge with the main beam for small cylinders.

The synthesis method first suggested by Sureau and Keeping [21], further developed by Dufort [22] and Olen and Compton [23], and described in Chapter 3 is readily applicable to conformal arrays on generalized surfaces. The technique uses adaptive optimization algorithms to form the array pattern in the presence of closely spaced sources of interference that are tailored or iterated [23] to achieve the desired pattern. Figure 4.18 shows results due to Sureau and Keeping [21] for a circular sector array of 32 identical elements disposed over a 120-deg sector of a cylinder and displaced 0.55λ between elements. The geometry shown in Figure 4.18(a) shows the 32 excited elements of a 96-element circular array. A total of 372 sources of interference were uniformly distributed outside of the main beam window to control sidelobes. The measured element pattern was used in the calculations. Figure 4.18(b) shows that increasing the window width produces a set of patterns with progressively lower sidelobes, but decreasing aperture efficiency. Sureau and Keeping also investigated varying the interference weights to control sidelobe decay and the use of asymmetric weights for monopulse pattern control. Chapter 3 gives other results due to Olen and Compton [23], who extended the results of Sureau and Keeping to produce a convergent iterative solution for detailed pattern control.

Unlike linear and planar arrays, when the pattern of a conformal array is scanned, the relative contributions of element patterns on either side of the array center are different, and the array pattern is distorted because of this asymmetry. Hannon et al. [24] and later Antonucci and Franchi [25] addressed this problem by superimposing an odd (monopulse) excitation along with the even array power

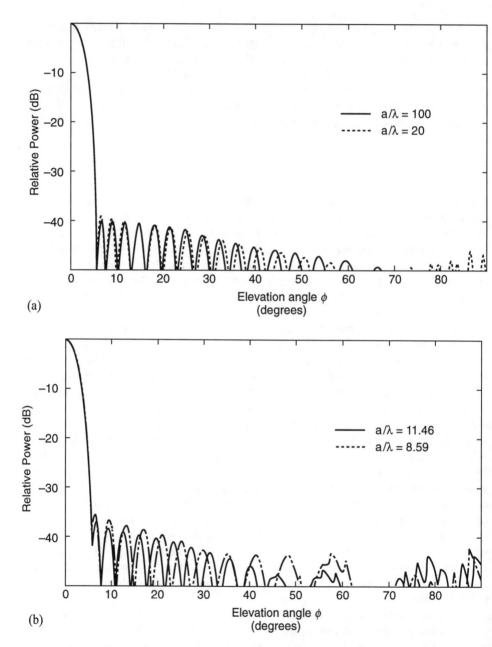

Figure 4.17 Synthesized patterns of circular sector array with projected Taylor distribution $\bar{n} = 8$, sidelobe level -40 dB: (a) $a/\lambda = 100$ (solid), 20 (dashed); (b) $a/\lambda = 11.46$ (90-deg sector) (solid), 8.59 (120-deg sector) (dashed).

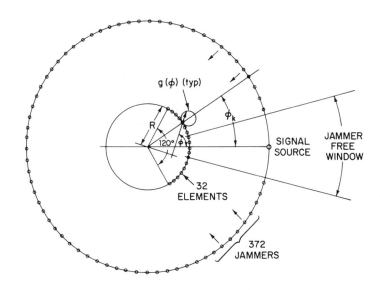

Figure 4.18(a) The synthesis procedure of Sureau and Keeping [21]: cylindrical array and signal environment.

distribution to produce the appropriate asymmetry to cancel the contribution from asymmetric element patterns.

Comparison Between Cylindrical and Multiface Planar Arrays

Cylindrical arrays are usually not phase-scanned in the azimuth plane because of the resulting pattern distortion, and so may rely entirely on the amplitude commutating network for azimuth scan. In systems where there are phase shifters at every element (perhaps for elevation scan or fine scanning), the requirements for the switching matrix can be relieved by scanning over limited angular sectors between commutated beam positions. One cannot scan more than a few degrees in this manner without incurring significant pattern distortion, so the phase shift option is balanced against cost for a particular system. In either case, one can assume that the circular array gain is nearly constant with scan angle. In comparison, each face of a four-faced array must scan to ± 45 deg to cover the 360 deg of azimuth field. A rough estimate of the relative number of elements in the cylindrical and four-faced planar arrays is obtained using the proportionality argument above and assuming a $\cos(\phi - \phi_n)$ element pattern for both arrays. From (4.42), the cylindrical array gain is proportional to $2a \sin \phi_{max}$, where ϕ_{max} again is one-half

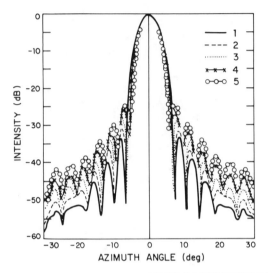

PATTERN	WINDOW WIDTH (deg)	BEAMWIDTH 3 dB (deg)	PEAK SL (dB)	APERTURE EFFICIENCY (dB)
1	13.06	5.1	-33.0	-0.924
2	12.16	5.0	-30.5	-0.787
3	11.25	4.76	-28.0	-0.644
4	10.34	4.20	-26.5	-0.508
5	9.42	4.36	-23.5	-0.377

Figure 4.18(b) The synthesis procedure of Sureau and Keeping [21]: optimum symmetric patterns.

the sector array subtended angle. The corresponding gain of one of the four-face planar faces of length L is proportional to $L \cos 45$, or $0.707L$. If, for the cylindrical sector array, a 90-deg sector angle is chosen ($\phi_{max} = 45$ deg), then the two gains are equal at $2a = L$, when the cylinder is tangent to the four faces of the square with sides L. In this case, the cylindrical array is required to have approximately $\pi/4$ or 79% of the elements of the four-faced array if the element spacing is taken as the same for both. However, as pointed out earlier, if high-quality sidelobe control is required, it may be necessary to space the cylinder elements near $\lambda/2$, while the linear array spacing for a 45-deg scan can be about 0.58λ. This means that the cylindrical array would have about 92% of the elements of the four-faced array. If a 120-deg sector is used, the relative size of the circular sector to four-faced array dimensions is given by setting the projection $2a$ (0.866) of the circular array equal to that of the four-faced array at 45 deg or $0.707L$. The ratio of the number of elements in the cylinder and the four faces is $(2\pi a/4L)$, or 64%, and so the cylindrical array can require substantially fewer elements than the four-faced array if minimum gain is the chosen criterion. This advantage is mostly lost if low sidelobes are required, since sector arrays that occupy up to 120 deg of the cylinder can have significant distortion, even at broadside, so elements need to be closely spaced.

Other practical concerns enter into the selection of cylinder or four-faced planar arrays. The cylindrical array is often required to have phase shifters at each element in order to provide elevation scanning, and these enable some simplification of the required commutating feed structure. In this configuration, the cylindrical array has nearly constant gain in azimuth. There is no need for a high-power switch to select the transmit array face as there is for the four-faced planar array. Moreover, if broadband radiation is required, then the cylindrical array has the advantage that time-delay cables can be built into the commutating matrix and the array may not require variable time-delay units. Alternatively, the cylindrical array commutating network is usually lossy and therefore substantially reduces gain or requires the use of amplification to overcome the loss. In addition, because of the need for commutation, the cylindrical array is usually organized into column subarrays, with the commutator switch exciting only one column input. This reduces the number of degrees of freedom, and the phase errors between columns become very significant and must be minimized if low sidelobes are important. This can lead to a requirement for phase comparator networks and the active real time correction of column phase.

The required scan sector also enters into the choice of cylindrical or multiface planar array. If it is required that the array scan to angles near zenith, a multiface array with faces tilted back or a truncated conical array must be chosen. References [17,26,27] give some details of tilted multiface planar arrays for wide sector coverage.

4.3 SPHERICAL AND HEMISPHERICAL ARRAYS

Spherical arrays are most often fed by exciting elements in groups or subarrays. The radiated pattern of an array of elements located at positions on the surface of a sphere or hemisphere is given by (4.8), with

$$\mathbf{r}'_{nk} = \hat{\mathbf{x}} x_{nk} + \hat{\mathbf{y}} y_{nk} + \hat{\mathbf{z}} z_{nk}$$
$$x_{nk} = a \sin \theta_k \cos \phi_n = a u_{nk} \qquad (4.46)$$
$$y_{nk} = a \sin \theta_k \sin \phi_n = a v_{nk}$$

and so

$$\mathbf{r}'_{nk} \cdot \hat{\mathbf{\rho}} = a[uu_{nk} + vv_{nk} + \cos \theta \cos \theta_k]$$
$$= a[\sin \theta_k \sin \theta \cos(\phi - \phi_n) + \cos \theta \cos \theta_k]$$

and

$$F(\theta, \phi) = \sum_n \sum_k I_{nk} f_{nk}(\theta, \phi) e^{jkr' \cdot \hat{\mathbf{\rho}}} \qquad (4.47)$$

Spherical and hemispherical arrays have the same limitations as cylindrical arrays [28] and must be fed by commutating an illuminated distribution to various points on the surface of the body. These have been fed by switch matrices to excite active sectors of the sphere. The largest hemispherical arrays have been lenses fed by scanning arrays because this is an effective and low-cost means of RF power commutation. In this configuration, called the DOME [29,30] antenna concept (Figure 4.19), the sphere is a passive lens with inserted phase shifters to collimate the distributed signal received from the array feed. The required nonlinear phase progression is selected to achieve a scan angle that is some fixed multiplier $K\theta$ ($K > 1$) times the scan angle of the feed array. The primary purpose of the DOME structure is to provide economical hemispherical coverage, and it can even provide coverage somewhat below the horizon with proper tailoring of the fixed phase shifts in the DOME geometry.

4.4 TRUNCATED CONICAL ARRAYS

Another important conformal array geometry is the truncated conical array. Shaped to suit missile and aircraft nose cones, the truncated conical geometry is nearly cylindrical if the cone angle is small and the array truncated far from the cone tip. Like the cylindrical array, the truncated conical array is usually fed by moving an illuminated region around the cone by means of switching matrices.

The field pattern of a truncated array is given below as the sum of patterns from each constituent circular loop array using (4.8):

$$F(\theta, \phi) = \sum F_k(\theta, \phi) \qquad (4.48)$$

The various radii a_k are given as

$$a_k = a_0 - z_k \sin \delta \qquad (4.49)$$

for cone half-angle δ. The current phases are given to collimate the radiated beam:

$$I_{nk} f_{nk} = |I_{nk} f_{nk}| \exp[-jk(a_k \sin \theta \cos(\phi - n\Delta\phi_k) + z_k \cos \theta)] \qquad (4.50)$$

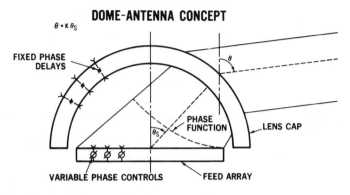

Figure 4.19 Hemispherical dome array for scan to $\theta = K\theta s$.

for element patterns f_{nk}. Mutual coupling and array active element patterns are determined by asymptotic methods, by approximate methods, or by full-wave expansions, as discussed in [1].

REFERENCES

[1] Borgiotti, G. V., "Conformal Arrays," Ch. 11 in *The Handbook of Antenna Design*, Vol. 2, A. Rudge et al., eds., Milne, Olver, Knight, Peter Peregrinus, 1987.

[2] Davies, D. E. N., "Circular Arrays," Ch. 12 in *The Handbook of Antenna Design*, Vol. 2, A. Rudge et al., eds., London: Peter Peregrinus, 1987.

[3] Hansen, R. C., ed., *Conformal Antenna Array Design Handbook*, Dept. of the Navy, Air Systems Command, Sept. 1981, AD A110091.

[4] Sheleg, B., "A Matrix-Fed Circular Array for Continuous Scanning," *IEEE Trans.*, Vol. AP-56, No. 11, Nov. 1968, pp. 2016–2027.

[5] Davies, D. E. N., "A Transformation Between the Phasing Techniques Required for Linear and Circular Aerial Arrays," *Proc. IEE*, Vol. 112, Nov. 1965, pp. 2041–2045.

[6] Rahim, T., and D. E. N. Davies, "Effect of Directional Elements on the Directional Response of Circular Arrays," *Proc. IEE*, Vol. 129, Part 11, No. 1, Feb. 1982, pp. 18–22.

[7] Provincher, J. H., "A Survey of Circular Symmetric Arrays," *Phased Array Antennas, Proc. 1970 Phased Array Antenna Symp.*, June 1972, Dedham, MA: Artech House, A. Oliner and G. Knittel, eds., 1972, pp. 292–300.

[8] Hill, R. J., "Phased Array Feed Systems, A Survey," *Phased Array Antennas*, A. Oliner and G. Knittel, eds., Dedham, MA: Artech House, 1972, pp. 197–211.

[9] Giannini, R. J., "An Electronically Scanned Cylindrical Array Based on a Switching and Phasing Technique," *IEEE Int. Symp. Antennas and Propagation Dig.*, Dec. 1969, pp. 199–207.

[10] Boyns, J. E., C. W. Gorham, A. D. Munger, J. H. Provencher, J. Reindel, and B. I. Small, "Step-Scanned Circular Array Antenna," *IEEE Trans.*, Vol. AP-18, No. 5, Sept. 1970, pp. 590–595.

[11] Holley, A. E., E. C. Dufort, R. A. Dell-Imagine, "An Electronically Scanned Beacon Antenna," *IEEE Trans.*, Vol. AP-22, No. 1, Jan. 1974, pp. 3–12.

[12] Honey, R. C., and E. M. T. Jones, "A Versatile Multiport Biconical Antenna," *Proc. IRE*, Vol. 45, Oct. 1957, pp. 1374–1383.

[13] Bogner, B. F., "Circularly Symmetric R.F. Commutator for Cylindrical Phased Arrays," *IEEE Trans.*, Vol. AP-22, No. 1, Jan. 1974, pp. 78–81.

[14] Irzinski, E. P., "A Coaxial Waveguide Commutator Feed for a Scanning Circular Phased Array," *IEEE Trans.*, Vol. MTT-29, No. 3, March 1981, pp. 266–270.

[15] Shelton, P., "Application of Hybrid Matrices to Various Multimode and Multi-beam Antenna Systems," *IEEE Washington Chapter PGAP Meeting*, March 1965.

[16] Skahil, G., and W. D. White, "A New Technique for Feeding a Cylindrical Array," *IEEE Trans.*, Vol. AP-23, March 1975, pp. 253–256.

[17] Mailloux, R. J., "Conformal and Low-Profile Arrays," Ch. 21 in *Antenna Engineering Handbook*, R. C. Johnson and H. Jasik, eds., New York: McGraw-Hill, 1984.

[18] Herper, J. C., A. Hessel, and B. Tomasic, "Element Pattern of an Axial Dipole in a Cylindrical Phased Array—Part I: Theory," *IEEE Trans.*, Vol. AP-33, No. 3, March 1983, pp. 259–272.

[19] Tomasic, B. and A. Hessel, "Linear Array of Coaxially Fed Monopole Elements in a Parallel Plate Waveguide—Part I: Theory," *IEEE Trans.*, Vol. AP-36, No. 4, April 1988, pp. 449–462.

[20] Hessel, A., "Mutual Coupling Effects in Circular Arrays on Cylindrical Surfaces—Aperture Design Implications and Analysis," *Phased Array Antennas*, A. Oliner and G. Knittel, eds., Dedham, MA: Artech House, 1972, pp. 273–291.

[21] Sureau, J. C., and K. J. Keeping, "Sidelobe Control in Cylindrical Arrays," *IEEE Trans.*, Vol. AP-30, No. 5, Sept. 1982, pp. 1027–1031.

[22] Dufort, E. C., "Pattern Synthesis Based on Adaptive Array Theory," *IEEE Trans.*, Vol. AP-37, 1989, pp. 1017–1018.

[23] Olen, C. A., and R. T. Compton, Jr., "A Numerical Pattern Synthesis Algorithm for Arrays," *IEEE Trans.*, Vol. AP-38, No. 10, Oct. 1990, pp. 1666–1676.

[24] Hannon, P., and E. Newmann, "Study and Design of a Cylindrical Lens Array Antenna for Wideband Electronic Scanning," RADC-TR-83-128.

[25] Antonucci, J., and P. Franchi, "A Simple Technique to Correct for Curvature Effects on Conformed Phased Arrays," *Proc. 1985 Antenna Applications Symposium*, RADC/TR-85-742, Vol. 2, Dec. 1985.

[26] Knittel, G. H., "Choosing the Number of Faces of a Phased Array for Antenna for Hemisphere Scan Coverage," *IEEE Trans.*, Vol. AP-13, Nov. 1965, pp. 878–882.

[27] Corey, L., "A Graphical Technique for Determining Optimal Array Antenna Geometry," *IEEE Trans.*, Vol. AP-33, No. 7, July 1985, pp. 719–726.

[28] Schrank, H. E., "Basic Theoretical Aspects of Spherical Phased Arrays," *Phased Array Antennas*, A. A. Oliner and G. H. Knittel, eds., Dedham, MA: Artech House, 1972.

[29] Schwartzman, L., and J. Stangel, "The Dome Antenna," *Microwave Journal*, Oct. 1975, pp. 31–34.

[30] Steyskal, H., A. Hessel, and J. Schmoys, "On the Gain-Versus-Scan Trade-Offs and the Phase Gradient Synthesis for a Cylindrical Dome Antenna," *IEEE Trans.*, Vol. AP-27, Nov. 1979, pp. 825–831.

Chapter 5
Elements, Transmission Lines, and Feed Architecture for Phased Arrays

5.1 ARRAY ELEMENTS

Of the many different kinds of elements used in array systems, most can be considered either as wire antennas, or slots, or a combination of these. Most arrays are designed with conducting ground screens, and so the potential functions introduced in Chapter 2 can be used to evaluate near-field coupling effects as well as far-field radiation. Recently, many arrays have been built using printed circuit dipoles or microstrip patch antennas, and the use of dielectric substrates above the metallic ground screen requires a more complex analytical formulation than that of Chapter 2. Similarly, arrays built of dielectric rods or other dielectric elements require a more generalized formulation.

The intent of this chapter is to catalog a body of technology that constitutes the hardware of phased arrays. Since the chapter deals with isolated elements, and the behavior of these elements is not directly relevant to their behavior in the array, the chapter presents only simple approximate equations for element impedance and radiation patterns. Many of the equations are for the resonant cases, even though there exists a vast body of technical literature for isolated elements of various resonant and nonresonant dimensions. Resonant data are given here to address the engineering problem of matching element impedance to the feed transmission line. This chapter will then help the reader choose transmission lines and element types to suit required pattern coverage and system bandwidth. The detailed evaluation of element behavior in a scanned array and the element patterns in an array environment will be discussed for some of these elements in Chapter 6.

5.1.1 Polarization Characteristics of Infinitesimal Elements in Free Space

The radiation pattern of any element is obtained from the integral over the currents or aperture fields of the given element. In many cases, the element can be considered as composed of straight, infinitesimal current carrying wire elements or filaments of tangential electric field in an aperture. The normalized element patterns from such isolated infinitesimal elements, as shown in Figure 5.1(a,b), are given below.

The radiation fields of electric current filaments in free space are readily derived from the vector potential. In the far zone, the normalized electric field radiated from an infinitesimal current source with components as shown in Figure 5.1(a) are given below, as obtained from (2.1) and (2.2).

Source Current	Normalized (Voltage) Element Pattern	Normalized Power Pattern	
I_x	$(\hat{\boldsymbol{\theta}} \cos\theta \cos\phi - \hat{\boldsymbol{\phi}} \sin\phi)/2^{1/2}$	$1 - \sin^2\theta \cos^2\phi$	
I_y	$(\hat{\boldsymbol{\theta}} \cos\theta \sin\phi + \hat{\boldsymbol{\phi}} \cos\phi)/2^{1/2}$	$1 - \sin^2\theta \sin^2\phi$	(5.1)
I_z	$\hat{\boldsymbol{\theta}} \sin\theta$	$\sin^2\theta$	

Due to the choice of coordinate system, the z-directed current results in a particularly simple form with a single component of polarization ($\hat{\boldsymbol{\theta}}$). Figure 5.1(c) shows a sketch of the electric fields radiated by an isolated infinitesimal vertical monopole, and Figure 1.1(d) shows the elevation power pattern ($\sin^2\theta$) of the infinitesimal monopole with current I_z. The dashed pattern is that of an isolated resonant half-wave dipole and has the form $[\cos(\pi/2 \cos\theta)/\sin\theta]^2$.

If the current is primarily in the x- or y-direction, it may be convenient to define a different coordinate system to correspond to the axis of the wire. For a dipole with its axis in the x-direction, redefining the coordinate system to that of Figure 5.1(e) results in the electric field normalized element pattern

$$f(\Phi) = \hat{\boldsymbol{\Phi}} \sin\Phi \tag{5.2}$$

In the case of radiation from an aperture in a ground screen, one must specify the plane of the ground screen in addition to the electric field component. Assuming a ground screen in the plane $z = 0$, with an infinitesimal slot aperture and electric field elements E_x, E_y as shown in Figure 5.1(b), the radiated far fields are given from (2.17).

Field	Normalized (Voltage) Element Pattern	Normalized Power Pattern	
E_x	$(\hat{\boldsymbol{\theta}} \cos\phi - \hat{\boldsymbol{\phi}} \cos\theta \sin\phi)/2^{1/2}$	$1 - \sin^2\theta \sin^2\phi$	(5.3)
E_y	$(\hat{\boldsymbol{\theta}} \sin\phi + \hat{\boldsymbol{\phi}} \cos\theta \cos\phi)/2^{1/2}$	$1 - \sin^2\theta \cos^2\phi$	

Actual elements have more complex polarization than these filaments, but their polarization can be considered as made up of contributions from the various filamentary currents and aperture fields.

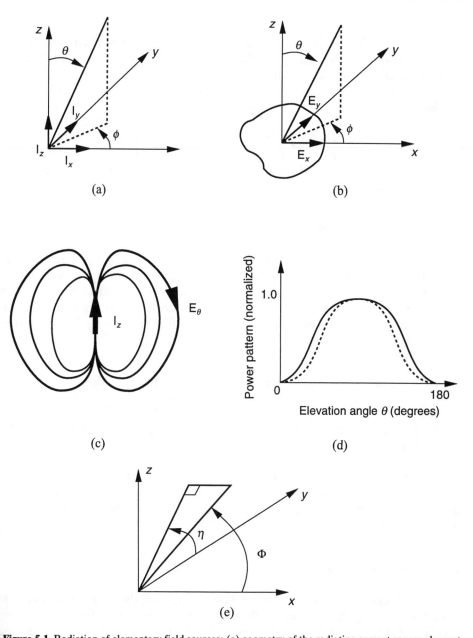

Figure 5.1 Radiation of elementary field sources: (a) geometry of the radiating current source element; (b) geometry of the radiating electric field (magnetic current) element; (c) electric field radiated from vertical dipole; (d) far-field of a dipole element with current I_z: ———— infinitesimal element ($\cos^2 \theta$), $\cdots\cdots$ resonant half-wave dipole $[\cos(\pi/2 \cos \theta)/\sin \theta]^2$; (e) alternate coordinate system for dipole with current I_x.

The radiated polarization of the element often changes when the array is scanned, because the interaction of other elements may cause currents or aperture fields to be excited which are not present in an isolated element.

Selection of coordinates to define primary polarization and crossed polarization for a given antenna can be handled in several ways. Ludwig [1] discusses three definitions of polarization and makes a convincing argument in the case of reflector systems for defining polarization coordinates that correspond to the natural coordinates for an azimuth/elevation measurement. This set of coordinates is also the one that corresponds to the polar system used throughout this text, with θ corresponding to the elevation tilt.

5.1.2 Electric Current (Wire) Antenna Elements

Most wire antenna elements used in arrays are variations of the dipole or the monopole. These elements are well understood when used separately or in the array environment. Since their radiation in an array is very different than when used as isolated elements, it is established practice to perform the full mutual coupling analysis to evaluate array performance before completing the design (see Chapter 6).

Effective Radius of Wire Structures With Noncircular Cross Section

The actual wire cross section does not significantly alter the radiation properties of the element. This is usually accounted for by defining an effective radius for the wire with a noncircular cross section. This effective radius is given below for a wire or group of wires of given cross sections.

Balanis [2, Table 8.3] lists most of the useful equivalent cross sections, and so is reproduced here as Figure 5.2. In addition, Tai [3] gives a table (Table 5.1 here) with equivalent radii a_{eq} of regular polygons in terms of the radius a of the outscribed circles.

The Dipole and the Monopole

Undoubtedly the most studied of any radiating structures, these basic wire antennas have been thoroughly analyzed as elements alone or in arrays. Figure 5.3 shows several orientations of dipoles and monopoles as commonly used in array antennas. In this figure, the monopole height is shown as the dimension A or ℓ interchangeably, in accordance with the references used throughout the chapter. The fundamental work of King and others [4–6] has led to convenient approximate results, which accurately describe not only the radiation patterns and radiation resistance,

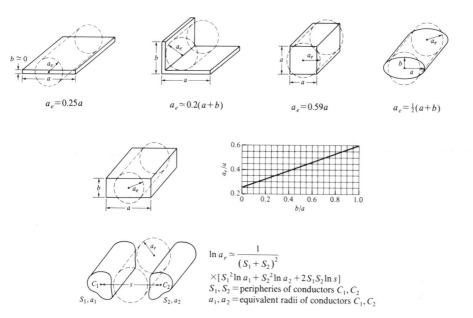

$a_e = 0.25a$ $a_e \simeq 0.2(a+b)$ $a_e = 0.59a$ $a_e = \frac{1}{2}(a+b)$

$$\ln a_e \simeq \frac{1}{(S_1 + S_2)^2}$$
$$\times [S_1{}^2 \ln a_1 + S_2{}^2 \ln a_2 + 2S_1 S_2 \ln s]$$
S_1, S_2 = peripheries of conductors C_1, C_2
a_1, a_2 = equivalent radii of conductors C_1, C_2

Figure 5.2 Conductor geometrical shape and their equivalent circular cylinder radii (after Balanis [2]).

Table 5.1
Equivalent Radii of Regular Polygons

n	a_{eq}/a
3	0.4214
4	0.5903
5	0.7563
6	0.9200

Note: n = the number of sides; a_{eq} = the equivalent radius; a = radius of oustscribed circle.
Source: [3].

but also near-field effects and mutual coupling. Even today, with accurate numerical procedures and convenient computer codes available, these approximate formulas still provide a valuable resource for handling large arrays. Chapter 6 briefly discusses this and other (largely numerical) procedures for calculating the array performance for a variety of elements.

A thin dipole at resonance [2] presents a input impedance of approximately

$$R_{\text{dipole}} = 73\Omega \qquad (5.4)$$

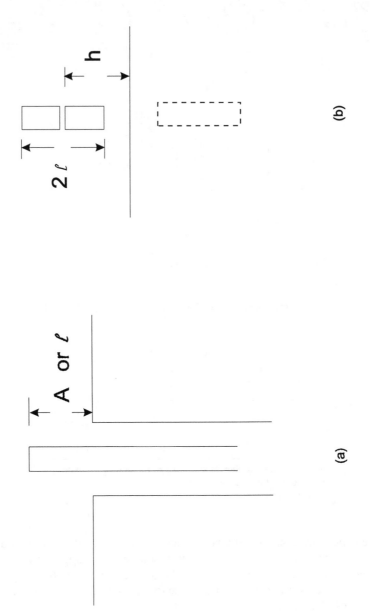

Figure 5.3 Basic wire radiating elements: (a) vertical monopole excited by coaxial feed; (b) vertical dipole and its image.

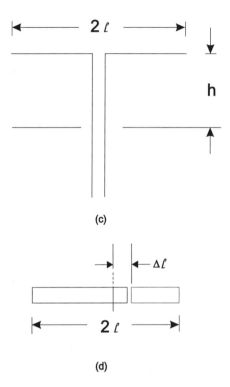

(c)

(d)

Figure 5.3 (cont.) Basic wire radiating elements: (c) horizontal dipole antenna; (d) off-center excitation of the dipole.

This impedance is not difficult to match to 50Ω transmission lines, and a number of convenient matching circuits have been designed to make the transition from various coaxial and other transmission lines.

At resonance, the thin vertical monopole (Figure 5.3(a)), used with a conducting screen, has half the input resistance of the dipole, or 36.5Ω. Figure 5.4(a) shows the measured resistance, and Figure 5.4(b) the reactance of a monopole of various electrical lengths (A) as a function of the element diameter D. The figure shows monopole reactance for elements with heights up to 240 electrical deg (with 360 deg representing one wavelength).

This range of monopole lengths extends beyond the first resonance and antiresonance. This experimental data, due to Brown and Woodward [7], describe monopole antennas over a large but finite ground screen and excited by a coaxial line, so it includes the effect of the impedance change at the junction between the coaxial feed and monopole antenna. For monopoles near a half wavelength, there

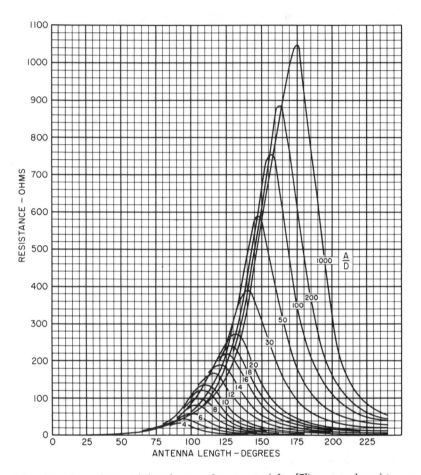

Figure 5.4(a) Impedance characteristics of monopole antennas (after [7]): monopole resistance versus length A, for diameter D.

is substantial difference in resistance with varying monopole thickness, but for heights less than the first resonant length (around 90 deg), there is relatively little dependence on monopole thickness.

The input impedance of an isolated dipole or monopole is found from an approximate solution to Hallen's or Pocklington's equation [8], but for dipoles of length 2ℓ to about a half wavelength, this simplified formula [3] is useful.

$$Z_i = R(k\ell) - j[120[\ln(2\ell/a) - 1] \cot(kh) - X(k\ell) \tag{5.5}$$

where $k = 2\pi/\lambda$.

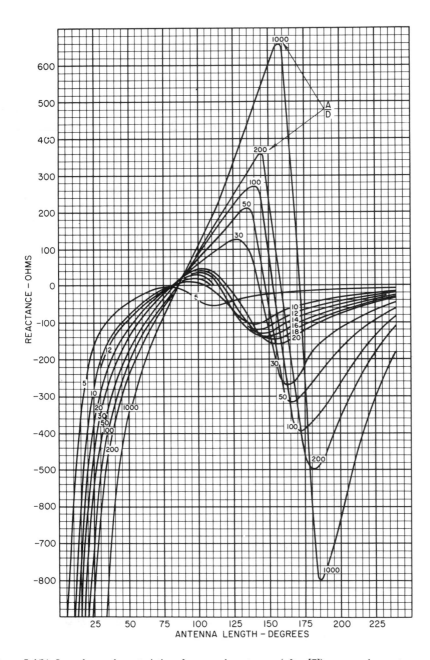

Figure 5.4(b) Impedance characteristics of monopole antennas (after [7]): monopole reactance versus length A.

The functions R and X are shown in Figure 5.5 [3] and are tabulated in the reference for $0 \le 2\ell \le \pi/2$. Elliott [9] has represented these functions with second-degree polynomials and so gives the following form, valid over the range $1.3 \le k\ell \le 1.7$ and $0.0016 \le a/\lambda \le 0.0095$.

$$Z = [122.65 - 204.1\, k\ell + 110(k\ell)^2] \\ - j[120(\ln(2\ell/a) - 1)\cot k\ell - 162.5 + 140k\ell - 40(k\ell)^2] \tag{5.6}$$

As noted earlier, the resistance is not significantly dependent on the dipole radius, and this is reflected in the form of the function R. Figure 5.5(b) shows the computed dipole resistance and reactance of an isolated dipole as computed by Elliott from (5.6). Values for the monopole are half those for the dipole.

The radiation pattern of a thin, vertical half-wave dipole in free space, or quarter-wave monopole (with ground screen), is approximately given by the following relationship. This expression is readily derived from (2.1), (2.2), and (2.15) by assuming a sinusoidal current distribution:

$$\mathbf{f}(\theta, \phi) = \hat{\mathbf{\theta}}\left\{\frac{\cos(\pi/2\,\cos\theta)}{\sin\theta}\right\} \tag{5.7}$$

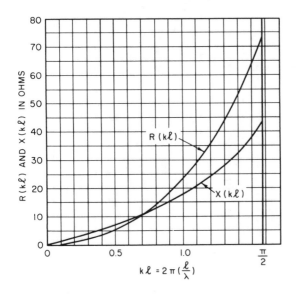

$k\ell$	$R(k\ell)$	$X(k\ell)$
0	0	0
0.1	0.1506	1.010
0.2	0.7980	2.302
0.3	1.821	3.818
0.4	3.264	5.584
0.5	5.171	7.141
0.6	7.563	8.829
0.7	10.48	10.68
0.8	13.99	12.73
0.9	18.16	15.01
1.0	23.07	17.59
1.1	28.83	20.54
1.2	35.60	23.93
1.3	43.55	27.88
1.4	52.92	32.20
1.5	64.01	38.00
$\pi/2$	73.12	42.46

Figure 5.5(a) Impedance functions and impedance of isolated dipole: impedance functions $R(k\ell)$ and $X(k\ell)$ (after [3]).

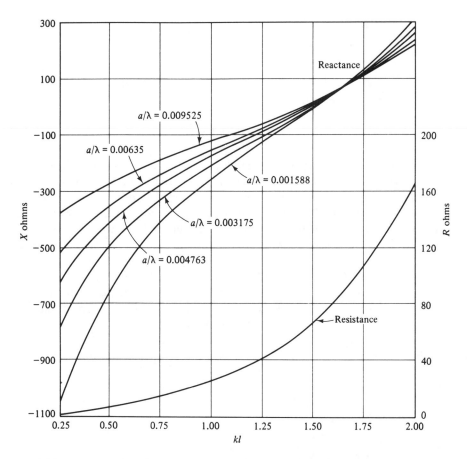

Figure 5.5(b) Impedance functions and impedance of isolated dipole: dipole resistance and reactance (after [9]).

The relationship shows the characteristic doughnut-shaped pattern, invariant in ϕ and with a null at $\theta = 0$ and a maximum at $\theta = \pi/2$. The pattern, shown for comparison (dashed) in Figure 5.1(d), is evidently much like the sin θ dependence of the infinitesimal dipole, but somewhat more directive, as befits the longer wire element.

The *vertical dipole* of Figure 5.3(b) has some application to arrays, but its use is restricted to situations where very restricted elevation coverage is required. The vertical dipole pattern is given by (5.7), modified by the contribution of the image with center at $z = -h$, as in the figure.

$$\hat{\mathbf{f}}(\theta, \phi) = \hat{\mathbf{\theta}} \left\{ \frac{\cos(\pi/2 \cos \theta)}{\sin \theta} \right\} \cos((2\pi h/\lambda) \cos \theta) \qquad (5.8)$$

Elevation coverage is restricted because the array ground screen image creates a zero in the elevation plane at

$$\theta = \cos^{-1}(\lambda/(4h)) \qquad (5.9)$$

Techniques for exciting vertical dipoles are summarized later in the chapter.

The *vertical monopole* has found use in a number of high-frequency ground radar systems, where low-angle coverage is at a premium and there is no coverage requirement near the zenith ($\theta = 0$), where the pattern (5.1) has a natural zero. As shown in Figure 5.3(b), the monopole is conveniently excited from a coaxial line beneath the ground screen. A recent treatment of the monopole array is given by Fenn [10].

The *horizontal dipole* (Figure 5.3(c)) is one of the most useful array elements. At resonance, the horizontal dipole, suspended over a conducting ground screen and with its axis oriented along the x-axis, has a pattern and input resistance that is strongly dependent on the height h above the ground screen.

Figure 5.6(a) shows the input resistance and reactance of a $\lambda/2$ dipole over a conducting plane. For height "h" very small, the input resistance and reactance

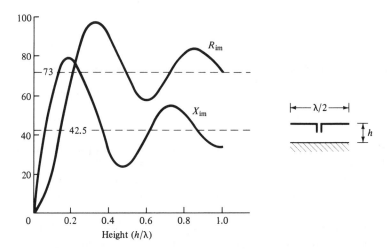

Figure 5.6(a) Impedance and elevation pattern of horizontal dipole over ground screen: impedance of horizontal $\lambda/2$ dipole versus height h above ground (after [11]).

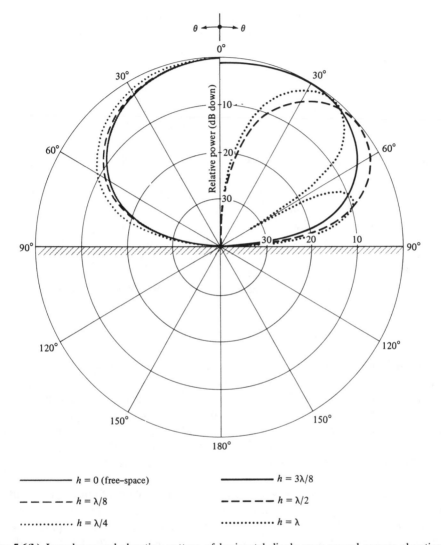

Figure 5.6(b) Impedance and elevation pattern of horizontal dipole over ground screen: elevation patterns of horizontal infinitesimal dipole versus height h (after [2, p. 145]).

both approach zero as the image and the direct radiation cancel. With increasing height, the radiation resistance increases until it reaches a peak value of over 90Ω at a height slightly greater than 0.3λ, and thereafter oscillates about the value 73Ω of the isolated resonant dipole.

The radiation pattern of an infinitesimal dipole with current I_x located a distance h above an infinite ground screen is given by the formula below (using the geometry of Figure 5.1(e)), showing the relationship in terms of both the (θ, ϕ) and (Φ, η) systems:

$$\mathbf{f}(\theta, \phi) = \hat{\mathbf{\Phi}} \sin \Phi \sin(kh \sin \Phi \sin \eta)$$

$$= \hat{\mathbf{\Phi}}(1 - \sin^2 \theta \cos^2 \phi)^{1/2} \sin(kh \cos \theta) \tag{5.10}$$

where here the unit vector $\hat{\mathbf{\Phi}}$ and coordinate system (Φ, η) is shown defined in Figure 5.1(e).

Figure 5.6(b) illustrates the contribution of the elevation pattern of the image in narrowing the elevation of an infinitesimal dipole over a ground screen. The pattern is shown along the plane $\phi = 0$. In practice, the height h is usually kept near a quarter wavelength in order that the elevation pattern not be narrowed by the presence of the image currents. For heights beyond $h \approx 3\lambda/8$, the gain at the zenith begins to decrease and becomes zero at $h = \lambda/2$. For heights above 0.75λ, the elevation pattern becomes multilobed and is not useful in general.

Special Feeds for Dipoles and Monopoles

A number of special feed arrangements have been developed for dipoles and monopoles intended to make the transition from various unbalanced lines and to match impedances. Several good surveys of the literature on baluns are available in [13–15], which describe some of the more important types. Figure 5.7(a) shows a dipole fed by a coaxial line "split tube balun" [13,15]. The impedance at the balanced output arms of the dipole are matched at four times that of the coaxial input $(Z_{AB} = 4Z_0)$. The coaxial outer conductor split is nominally a quarter wavelength.

Figure 5.7 also shows several printed circuit dipole configurations suitable for low-cost arrays. Figure 5.7(b) shows a printed circuit dipole due to Wilkinson [16]. A printed circuit distribution network is fabricated by two photographic exposures using a two-sided printed circuit board. The network and dipoles are configured as a printed circuit two-wire line, with the intervening substrate serving to support both printed conductors, as shown in the figure. This unconventional transmission line medium was first analyzed by Wheeler [17] and is particularly appropriate for unscanned "flat plate" arrays. The two conductors, with dipoles and substrate, are mounted a quarter wavelength above the ground screen. The entire network was fed by a special wide-band split tube balun [18].

Figure 5.7(c) shows a printed circuit element, described by Edward and Rees [19], but referenced earlier without description [20]. The element consists of a

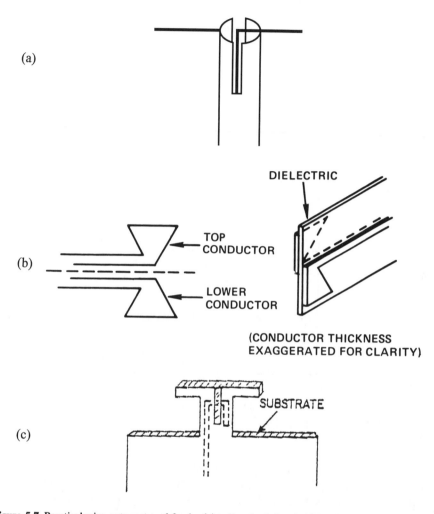

(a)

(b)

DIELECTRIC

TOP
CONDUCTOR

LOWER
CONDUCTOR

(CONDUCTOR THICKNESS
EXAGGERATED FOR CLARITY)

(c)

SUBSTRATE

Figure 5.7 Practical wire antennas and feeds: (a) split tube balun feed for dipole antenna; (b) bowtie dipole excited by coplanar strips transmission line (after [16]); (c) microstrip-fed dipole and balun for radiation endfire to substrate (after [19]).

dipole etched into the ground screen side of a printed circuit board and capacitively coupled to a loop feed formed by the printed microstrip line. The element is very compatible with monolithic fabrication techniques and makes effective use of the substrate ground screen, which is mounted normal to the plane of the array. Details of the design are given in the paper by Edward and Rees and in a later report by Proudfoot [21]. The feed uses a printed circuit balun due to Roberts [22] and

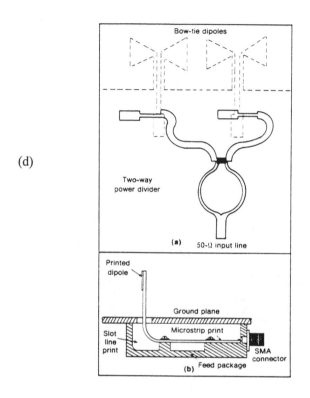

Figure 5.7 (cont.) Practical wire antennas and feeds: (d) bowtie dipole excited by microstrip, slot line, and coplanar strip transitions (after [24]).

adapted by Bawer and Wolfe [23] to a printed circuit configuration. The balun matches a balanced coplanar microstrip dipole feed section to a microstrip line on the opposite side of the substrate. This balun provides no impedance transformation at center frequency, which must be accomplished separately if necessary. Since it is built of microstrip, it is nearly ideal for monolithic printed circuit integration. Since the dipole and balun combination can be "double tuned" to produce a broadband impedance match, the antenna can operate over a 40% bandwidth with a 2:1 voltage standing wave ratio (VSWR). One disadvantage of this element is cross-polarized radiation [21], which appears to be caused by unbalanced orthogonal currents in the feed region. The cross-polarized radiation can be as large as −15 dB relative to the copolarized signal, and so can be a significant limitation to performance for certain applications.

Another new dipole feed configuration [24] provides a very wide-band feed from microstrip to slot line balun, and then makes a transition to a printed coplanar strip transmission line feed for the bowtie dipole. This geometry is shown in Figure 5.7(d). The balun [25] is a printed circuit version of a Marchand [26] balun. The balun alone has a 4:1 bandwidth for 11-dB return loss and octave bandwidth for 15-dB return loss. The combination balun and dipole has better than a 10-dB return loss over an octave. Although this dipole, like that of Edward and Rees, has a balanced coplanar strips feed, the balun is quite different in that it makes a transition from the microstrip line (shown solid) on the top surface of the substrate to a slot line (shown dashed) of the other surface. The slot line then makes a transition to a coplanar strip transmission line, which feeds the dipole. The dashed horizontal line indicates the end of the metallized region (the slot line region) on the bottom of the substrate. In addition, the balun transition as shown includes a quarter-wave impedance transforming region.

One problem that can grow to be severe in dipole arrays is that the vertical feed wires can themselves be the source of radiation. This is not a significant problem with individual dipoles because the vertical wires contain balanced opposite currents. However, when the array is scanned, the vertical pair is excited by the coupled signal from other elements and radiates an unwanted θ polarization. In addition, the coupling can even cause "array blindness" effects [27], as described in Chapter 6.

Dipoles Fed Off-Center

The radiation pattern and the current distribution of a dipole of length less than $\lambda/2$ are relatively independent of the location of the driving source. The approximate input impedance of an off-center-fed dipole (Figure 5.3(d)) is given in terms of its input impedance at the center Z_c by

$$Z_{in} = \frac{Z_c}{\cos^2(k\Delta l)} \tag{5.10}$$

where Δl is the displacement from the center. Assuming the displacement is small, this relationship shows that the impedance increases with the displacement Δl. This technique can be used for impedance matching without modifying the radiation characteristics. A more detailed analysis, obtained by an approximate solution of the integral equation, is given by King and Wu [28].

The Sleeve Dipole and Monopole

Sleeve antennas incorporate a tubular conductor (sleeve) such that the exterior of the sleeve is a radiating element, while the sleeve interior is used as the outer conductor of the coaxial transmission line that feeds the antenna. The monopole or dipole protrudes out of the enclosing sleeve and is an extension of the center conductor of the feed coaxial line, whose outer conductor is terminated at the monopole feed point, a distance L from the ground screen. The coaxial line outer conductor is shorted to the ground screen. The entire structure is enclosed by a cylindrical shell, or a sleeve of length d, which is also shorted to the ground screen. Other dimensions are indicated in Figure 5.8. The sleeve dipole, not shown, includes the mirror image of the structure in Figure 5.8(a) and so has length $2H$.

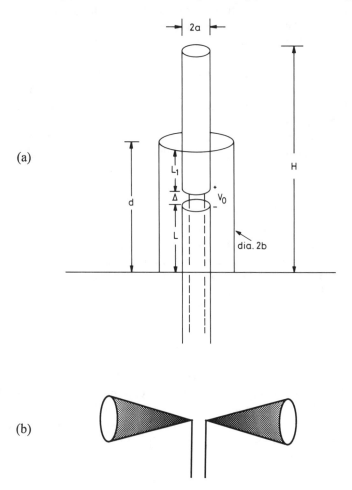

Figure 5.8 Practical wire antennas and feeds: (a) sleeve monopole antenna. (b) biconical dipole.

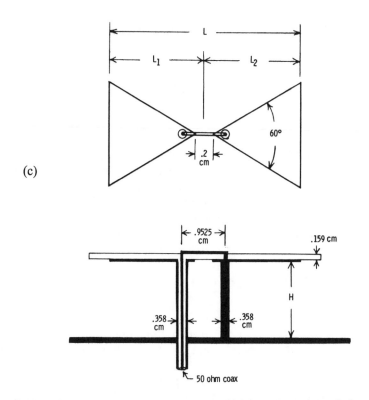

Figure 5.8 (cont.) Practical wire antennas and feeds: (c) bowtie dipole (after [37]).

Very High Frequency Techniques [29] devotes a chapter to an excellent and comprehensive discussion of the variety of sleeve antennas. Figure 5.8(a) depicts a very generalized version of the sleeve monopole.

Sleeve antennas have certain advantages in terms of ruggedized construction, but they are primarily important because of their excellent broadband impedance characteristics. Since the diameters of the sleeve and inner conductor, as well as the lengths L, H, and d, can all be varied, this double tuned dipole (or monopole) has been shown [30,31] to have excellent broadband characteristics.

Poggio and Mayes [32] presented a method for optimizing the pattern bandwidth of the sleeve monopole. They observed that neither the sleeve diameter nor the monopole diameter has any significant effect on the radiation patterns, although increasing the monopole diameter does lower the antenna Q and broaden the bandwidth. The total height H is set to resonate at approximately one-quarter wavelength at the lowest frequency. This done, only the length d remains to control the pattern bandwidth through control of the current distribution at the higher frequencies within the band.

Within the sleeve region, the feed sees two impedances in parallel. The impedance seen looking vertically up from the feed point (see Figure 5.8(a)) is the antenna impedance Z_A seen at the top of the sleeve and transformed by a coaxial line transformer of length L_1, inner diameter $2a$, outer diameter (the sleeve diameter) $2b$, and characteristic impedance Z_{01}. The second impedance is that looking down from the feed point. The monopole impedance Z_A can be obtained from (5.5) or (5.6) from published curves or available software. From the figure, this is obviously the impedance of a shorted section of transmission line of length L with characteristic impedance Z_{02}, whose inner conductor is the feed transmission line (diameter $2a$) and whose outer conductor is the sleeve. Note that the figure shows the same transmission lines ($Z_{01} = Z_{02}$), but in the general case this may not be so. For the more general geometry, Poggio and Mayes give the feed point impedance as

$$Z_{in} = Z_{01} \frac{Z_A + jZ_{01} \tan kL_1}{Z_{01} + jZ_A \tan kL_1} + jZ_{02} \tan kL \qquad (5.12)$$

Poggio and Mayes present data for a sleeve monopole operating over a 4:1 bandwidth and demonstrate pattern optimization by proper choice of the ratio $(H - d)/d$.

The recent paper by Wunsch [33] presents an accurate numerical model of the isolated sleeve monopole. Wunsch's results show the relatively complex current distribution that must be properly modeled to predict behavior at the high end of the band. In addition, Wunsch shows excellent correlation between the measured data of Poggio and Mayes and computed impedance over a 4:1 bandwidth for the isolated antenna.

Although the characteristics of isolated sleeve dipoles and monopoles are now well known, there are relatively little data on scanned arrays. Certainly, the bandwidth of sleeve antennas in arrays is far less than that of isolated antennas. Specific design data for arrays are given in the article by Wong and King [31], showing how bandwidth can be improved by the proper selection of dimensions. Typical achievable bandwidth ratios are up to 1.8.

The Bowtie and Other Wide-Band Dipoles

Thin dipole and monopole elements have reasonable bandwidths for many phased array applications. However, the dipole bandwidth can be increased substantially by using fatter conductors. Balanis [34] quotes the narrow bandwidth of 3% for a very thin dipole with $\ell/a = 5,000$, but with a fatter dipole ($\ell/a = 260$) the dipole bandwidth is approximately 30%. It is common practice to use fatter dipoles and specially shaped antennas to increase bandwidth. The biconical dipole [35] of Figure 5.8(b), the conical monopole [36], and the bowtie [37] element shown in Figure 5.8(c) are examples of wide-band elements that are treated in some detail in the literature and that offer significantly improved wide-band operation when used as isolated elements. A broadband version of monopole or dipole, the bowtie element (Figure 5.8(c)) has a flat triangular shape and is lightweight compared to the biconical structures, but still retains some of the wide-band properties. The bowtie element of Figure 5.8(c) described by Bailey [37] includes a balanced to unbalanced (balun) impedance matching transformer, which is formed using the cylindrical conductor (solid left vertical line), and the coaxial line outer conductor to form a short-circuited transmission line. This element operates over a 37% bandwidth with VSWR < 2.0 and has cross-polarized components of radiation suppressed below −25 dB. Although the element bandwidth in a scanned array will be significantly less than that of the isolated element, these numbers still serve as guides to estimate the maximum that can be expected in the array environment.

The Folded Dipole

Figure 5.9(a,c) show a folded dipole antenna excited by a two-wire line. The folded dipole has an impedance transforming feature that multiplies the antenna impedance by a number related to the diameter and spacing of the wires in the folded dipole. The structure was first analyzed by Mushiake [38] as the combination of symmetrically and antisymmetrically driven modes: a transmission line mode and an antenna mode. A more recent full-wave numerical solution [39] confirms the accuracy of this method.

Design equations for selecting wire diameters and other dimensions are given below for a folded dipole with different radii, as shown in the figure. The equations are as cited by Tai [40] from the symmetric-antisymmetric solution of Mushiake. The input impedance of the folded dipole is given by

$$Z = \frac{2(1 + a)^2 Z_r Z_F}{(1 + a^2)Z_r + 2Z_F} \tag{5.13}$$

where the impedances Z_r and Z_F are the input impedances of symmetrically and asymmetrically fed lines. The asymmetrically fed line with impedance Z_0 has equal and opposite currents in the two arms and has as input impedance the impedance of a shorted transmission line of length L.

$$Z_F = jZ_0 \tan(kL/2) \tag{5.14}$$

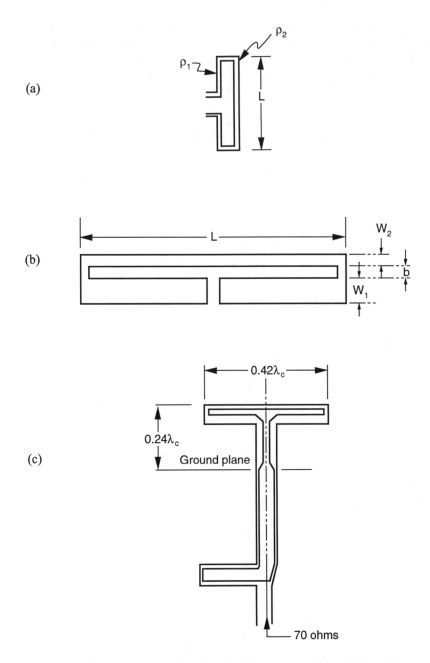

Figure 5.9 Folded dipole antennas: (a) basic folded dipole antenna; (b) strip folded dipole geometry; (c) strip folded dipole (after [41]).

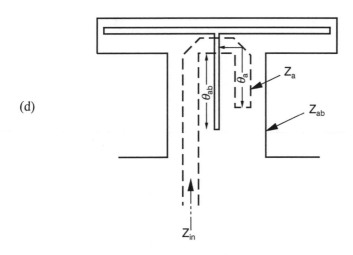

Figure 5.9 (cont.) Folded dipole antennas: (d) folded dipole excited by microstrip balun (after [44]).

The impedance Z_r is that of a cylindrical dipole with equivalent radius a_{eq} for the case of two parallel conductors with radii ρ_1 and ρ_2, as given in the figure.

The impedance step-up ratio $(1 + a)^2$ is always greater than 1. The parameter a is given by the equation

$$a = \frac{\cosh^{-1}[(\nu^2 - \mu^2 + 1)/2\nu]}{\cosh^{-1}[(\nu^2 + \mu^2 - 1)/2\nu\mu]} \tag{5.15}$$

where $\mu = \rho_2/\rho_1$ and $\nu = d/\rho_1$.

Figure 5.10 shows the step-up ratio $(1 + a)^2$, as computed by Hansen [40]. If both wires have the same radius, then the step-up ratio is 4, and so dipole impedances of approximately 73Ω are transformed to closely match the 300Ω transmission line. However, a wide range of step-up ratios can be obtained through the proper selection of the spacing and relative dimensions. Figure 5.10(b) shows the (ρ_1/ρ_2) ratio of conductor diameters versus the transformation ratio and spacing parameter d.

A particularly convenient folded dipole circuit, shown in Figure 5.9(c), with dimensions given in the paper by Herper et al. [41] is a printed strip line folded dipole with a Schiffman balun. The dipole length was 0.42λ and the top of the dipole was located 0.24λ above the ground screen at center frequency. This element was fed by strip line, but could also be fed by a microstrip transmission line. One major advantage of this element is that it can be printed in a single process all on

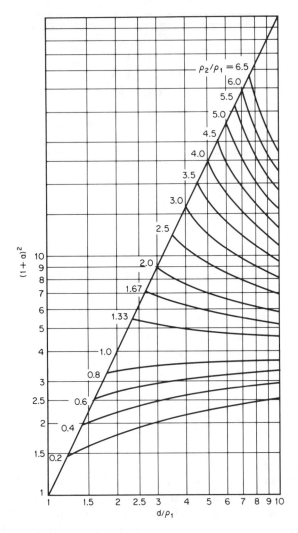

Figure 5.10(a) Folded dipole parameters (after [40]): step-up ratio $(1 + a)^2$ versus ρ_2/ρ_1 and d/ρ_1.

one side of a circuit board, and so is relatively inexpensive to produce. A recent paper by Lampe [42] gives some design formulas for such a printed folded dipole designed from the equations given above, using the transmission line parameters of an asymmetrical coplanar strip transmission line with dimensions given in the figure.

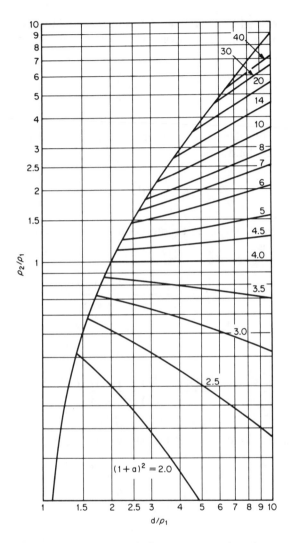

Figure 5.10(b) Folded dipole parameters (after [40]): conductor ratio ρ_2/ρ_1 versus step-up ratio $(1 + a)$ and d/ρ_1.

Lampe obtained the characteristic impedance of the asymmetrical coplanar strip transmission line from a Schwarz-Christoffel transformation:

$$Z_0 = \frac{120}{\epsilon^{1/2}} \frac{K(k)}{K'(k)} \tag{5.16}$$

where ϵ is the relative dielectric constant of the substrate, K is the complete elliptical integral of the first kind, and

$$K' = K([1 - k^2]^{1/2}) \tag{5.17}$$

For coplanar strips, and using the notation of Figure 5.9(b), the parameter k is given by Lampe as

$$k = \frac{b/2[1 + e(b/2 + W_1)]}{b/2 + W_1 + e(b/2)^2} \tag{5.18}$$

where

$$e = \frac{W_1 W_2 + (b/2)(W_1 + W_2) - [W_1 W_2 (b + W_1)(b + W_2)]^{1/2}}{(b/2)^2 (W_1 - W_2)}$$

In the particular case of symmetrical strips ($W_1 = W_2 = W$), the parameter k reduces to

$$k = \frac{b}{b + 2W} \tag{5.19}$$

Efficient methods for evaluating the complete elliptical function K and its complement are available in the literature. For example, Lampe cites the simple formulas of Hilberg [43]. The parameter a for the strip folded dipole is given by Lampe as

$$a = \frac{\ln\{4c + 2[(2c)^2 - (W_1/2)^2]^{1/2}\} - \ln(W_1)}{\ln\{4c + 2[(2c)^2 - (W_2/2)^2]^{1/2}\} - \ln(W_2)} \tag{5.20}$$

The remaining parameter required for computing the input impedance of the folded dipole is the impedance Z_r for the dipole of the equivalent radius. Lampe gives a relationship for this parameter in terms of integrals. In the case of equal-width strips, the equivalent radius is given as

$$\rho_e = \{(W/4)(c + [c^2 - (W/4)^2]^{1/2}\}^{1/2} \tag{5.21}$$

If the strips are narrow relative to spacing $2c$, then the equivalent radius becomes

$$\rho_e = [We/2]^{1/2} \tag{5.22}$$

Using these relationships, Proudfoot [44] constructed a balun-fed folded dipole using the feed that he previously described in earlier studies [19] and [21] (see Figure 5.9(d)). The demonstrated 36% bandwidth for the isolated element. Cross-polarized radiation levels achieved were similar to those of the balun-fed dipole of [21], thus confirming the likely cause of the cross polarization as the hook balun.

Microstrip Dipoles

Printed circuit dipoles are strips of printed conductor cut to resonant lengths and excited by various means, including electromagnetic coupling to nearby transmission lines or direct coupling to feed lines or probes. Figure 5.11 shows several resonant microstrip dipole elements excited by direct coupling and proximity coupled to an open-circuited line in the same plane or below the dipole. Proximity-coupled arrays developed by James and Wilson [45] (Figure 5.11(a,b)) and by Mise [46] center each element at the voltage minimum of a reactively terminated feed line. Mise shows the dependence of normalized line resistance on radiator position for a single radiator, and indicates how various dimensional parameters influence impedance and resonant frequency for a variety of element positions. A similar radiator (Figure 5.11(c)) was developed independently by Oltman [47] and places the resonant radiator about halfway beyond the end of the feed line. A paper by Oltman and Huebner [48] is an application of Oltman's element to a corporate fed array.

Recent studies of printed microstrip dipoles have been conducted using full-wave integral equation solutions and have considered a number of direct and proximity-coupled configurations [49–52]. Single and dual parasitic elements [52] have been shown to give bandwidths in excess of 11% for isolated elements. In comparison with the microstrip patch antenna, the microstrip dipole is more narrow-band unless used with relatively thick substrates.

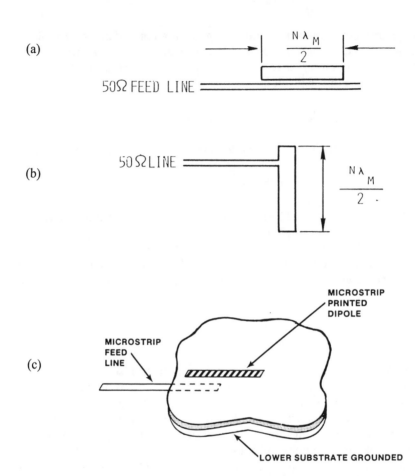

Figure 5.11 Microstrip dipoles for radiation perpendicular to substrate: (a) electromagnetic coupled microstrip dipole with coplanar feed; (b) microstrip dipole fed directly by microstrip line; (c) electromagnetic coupled microstrip dipole with feed below substrate.

Other Wire Antenna Structures

The wire elements listed above have been used in arrays and are clearly suited for such use. However, they comprise only a very small subset of all of the wire antennas developed, some of which may have unique qualities for arrays. Since many of these were developed for use with coaxial or parallel wire transmission lines, their application to a variety of printed transmission lines has not been studied, and could prove a new and potentially fruitful area for future developments. The remarkable variety of these creative endeavors is chronicled in Chapter 3 of the text by King [4], a chapter that bears rereading from the perspective of modern transmission circuits.

Broadband Flared-Notch, Vivaldi, and Cavity-Backed Antennas

Various flared transitions have been used as transmission line interconnects for many years, and flared antennas [53] are described in a number of texts. Kerr [54] developed a series of flared-ridge-loaded horn antennas that have found utility as a wide-band feed for reflectors and anechoic chambers.

Variations of these basic elements (Figure 5.12) have been developed and several incorporated into full phased arrays. The early work reported by Lewis [55] described the flared notch (Figure 5.12(a,b)), in which notches in the outer strip line conductors are excited by an open-circuited orthogonal center conductor. In an array environment, the element showed acceptable scan characteristics with a linearly polarized radiation pattern. Broadside arrays of the same elements had previously demonstrated greater than 4:1 bandwidth [56,57], but in the study of Lewis et al. [55], the scanned array bandwidth was limited to about one octave. The flared notch, as it was first investigated, is a strip line element excited by proximity coupling to the open-ended strip line center conductor. More recent investigations of abrupt transitions to notch antennas have also shown substantial bandwidths on the order of 6:1 [57] and included some data on the strip line capacitively coupled feed that excites the notch.

A similar antenna, the Vivaldi element (Figure 5.12)(c) [58] is actually a flared, truncated slot line. The paper by Gibson [58] popularized the name *Vivaldi* for the exponentially flared slot line antenna and expressed bandwidths of 6:1. A linearly tapered slot line antenna is described in the work of Prasad and Mahapatra [59]. More recent studies by Yngvesson [60] have included design data for obtaining increased bandwidth. Franz and Mayes [61] describe a microstrip-fed Vivaldi antenna using several different feeds, which has demonstrated good impedance match over a 4:1 bandwidth, and good pattern bandwidth over an 11:1 bandwidth.

The Vivaldi flare is given in the following form by Franz and Mayes [61]:

$$Y(x) = (w_0/2) \exp(\alpha x) \tag{5.23}$$

where Y is the half separation of the Vivaldi radiator conductor; x is the longitudinal coordinate; α is the flare scaling factor; and w_0 is the width of the uniform slot line.

Franz and Mayes give data describing the radiation length and reflection coefficient, and show that the length L_R should exceed 0.5λ, with 0.9λ being nearly optimum. Similarly, the dimension W_{max} must exceed about 0.5λ.

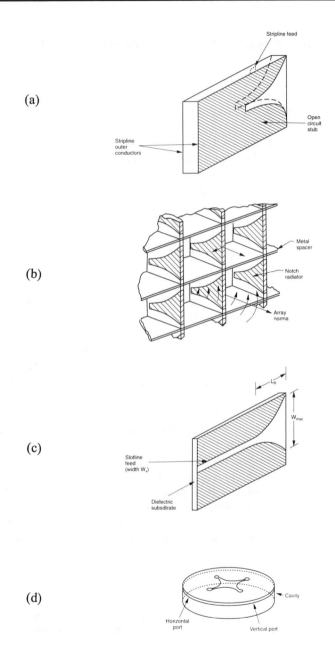

Figure 5.12 Notch type radiators: (a) flared notch strip line-fed radiator; (b) two-dimensional flared notch array; (c) Vivaldi slot line element (after [58]); (d) cavity-backed arrangement of four microstrip-fed flared slot lines (after [64]).

Recent published developments have extended the practical implementation of these elements. Povinelli [62] describes a scanned array of strip line flared notches that maintained an average VSWR under 2:1 over the band 6 to 18 GHz for a 60-deg scan in all planes. To achieve good performance over this scan range without grating lobes, the elements were very closely spaced (0.19λ) at the lowest frequency. The flare was designed using a modified Dolph-Chebyshev taper given by Klopfenstein [63], and this resulted in an extremely short flared region (approximately 0.15λ long at the lowest frequency).

A new wide-band flared element due to Povinelli [64] is a microstrip-fed slot line exciter for a cavity-backed slot. Figure 5.12(d) shows a circularly polarized crossed-slot version of this element. In this case, the radiation is orthogonal to the array slot aperture. The circularly polarized cavity-backed antenna had less than 2:1 (average) VSWR over most of the 4- to 18-GHz frequency band. Without dielectric loading, the circularly polarized element required a cavity more than one wavelength across, and so, at least in this form, the array applications are limited.

5.1.3 Aperture Antenna Elements

The variety of aperture antennas is far less than that of wire antenna elements and consists primarily of slot antennas, waveguide antennas, and horns.

Slot Elements

Slot antenna elements are among the oldest radiators. They are the well-known complement of the dipole, and the impedance of a slot in a ground screen in free space (Figure 5.13(a,b)) is obtained from Babinet's principle as [65]

$$Z_S = Z_0^2/(4Z_c) \tag{5.24}$$

Here, Z_c is the dipole impedance at the corresponding point on the complementary dipole, and Z_0 is the free space impedance ($Z_0 = 120\pi$ ohms). The slot "equivalent radius" is usually taken as one-fourth the slot thickness in accordance with the analogy of the strip dipole of Figure 5.2.

Since the microstrip slot antenna is bidirectional, it is necessary to use a ground screen or cavity behind each slot to restrict radiation to the front hemisphere. Typically, the required back plate spacing is about 0.25λ between slot and reflector. In an array, it is usually necessary to have separate cavities [66] behind each element instead of just a reflecting plate, because coupling into parallel plate modes in the region below the slots can lead to serious pattern degradation in the array scan behavior [67].

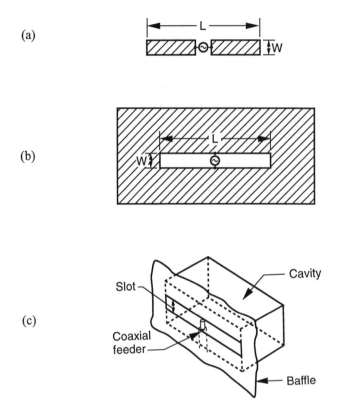

Figure 5.13 Fundamental slot and dipole strip antennas: (a) strip dipole; (b) equivalent slot in ground screen; (c) cavity-backed slot antenna.

If the slot is instead in a cavity, as shown in Figure 5.13(c), the radiation resistance is approximately double the free space resistance. This relationship is not exact because of the presence of higher order modes in the cavity behind the slot, which alter the element field distribution and change the radiating impedance. The cavity-backed slot [66] is an excellent element for scanned or unscanned arrays because of its polarization purity and its good scanning characteristics [67] and relatively broadband radiation characteristics. The techniques for exciting the slot using strip line are shown in Figures 5.14(a,b). In Figure 5.14(a), a strip line feed is shorted to the slot ground plane after passing over the slot [68]. Figure 5.14(b) shows the slot and a cavity formed by soldered pins or plated through holes in a dielectric. For this line, the strip line center conductor is terminated in an open circuit about a quarter wavelength beyond the slot [69]. The impedance of narrow slot radiators is very high and may be on the order of 400Ω to 500Ω without a

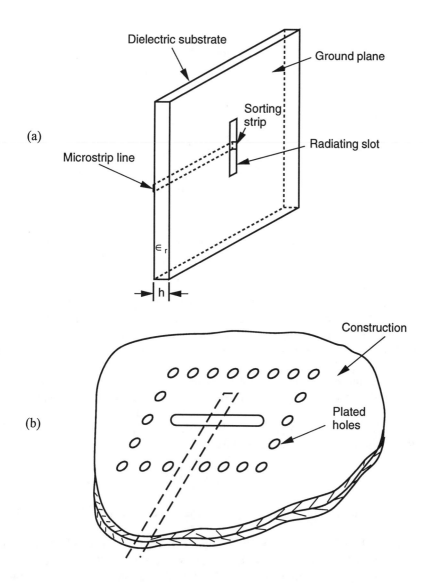

Figure 5.14 Practical feeds for slot antennas: (a) strip line shorted beyond slot; (b) open-circuited strip line beneath cavity-backed slot.

cavity and double that with a cavity, and so there is some engineering necessary to design slot feed networks. The use of a T-bar feed (Figure 5.14(c)) [70,71] compensates the impedance characteristics to provide a broadband impedance

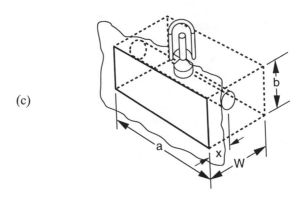

Figure 5.14 (cont.) Practical feeds for slot antennas: (c) tee-bar feed for slot antenna (after [72]).

match. Detailed design data for slot antennas coupled by T-bar feeds are given in the paper by Newman and Thiele [72]. A recent application of a T-bar (Figure 5.14(c)) feed located in the plane of the slot is readily adapted to printed circuit technology [73]. This technology has proven successful at frequencies up to 45 GHz.

The input impedance of the slot is maximum at the slot center, and so it is often convenient to excite the slot off center. As in the dipole case, this does not change the pattern measurably. The approximate impedance of the off-center-fed slot is given by the complementarity relationship (5.24), and so from (5.11) for relatively small displacement Δl, one obtains

$$Z_{in} = Z_c \cos^2(k\Delta l) \tag{5.25}$$

where Z_{in} is the input impedance and Z_c is the impedance of the center-fed slot ($k = 2\pi/\lambda$).

Bahl and Bhartia [74] summarize the literature of microstrip-fed slot antennas and quote the analysis of Nakaoka et al. [75] for the slot resistance when excited off center by a microstrip line shorted to the slot edge, as shown in Figure 5.14(a).

Waveguide Radiators

Still the most important element for high-power radar and communication arrays, the rectangular or cylindrical waveguide radiating element (Figure 5.15(a,b)) has been investigated in detail and optimized to develop excellent scanning properties.

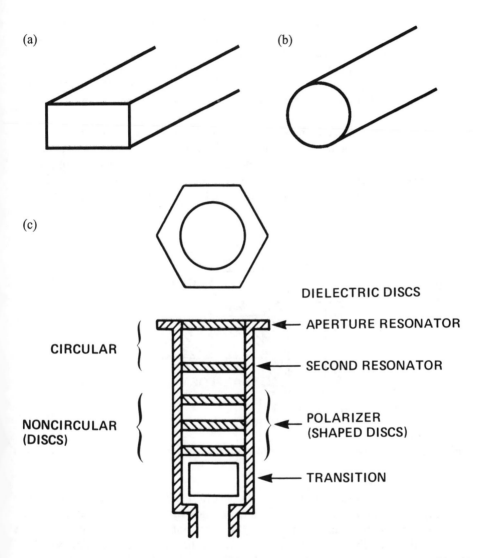

Figure 5.15 Waveguide radiating elements: (a) rectangular waveguide; (b) circular waveguide; (c) circularly polarized waveguide element for triangular grid array (after [77]).

Waveguide arrays, though heavy, tend to have low loss, a bandwidth exceeding 50%, and graceful scan degradation. Impedance matching at broadside is usually not difficult because the impedance of an unloaded waveguide is close to the free-space impedance.

Detailed considerations of rectangular and circular waveguide antennas as elements and in arrays are given in a number of texts [76]. For waveguides radiating as apertures in conducting screens (as in Figure 5.15(a,b)), the published numerical results are quite accurate, since waveguides can usually be assumed to operate with a single incident mode with all other modes cut off. This separates the feed and radiation properties and results in well-defined boundary value problems.

Often the waveguide element is dielectrically loaded to make a transition to a ferrite phase shift section. Examples of specific waveguide element design are the studies of Wheeler [77], wherein matching networks were derived using waveguide transmission circuits like that shown in Figure 5.15(c), consisting of dielectric slabs mounted in and above the waveguide. The later studies of McGill and Wheeler

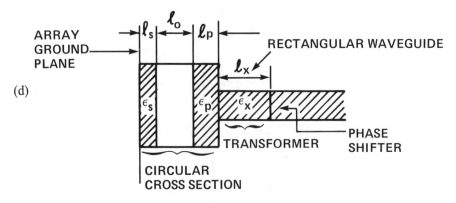

Figure 5.15 (cont.) Waveguide radiating elements: (d) doubly tuned waveguide element (after [79]).

[78] introduced the use of a dielectric sheet, often called a *wide-angle impedance matching* (WAIM) sheet, to produce a susceptance variation with a scan angle that partially cancels the scan mismatch of the array face. Figure 5.15(d) shows an example of a broadband waveguide element that has doubly tuned response characteristics synthesized using dielectric loading and a section of waveguide beyond cutoff as an impedance transformer.

References to the scan matching and performance of rectangular and circular waveguides are given in Chapter 6.

In general, it is now possible to predict wide-band waveguide scan characteristics for large arrays with such accuracy that, using available transmission line software, one can readily synthesize appropriate matching networks for wide-angle, wide-band performance.

Ridged Waveguide Elements

Ridged waveguides are broadband transmission lines that can be used as efficient, high-power, broadband array elements. Single-, double-, and quad-ridged waveguides are shown in Figure 5.16. Quad-ridged waveguides (Figure 5.16(c)) extend these features to circularly polarized arrays. They are, of course, more expensive to build than conventional waveguides, and so have application only when the specifications require these features.

The design of ridged-waveguide arrays involves a tradeoff between waveguide bandwidth and scan matching of the array aperture. Figure 5.16(d,e) shows the

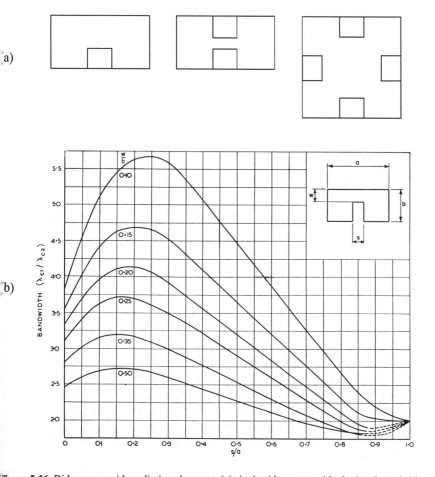

Figure 5.16 Ridge waveguide radiating elements: (a) single-ridge waveguide dual and quad-ridge waveguide; (b) dual single-ridge waveguide.

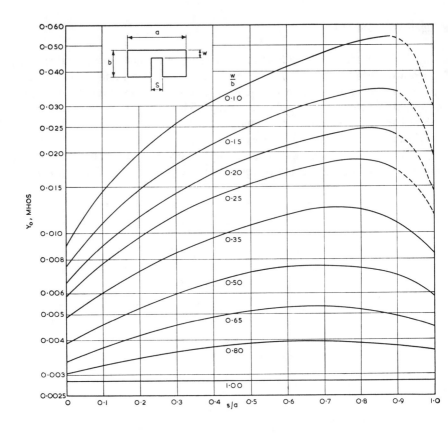

Figure 5.16(c) Impedance of single-ridge waveguide (after [80]) with $b/a = 0.45$ and infinite frequency.

bandwidth of a single-ridged waveguide as a function of ridge and waveguide dimensions. The plotted bandwidth is defined as the ratio of the cutoff wavelengths of the TE_{10} mode and the next higher mode. Clearly, the highest bandwidth is obtained for a heavily loaded waveguide, with ridges that extend most of the way across the guide. Figure 5.16(e) shows that this case corresponds to low characteristic impedance, and so is a poor match to free space ($Y_0 = 0.0027$). Matching for a scanned array is possible by varying the ridge parameters or through the use of dielectric layers.

Other references to ridged waveguide characteristics are included in the texts [81,82]. Scan parameters of ridge guide arrays are given by Chen and Tsandoulas

[83], Montgomery [84], Chen [85], and by Wang and Hessel [86] for double-ridged elements. The results of this last detailed study of a two-dimensional scanning array confirmed that by properly selecting ridge parameters, one can design elements to scan a quarter hemisphere and have between 40% and 58% bandwidth. The maximum VSWR of such elements without broadband matching circuits and for any scan angle was about 7:1 for the 40% bandwidth case and 10:1 for the 58% bandwidth case. An attempt to design a 75% bandwidth element led to maximum mismatch of about 16:1 within the scan volume.

Horn Elements

In distinguishing horn from waveguide elements, we mean to restrict the consideration of horns to apertures that are generally more than a wavelength or so on a side, and often some number of wavelengths. These generally have little application to scanning arrays, except those that scan only a few degrees. In general, horn arrays can scan approximately to the horn 3-dB point, or roughly to

$$\sin \theta_{max} = 0.443\lambda/D \tag{5.26}$$

where D is the horn aperture length in the scan plane. Such a scan results in potentially serious pattern distortion in the form of grating lobes. These and other antennas for restricted sector scan coverage are discussed in Chapter 8.

5.1.4 Microstrip Patch Elements

Microstrip Patch

The microstrip element is a most interesting and useful element. Figure 5.17(a,b) show the basic rectangular patch first described by Munson [87] and the circular disk radiator of Howell [88]. The key to its utility has been that it can be fabricated with low-cost lithographic techniques on printed circuit boards. It can also be produced by monolithic integrated circuit techniques that fabricate controls, phase shifters, amplifiers, and other necessary devices, all on the same substrate and all by automated processes. Several recent books and numerous technical papers present design data for these elements [89–92]. Figure 5.17(a,b) also illustrate the two most common feed structures: the inline microstrip feed and the coaxial probe feed.

The microstrip patch was not included in the sections on generic wire elements or aperture elements because its radiation, and in fact all of its properties, can be formulated from either perspective. This is illustrated in Figure 5.18, which shows microstrip patches above a ground screen. In this case, we show no dielectric under

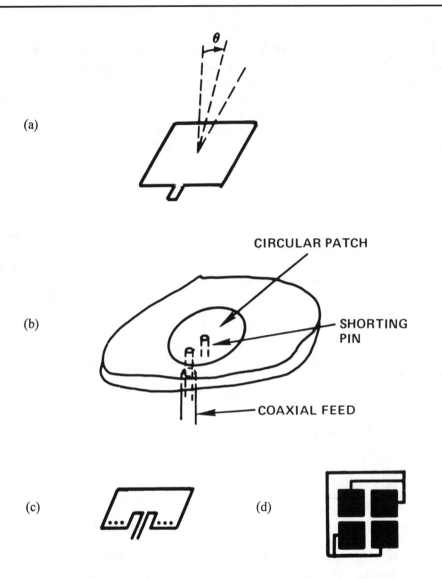

Figure 5.17 Basic microstrip patch radiating elements: (a) rectangular patch radiating elements; (b) circular disk excited by probe feed; (c) shorted patch; (d) crossed slot of four shorted patches.

the patches in order to simplify the exposition. In Figure 5.18(a), the currents are shown as solid lines on the patch. In the case of the patch current model (Figure 5.18(a)), the field for $z > 0$ can be rigorously expanded using (5.2) and (5.3) for

(e)

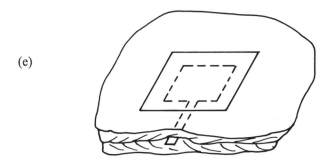

Figure 5.17 (cont.) Basic microstrip patch radiating elements: (e) electromagnetically coupled patch.

current sources over a ground screen, and including the image currents as directed by (2.9) for the vector potential. This is an antenna current formulation, such as was done for the wire antennas discussed earlier. The currents are not known and remain the subject of a more complex analysis that includes the source excitation. Alternatively, from the perspective of Figure 5.18(b,d), one could assume that the ground screen, a perfect electric conductor, is at the plane of the patches (see dotted line in Figure 5.18(b)), and this allows the fields to be expressed using the half-space potential function. In this case, the integral in (2.9) is strictly taken over the (unknown) tangential electric field all along the dotted surface. The tangential aperture fields are obtained from the solution of a boundary value problem beneath and at the edges of the patch, and in the array case includes coupling to other patches, and the periodic array at $z = 0$, in addition to the exciting source conditions. In practice, the most intense fields are usually confined very close to the patch edges, and this is the basis for a convenient two-slot approximation to the patch radiation patterns. In the two-slot approximation of Figure 5.18(d), the patch is assumed to radiate like two slots of width equal to the substrate thickness. This representation does not account for cross-polarized components of radiation, although these can be included by integrating the antisymmetrical fringing fields along the side edges of the patch.

Still, other formulations based on alternative combinations and locations of electric and magnetic sources are equally valid, and a number of these are cataloged in the text by Bahl and Bhartia [90].

Before leaving the subject of theoretical models to present specific engineering results of the two-slot model, it is worthwhile noting that the above statements are strictly true only for patches without a dielectric substrate. In general, for a patch over a dielectric substrate, one cannot use the half-space Green's function rigorously, and the analytical problem is significantly complicated. If the substrate is not air, the more complex Sommerfeld Green's function (or the so-called spectral

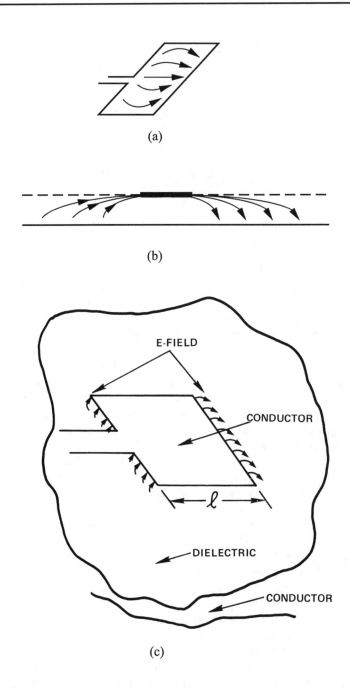

(a)

(b)

(c)

Figure 5.18 Microstrip patch models: (a) electric current model; (b) aperture field model; (c) microstrip patch radiator showing fringing fields.

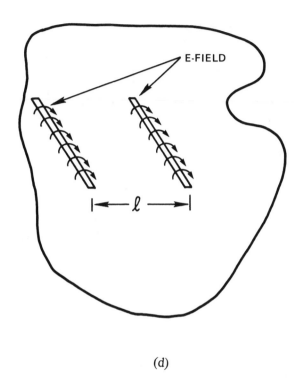

(d)

Figure 5.18 (cont.) Microstrip patch models: (d) simplified two-slot radiator model.

form of the Green's function) is required in the patch current formulation, and (2.9) is no longer used. Nevertheless, the patch current formulation has proven to be of great utility in computing the detailed mutual coupling analysis, as will be discussed in Chapter 6.

The aperture field approach has proven especially useful and intuitive. From this perspective, the element is viewed as two radiating slot apertures with electric fields in the plane of the patches. The slots are spaced b apart and have thickness t equal to the substrate thickness. Radiation from the edges that run parallel to the currents is orthogonally polarized, but is often neglected for nonscanning broad-

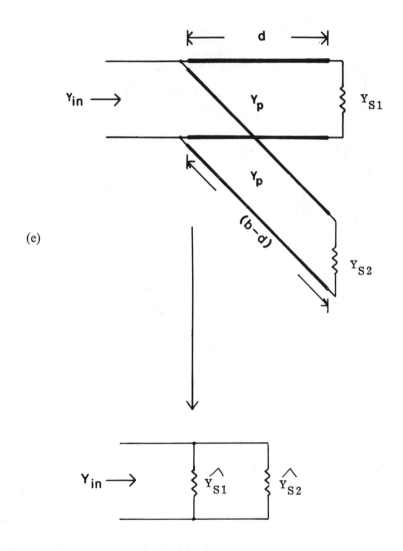

(e)

Figure 5.18 (cont.) Microstrip patch models: (e) equivalent circuit for transmission line approximation

side arrays because the field is asymmetrical along the patch and its radiation tends to cancel in the broadside direction. Early studies obtained resonant frequencies by modeling the patch as a resonator made of a parallel plate transmission line with susceptances to represent the discontinuity at each end. The success of this transmission line model has been in producing convenient and reasonably accurate formulas for rectangular patch resonant dimensions and pattern and radiation resistance. These are presented below as given in the model presented in a number of texts and references [89–96] for a patch with an arbitrary feed point, as used for the inset feed of Figure 5.19(b).

The transmission line model for the patch of Figure 5.18(d), with arbitrary feed point, represents the rectangular patch as two slots separated by a distance b, which is usually very nearly one-half wavelength in the dielectric, and width a and thickness h. The slot thickness h is usually taken as the substrate thickness. Assuming uniform fields across the patch, the normalized element pattern of this combination is approximately

$$F(\theta, \phi) = \frac{\sin(khu/2)}{khu/2} \frac{\sin(kav/2)}{kav/2} \cos(kbu/2) \tag{5.27}$$

for direction cosines u and v. This pattern can be integrated to give the patch directivity [90].

An approximation of the radiation resistance is also found from the transmission line model by considering that the resonant element radiates as two slots of length equal to the patch width a radiating in parallel [95].

$$R_{\text{in}} = \frac{60/\lambda}{a} \tag{5.28}$$

Width a is chosen to be about

$$a = \frac{c}{2f} \{(\epsilon_r + 1)/2\}^{-1/2} \tag{5.29}$$

for c, the velocity of light. Since this width is usually about a half wavelength or less, this input resistance is often from 100Ω to 200Ω. A match to nominal 50Ω transmission lines can be accomplished either by a matching network in the feed line or by modifying the patch. The most convenient solution to date has been to utilize the transmission line model circuit to transform the impedance to a feed point inside (or beneath) or along a nonradiating edge of the patch. In practice, this is most often done using a probe feed from beneath the patch or an inset feed using a coplanar waveguide within the patch, as shown in Figure 5.19.

The input impedance at an arbitrary feed point beneath the patch is obtained using the transmission line model by assuming the patch is a transmission line terminated in an open circuit at both ends. The patch is modeled as two slots

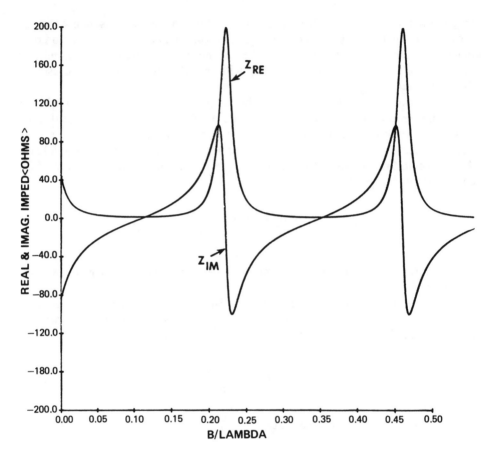

Figure 5.19(a) Microstrip patch antenna with inset feed: edge input impedance versus patch length b.

centered at the edges of the patch. Harrington [96] gives the admittance of a narrow H-plane slot (for $h/\lambda < 0.1$) as

$$Y_s = (\pi a/(\lambda_G \eta_0))(1 - k^2 h^2/24) + j(a/(\lambda_G \eta_0))[3.135 - 2 \ln(kh)] \quad (5.30)$$
$$= G_s + jB_s$$

where h is the height above ground; a is the width of strip line; and η_0 is the impedance of free space, which equals 377Ω. In this expression, $k = 2\pi/\lambda$ for free space wavelength λ, and λ_G is the microstrip guide wavelength.

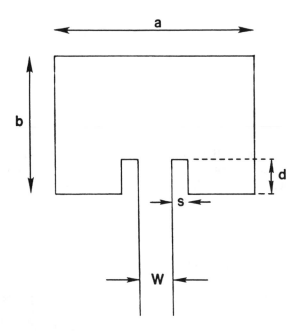

Figure 5.19(b) Microstrip patch antenna with inset feed: inset feed geometry (substrate thickness h not shown).

The admittance of the microstrip line of width A is given by Schaubert et al. in terms of the effective dielectric constant ϵ_E as [89,97]

$$Y_P = [(\epsilon_E)^{1/2}/\eta_0][a/h + 1.393 + 0.667 \ln(a/h + 1.444)] \qquad (5.31)$$

where

$$\epsilon_E = (\epsilon_r + 1)/2 + (\epsilon_r - 1)/(2[1 + 10h/a]^{1/2}) \qquad (5.32)$$

The input impedance is obtained using the equivalent circuit in Figure 5.18(e) to sum the admittances Y_0 (seen looking toward the far end of the patch) and Y_F (looking toward the feed side), both referred to the feed point. In these expressions, the dimensions l_1 and l_2 are electrical lengths, not actual lengths. They are used in the transmission line formulas and must be related to physical dimensions later. At the feed point, the admittance of the transmission line that terminates at the far end of the patch is

$$Y_0 = Y_P \frac{e^{jkl_2} + \Gamma_1 e^{-jkl_2}}{e^{jkl_2} - \Gamma_1 e^{-jkl_2}} \tag{5.33}$$

where

$$\Gamma_1 = \frac{Y_S - Y_P}{Y_S + Y_P}$$

and Y_S is the admittance of the slot at the far end of the patch.

The admittance looking toward the feed side of the patch is denoted by Y_F and is given by the expression above for Y_0 but with l_1 substituted for l_2, since the slot impedance is the same at the feed side. The total admittance is thus

$$Y_{in} = Y_0 + Y_F \tag{5.34}$$

The electrical lengths l_1 and l_2 define the conditions of resonance and the input patch resistance using the above ideal transmission line theory. They are related to the physical lengths d and $(b - d)$ by the following relations:

$$l_1 = d(\epsilon_E)^{1/2} + \Delta L \qquad l_2 = (b - d)(\epsilon_E)^{1/2} + \Delta L \tag{5.35}$$

where it is shown that the physical lengths are multiplied by the square root of the effective dielectric constant, but then increased by a correction factor that accounts for fringing fields. Since the fringing fields cause the radiating slots of the patch to appear electrically some small distance beyond the patch edges, there is a need to include a length extension to the electrical length. Hammerstadt [98] gives an approximation of the length extension of an open-circuited microstrip transmission line as

$$\Delta L = 0.412h \frac{(\epsilon_E + 0.3)(a/h + 0.262)}{(\epsilon_E - 0.258)(a/h + 0.813)} \tag{5.36}$$

Equation (5.33) gives the input admittance of the patch with an arbitrary feed point. The model of [94] proceeds by setting $d = 0$ (no inset) and making successive guesses at the patch length b until the imaginary part of the admittance is zero. Figure 5.19(a) shows a typical plot of the patch edge input impedance versus the length b and shows resonant impedance peaks on the order of 200Ω, with the imaginary part of the impedance zero at the peaks.

A good initial guess at the dimension b is

$$b = 0.49\lambda_0/(\epsilon_E)^{1/2} \tag{5.37}$$

Once this resonant length is selected by iteration, the inset feed dimension d is increased from zero to some number at which the real part of the patch impedance equals the feed line impedance. An initial guess at d can be obtained using the approximate expression of Carver and Mink [89] for the impedance R_0, a distance d from the edge of a rectangular patch:

$$R_0 = R_E \cos^2(\pi d/b) \tag{5.38}$$

This leads to an expression for d:

$$d \approx (b/\pi) \cos^{-1}[(R_0/R_E)^{1/2}] \tag{5.39}$$

where R_0 is the feedline impedance and R_E is the patch impedance at the edge (given by (5.33)). Since the patch is at resonance, the input impedance remains real as the feed point is moved from the edge. The resonant frequency remains unchanged.

Solving (5.39) for a 50Ω feed point leads to a feed location some distance in from the radiating edge of the patch. Locating a feed at this point is achieved by the several means shown in Figures 5.17(b,c) and 5.19. Coaxial probe excitation (Figure 5.17(b)) has proven very successful and practical, but the susceptance of the coaxial probe can alter determination of the patch resonant frequency. The inset feed of Figures 5.17(b) and 5.19 has been particularly successful, and since the inset feed does not add an additional susceptance, one can get quite accurate design dimensions from the transmission line theory. Recent work with inset feeds has shown good correlation between theory and experiment. Typical dimensions for the grounded coplanar feed region use the coplanar slot width equal to two times the microstrip width.

Although a number of accurate numerical models are now available for thin patches, the transmission line model is still very useful and has been extended to give the resonant frequencies of even relatively complex shapes. Unfortunately, neither this method nor most of the numerical methods are accurate for thick patches (especially probe-fed patches) much beyond 0.05λ unless special care is taken to accurately include the electromagnetic characteristics of the feed. The primary limitation of the transmission line model is its inability to account for coupling between patches in an array environment. The text by Bahl and Bhartia [90] gives a summary and comparison of a number of different approaches to microstrip antenna analysis. Until recently, the main disadvantage of microstrip elements has been that they are quite narrow-band. The bandwidth of an isolated element with a probe or inset feed and without a broadband matching network can be modeled as a simple tuned R-L-C circuit, and is given below in terms of the band edge standing wave ratio s [92]:

$$\frac{\Delta f}{f} = \frac{s - 1}{Q_T(s)^{1/2}} \tag{5.40}$$

where the total quality factor is given as a function of conductor and dielectric loss and the loss associated with the radiation resistance. In the limit when the radiation resistance dominates, for dielectric constants ϵ_E greater than 2 and equivalent patch width $\lambda/2$, the total quality factor is given by

$$Q_T = \frac{3\epsilon_E}{8} \frac{\lambda}{h} \quad \text{for} \quad \epsilon_E > 2 \tag{5.41}$$

James et al. [92] give other relationships for a lower dielectric constant.

The fractional bandwidth is

$$\frac{\Delta f}{f} = \frac{(s - 1)}{\epsilon_E(s)^{1/2}} \frac{8h}{3\lambda} \tag{5.42}$$

This formula shows the bandwidth as decreasing with the dielectric constant, but is still somewhat deceptive because the microstrip thickness is ideally kept at some fraction of the wavelength λ_ϵ in the dielectric line in order to avoid surface waves. Since $\lambda_\epsilon = \lambda/(\epsilon_E)^{1/2}$, the resulting bandwidth for normalized patches that have a substrate thickness h/λ_E constant is given by the expression below:

$$\frac{\Delta f}{f} = \frac{(s - 1)}{s^{1/2}} \frac{8}{3(\epsilon_E)^{3/2}} (h/\lambda_\epsilon) \tag{5.43}$$

and portrays the bandwidth as proportional to the inverse of the relative dielectric constant raised to the three-halves power.

More sophisticated than the transmission line model [99,100] is the cavity model [101]. The cavity model is again used in conjunction with the aperture field integration method. This model assumes a perfect magnetic conductor around the perimeter of the antenna and uses a modal description of the internal fields. Some other cavity-type solutions assume impedance boundary conditions [89] at the patch edges. Like the transmission line model, the cavity model is not one that is ideally suited to solving the scanning array problem, but it has been used with great success for single elements. Since this model can accurately account for probe excitation, it has yielded excellent resonant frequency data for the coaxial probe-fed patches in rectangular and circular geometries. A comprehensive description of the range of geometries that have been investigated using the cavity model is included in [93]. Recent results have extended the model to treat slot line and coplanar line feeds [102].

The electric current model Fig. 5.18(a) has been used very successfully by several authors [103,104]. This model readily accounts for mutual coupling, as indicated in Chapter 6, and has led to useful modeling of array scan dependence. The model has also been used extensively to study a variety of electromagnetically coupled feeds, including aperture coupling through slots in the ground screen [105] and various wire and capacitively coupled metallic patch feeds [106]. A recent paper by Chew and Liu [107] includes an accurate polynomial approximation of the resonant frequency for rectangular and circular patch models.

Variants of the conventional rectangular patch are shown in Figure 5.17 and include shorted patches (Figure 5.17(c)) [108] that resonate when the element size is approximately one-quarter wavelength long in the dielectric medium. Although these elements do save space and have a broader radiation pattern because they radiate like a single slot instead of a slot pair, they have not found extensive use because the need to use plated-through holes or soldered pins is an expensive and not always reliable fabrication procedure. This element has somewhat poorer polarization characteristics than the conventional microstrip element because the field along the nonradiating edge is not asymmetrical (as it is for the conventional rectangular patch), and so contributes a significant cross-polarized component of radiation [109,110].

Other patch elements have been used because of (1) their polarization characteristics, (2) their ability to sustain dual resonant frequencies, or (3) their enhanced bandwidth properties. Special element configurations for exciting either dual or circular polarization will be addressed in a later section. Among the dual-frequency elements developed, the details of several vertically oriented elements and elements using stubs or asymmetries to produce a second resonance have been published [111–113].

Broadbanding of microstrip elements has met with some success, though not without complicating the element design. Paschen [113] developed several networks for double tuning elements and showed an increased bandwidth that exceeded 20% in single elements. Other authors [114–116] have reported very wide-band characteristics of capacitively excited elements (such as that shown in Figure 5.18(d)). Recent studies of infinite arrays of various electromagnetically coupled patches [106], using idealized feeds, indicate that one can indeed choose dimensions, substrate thicknesses, and dielectrics to provide bandwidth exceeding 15% without blindness.

The Balanced Fed Radiator of Collings

A number of useful antenna designs have used patches fed with two probes excited 180 deg out of phase. These antennas are derived from the Collings radiator [117,118], which was invented about the same time as the microstrip patch. The

Collings radiator, as shown in Figure 5.20, consists of a disk radiator excited by the center conductors of two coaxial lines with a 180-deg phase reversal.

Recent studies and developments of balanced fed microstrip patch antennas [119,120] show the advantage of this design for reducing cross-polarized radiation [119] and for producing an improved axial ratio in circularly polarized arrays [120]. The scan performance and bandwidth of infinite arrays of these elements has also been studied [121,122].

5.1.5 Elements for Alternative Transmission Lines

The microstrip transmission line is a very practical one for a number of applications, but there are several other transmission lines that are also amenable to monolithic fabrication. Figure 5.21 indicates examples [123] of slot line and coplanar strip line antennas that may have advantages for array use in a variety of applications. Figure 5.21(c) shows a wire antenna array [124] used for a flat-plate (nonscanning) antenna. This antenna is a printed circuit implementation of an antenna often called a *Kraus grid array*, developed by Kraus [125]. The new feature provided by the printed circuit microstrip transmission line medium is that it allows tailoring of wire impedances by varying the microstrip conductor width. This makes it possible to design

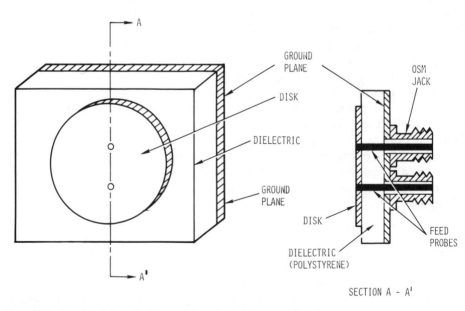

Figure 5.20 Circular disk excited by antiphase feed (Collings radiator).

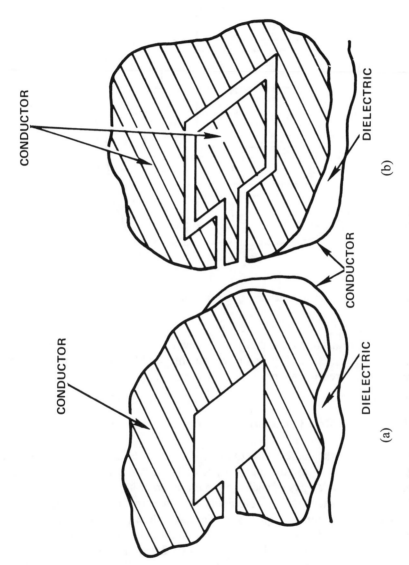

Figure 5.21 Other printed circuit radiators: (a) slot line antenna; (b) coplanar strip line antenna.

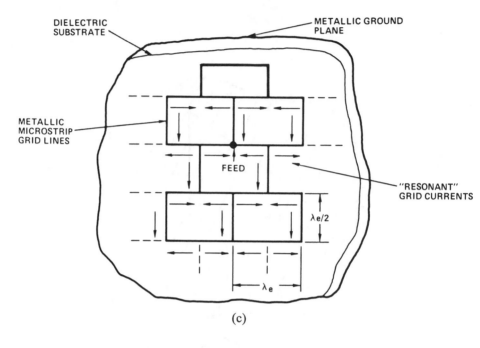

Figure 5.21 Other printed circuit radiators: (c) wire grid antenna (after [124]).

and fabricate appropriate aperture tapers for low-sidelobe radiation. Concepts such as these may have great utility in future reconfigurable antennas that can offer agility in resonant frequency and in system function.

5.1.6 Elements and Row (Column) Arrays for One-Dimensional Scan

If scanning in a single plane is adequate, the elements can be small individual elements like dipoles, slots, or microstrip elements, while for higher gain arrays the elements might be arrays of such simple elements. Such linear arrays are often called *line sources* when used as elements for an array that scans in the plane orthogonal to the elements. Figure 5.22 depicts column arrays excited by equal path corporate power dividers and in series-fed configurations. Column arrays fed by corporate power dividers, as shown in Figure 5.22(a), have far wider bandwidth and better power handling capacity than series arrays (Figure 5.22(b)), but are more bulky and expensive to construct. The series-fed geometry must be carefully designed to provide well-collimated radiation from each element. In general, one has a choice between waveguide and coaxial line corporate-fed power dividers or

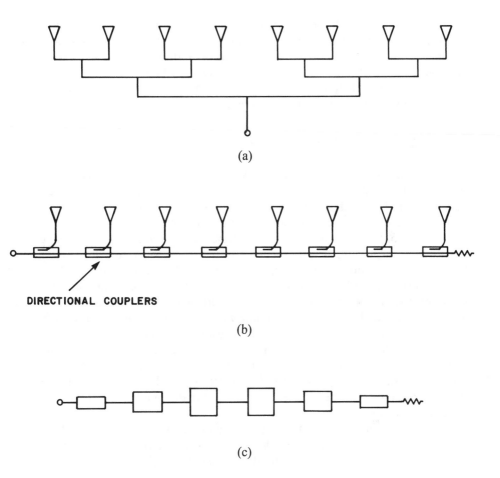

(a)

DIRECTIONAL COUPLERS

(b)

(c)

Figure 5.22 Column array configurations: (a) parallel (equal line length) corporate feed; (b) series-fed array; (c) series-fed microstrip patch array.

lower cost strip line or microstrip power dividers, which are lossy at higher frequencies and power-limited. The technology has become sufficiently advanced so that close tolerance control and sidelobes at the −40- and −50-dB level is possible with coaxial line, waveguide, or strip line power dividers. Microstrip power dividers, on the other hand, provide limited sidelobe control because their open surface allows radiation from bends and junctions and parasitic interactions between feed paths. These effects limit the achievable sidelobe level with open microstrip power dividers.

Variants of both the microstrip and the Collings radiator have been used as elements for arrays that scan in one dimension [118]. The strip version of the Collings radiator was scanned in one dimension and one version was excited with multiple feeds and used plated-through holes to divide the strip into cavities. Although the demonstration model was not scanned, element patterns indicated that scanning over a wide scan angle was possible.

Waveguide Slot Array Line Source Elements

Among the most important and well-understood line source elements are waveguide slot arrays [126–128]. Waveguide series slot arrays are simpler to construct but of narrower bandwidth than equal-path corporate-fed slot arrays, but can also have several undesirable pattern characteristics that will be described later.

Figure 5.23 shows three useful slot configurations for waveguide: edge slots, longitudinal displaced slots, and inclined series slots. Since the slot spacing must be restricted to avoid grating lobes, slot angles or locations are alternated as shown in the figure to introduce the extra 180-deg phase shifts to collimate radiation from slots approximately one-half wavelength apart in the transmission line. The slot arrays are designed as *resonant* (standing wave) or *traveling wave* arrays. Resonant arrays are terminated in short circuits to establish a standing wave in the feed waveguide, while traveling wave slot arrays are terminated by matched loads. Traveling wave arrays usually operate over broader bandwidths than resonant arrays, and the traveling wave array has an off-broadside (squinted) pointing angle that is a function of frequency. Resonant arrays are designed to radiate an "inphase" broadside pattern.

For traveling wave arrays with waveguide propagation constant $\beta = 2\pi/\lambda_g$, slot spacing d_y near half wavelength, and added phase shift π between each successive slot, the slots are excited by the exponential

$$\exp[-j2\pi v_0 n d_y/\lambda] = \exp\{-j[\beta n d_y - n\pi]\} = \exp[-jn\alpha] \qquad (5.44)$$

where v_0 is the usual direction cosine expression.

Using (5.44), the main beam occurs at the angle

$$v_0 = \frac{\alpha}{2\pi}(\lambda/d_y) = \frac{\lambda}{\lambda_g}\frac{-\lambda}{2d_y} \qquad (5.45)$$

or at

$$\phi = \pi/2 \qquad \theta = \sin^{-1}\left[\frac{\lambda}{\lambda_g} - \frac{\lambda}{\lambda d_y}\right] \qquad (5.46)$$

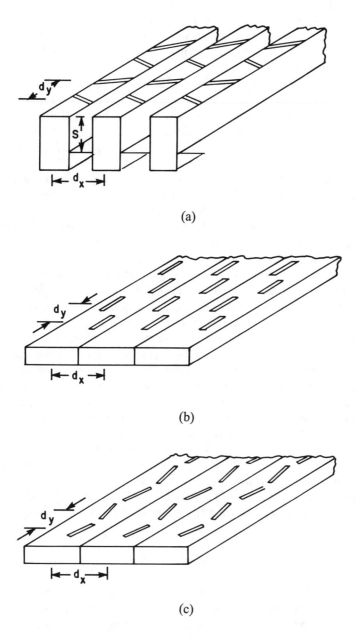

Figure 5.23 Waveguide slot array geometries: (a) edge slot array; (b) displaced longitudinal slot array; (c) inclined series slot array.

Since the guide wavelength λ_g is greater than the free space wavelength λ, the spacing d_y is usually chosen to be greater than $\lambda/2$ to bring the beam angle θ near broadside. Traveling wave arrays are not designed for broadside radiation ($\theta = 0$) because at that angle the various slot reflections add coherently and result in severe mismatch at the input port.

With precision fabrication, waveguide arrays can provide excellent pattern control, even at millimeter-wave frequencies. In a recent development, Rama Rao [129] used photolithographic technology to build waveguide longitudinal shunt slot and inclined series slot arrays at 94 GHz. The longitudinal shunt slot array was designed according to the formulas of Yee [130], while the inclined series slot array was designed following the analysis of Oliner [131].

One other important characteristic of traveling wave arrays is that each slot radiates only a fraction of the power incident upon it, and it is necessary to dissipate some of the power (often 5% to 10%) in matched loads in each waveguide. Unless the loads can be well matched over the frequency band, some power will be reflected by each array and radiated as an unwanted sidelobe at the angle $-\theta$. This limitation can be overcome by careful design, and traveling wave arrays have been incorporated into some of the lowest sidelobe antennas ever built.

In addition to these distinctions between resonant and traveling wave arrays, there are major differences between the performances of arrays that use different slot types. The displaced longitudinal slots of Figure 5.23 achieve the required 180-deg phase change by virtue of their displacement on either side of the broad-wall center line, while the various tilted slots (Figure 5.23) achieve the same phase change by alternating tilt angles. Tilted slots, however, radiate cross-polarized fields in addition to the fields of the principal polarization, and this is usually undesirable.

Unless they are loaded with dielectric to reduce interslot dimensions, all slot arrays produce unwanted lobes in the plane of the waveguide axis [132,133]. The lobes are of two types, and both result from either displacing or tilting alternate slots in different directions to produce the required 180-deg phase increment for the main beam. For tilted slots, the lobes with polarization orthogonal to the principal polarization are not grating lobes of the main beam, but result from a more rapid phase variation due to the added phase shifts. This added phase shift produces two lobes displaced from the main beam by (π/kd_y) in v-space, or appear at angles given by

$$v = v_0 \pm (\lambda/2d_y) \tag{5.47}$$

Since v_0 is usually small and since the spacing d_y is larger than a half wavelength for waveguides without dielectric loading, both of these lobes usually appear in real space. The magnitude of the lobes is zero along the y-axis at $x = 0$ ($u = 0$).

A second type of unwanted lobe is actually a grating lobe caused by the fact that the periodic cell of the line source array is two elements, not a single element.

For the longitudinal displaced slot array, this results in copolarized lobes, displaced a distance ($\lambda/2d_y$) from the main beam, but again the magnitude is zero along the principal plane $u = 0$. For a displaced slot array with slot displacement Δ, Derneryd [134] shows that the lobe magnitude varies like the product $u\Delta$ for small Δ, and gives convenient curves for estimating the magnitude of the lobes for small slot displacement. Tilted-slot arrays also have an asymmetry in the fields that radiate their principal polarized component (fields in the slots tilted clockwise have a different symmetry than those for slots tilted counterclockwise). Since this asymmetry repeats every two elements, this higher order principal plane radiation contributes to a set of grating lobes with the principal polarization and located at the same points ($\lambda/2d_y$) away from the main beam.

Since the lobe amplitudes are zero at $u = 0$, they do not radiate from two-dimensional flat-plate arrays composed of waveguide line sources but not scanned in the u-plane. However, these lobes can become very significant if the array is scanned to large angles in the u-plane. In a recent development [135], Green and Schnitkin presented a periodically ridge-loaded broad-wall waveguide low-sidelobe line source array. The system used an asymmetrical periodic ridge loading to introduce the required additional 180-deg phase shift every half wavelength along the waveguide. This allowed all slots to be on the waveguide center line and so the design radiates only principal plane radiation and does not suffer the first-order effects of slot displacement.

The edge slot arrays of Figure 5.23 are among the most commonly used elements for arrays that scan in one plane because the element spacing d_x can be made one-half wavelength or as appropriate for wide-angle scanning. Each waveguide slot array is excited by a different progressive phase for scan in the plane orthogonal to the waveguide axis. Low-sidelobe aperture distributions in the plane including the waveguide axis can be synthesized by varying the tilt angle of each slot, which changes its conductance. In order to maintain the high degree of aperture control necessary for low-sidelobe illuminations, mutual coupling effects must be included in the array design. A detailed treatment of the synthesis procedure including coupling is given in [136]. Edge slot arrays have several disadvantages. They are narrow-band and, like the inclined series broad-wall slots, they radiate cross-polarized lobes. The primary radiated beam is due to the electric field E_y in each slot, but the inclined slot produces a radiated component derived from the cross-polarized E_x field. Since the array is scanned only a small angle from broadside in the v-plane, these lobes are far from broadside and are suppressed by the cross-polarized element pattern. It is common practice to partially suppress the cross-polarized grating lobes by adjusting the depth S between the plane of the slots and the ground screen (see Figure 5.23).

Waveguides with displaced longitudinal or inclined series slots in the broad wall of the waveguides cannot be placed close enough to suppress the principal plane grating lobes for wide-angle scan (in the u-plane), and so these arrays are most commonly used unscanned. Dielectric loading the waveguide reduces this dimension to one appropriate for scanning, but there remain grating lobes due to the periodic displacement of the longitudinal slots or the periodic tilt of the inclined slots. In addition, unless the dielectric is extremely homogeneous, the variation in propagation constant along the waveguide can lead to high sidelobes.

Printed Circuit Series-Fed Arrays

Other narrow-band series arrays for one-dimensional scan include the series-fed microstrip patch arrays (Figure 5.22) and a variety of strip line and microstrip dipole arrays. Since these elements are more symmetrical than the waveguide slot arrays, they do not produce the second-order beams radiated by the slot configurations. Microstrip transmission line is, however, more lossy and cannot handle as much power as waveguide.

Comb line arrays use the half-wave section of microstrip transmission line as a microstrip dipole [137], but in this case the line is open-circuited at only one end, with the remaining end excited directly by the feed transmission circuit, as shown in Figure 5.24(a). The input impedance of each half-wavelength line, properly trimmed to account for reactive contributions, is the radiation resistance of the open-circuited stub element. The elements are placed a dielectric wavelength apart, and so the array radiates broadside in a manner directly analogous to a resonant waveguide slot array. Control of the radiation resistances is afforded by tailoring the width of the lines.

A planar array of comb line arrays, as shown in Figure 5.24, could provide the proper spacing for a scanned array, since the dielectric substrate allows spacing between line sources to be reduced to less than half wavelength.

5.1.7 Elements and Polarizers for Polarization Diversity

The elements described in previous sections are linearly polarized. However, in many cases there is a need to radiate several orthogonal components of radiation. In airborne and space communication systems, it is most common to radiate circularly polarized waves in order that transmit and receive antennas can never be completely orthogonal. Some radar systems have circular polarization (for removal of rain clutter), but most are linearly polarized. Recently, there has been an increasing interest in radar systems with variable polarization for target classification and the suppression of jamming signals within the main beam region. The subject of polarization synthesis is discussed in a number of technical papers and will not be treated here. A more detailed discussion is found in [138].

Most of the elements described in previous sections can be paired with identical elements positioned orthogonally to radiate two components of polarization.

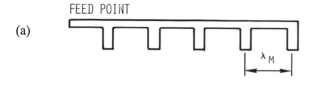

(a)

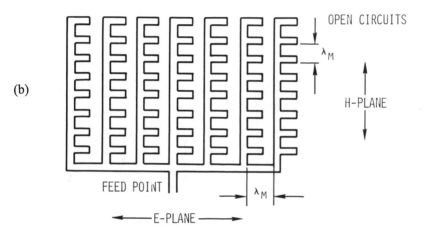

(b)

Figure 5.24 Comb line arrays (after [137]): (a) series-fed comb line array; (b) planar array of series-fed comb line arrays.

One antenna configuration commonly used for polarization diversity is a pair of orthogonal dipoles (Figure 5.25(a)) mounted over a ground screen. Arrays with waveguides of square, circular, or quad-ridged cross section or crossed-slot arrays are also commonly used. Microstrip patches can be excited with orthogonal feed lines (Figure 5.25(b)), and even the Vivaldi and flared-notch antennas can be combined in pairs to produce two orthogonal linear polarizations.

It is very expensive to excite an array to radiate arbitrary polarization. To do this requires a power divider and an extra phase shifter behind each element, or one must duplicate the entire array feed to excite two coincident, separate arrays with orthogonally polarized elements. In addition to these added components, there is usually not enough room within the array aperture to contain this extra circuitry. In the case of microstrip arrays, with low dielectric substrates for example, there is usually not enough room on one surface to provide power dividers and phasing

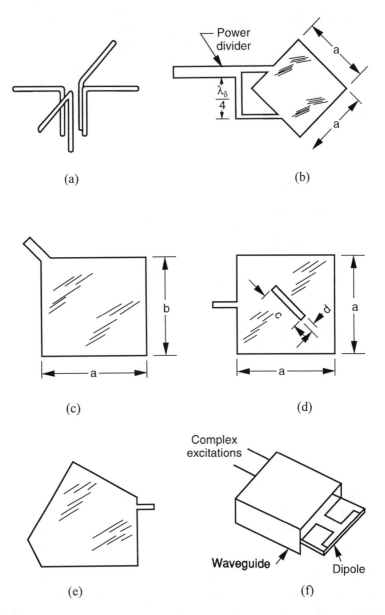

Figure 5.25 Elements for radiating circular polarization: (a) crossed dipole element for radiating both circular polarizations; (b) circularly polarized microstrip patch, orthogonal feeds; (c) circularly polarized microstrip patch (after [89]); (d) circularly polarized microstrip patch (after [139]); (e) circularly polarized microstrip patch (after [140]); (f) dipole slot circularly polarized element (after [142]).

controls for a two-dimensional scanning array, so exciting arbitrary polarization necessitates multilayer feed circuits. Nevertheless, the requirement to radiate independent polarizations is sometimes justified, and so a number of special techniques have been developed to provide this option.

If, however, one or both components of circular polarization are desired, the array feed can be simplified and be only slightly more complex than the corresponding linear (or dual linear) feed. This is accomplished using either circularly polarized elements or linearly polarized elements and a wave polarizing panel in front of the array. Figure 5.25(a) shows a pair of crossed dipoles. An added phase delay of 90 deg in series with one of the dipoles will produce a right- or left-handed circularly polarized radiated signal. This same antenna will receive a left- or right-handed circularly polarized wave at its input port. If the power split is accomplished using a four-port hybrid, an incident signal with the opposite circular polarization would be delivered to the port terminated in the load.

Figure 5.25(b) shows an accepted method for exciting patch antennas for circular polarization. This straightforward configuration requires considerable space on the patch array surface and so is not always the selected geometry. Sanford and Klein [108] used a combination of four shorted quarter-wave patch antennas, shown in Figure 5.17(e), to form a crossed slot with circular polarization, but this geometry also requires considerable space at the aperture. Figure 5.25(c–e) shows other means of exciting patch antennas using a single feed and producing the circular polarization by exciting asymmetrical current distributions. Figure 5.25(c) [89] shows a corner feed point used to execute an asymmetrical patch. Varying the dimensions a and b can produce equal orthogonal radiation components with the proper 90-deg phase for circular polarization. The geometry of Figure 5.25(d) [139] uses a symmetrical patch and a slot at a 45-deg angle centered in the patch, and 5.25(e) shows a pentagon-shaped patch [140] due to Weinschel. Each of these geometries includes an asymmetry, and by judiciously selecting dimensions can produce circularly polarized radiation. Schaubert et al. [141] discuss the use of shorting posts and asymmetrical feed locations to produce polarization diversity with single feed points.

Arrays of circular waveguides are usually excited for circular polarization using waveguide circular polarizers. These operate by introducing an asymmetry to produce orthogonal linear polarizations and then delaying or advancing one component an extra 90 deg. These components are used after the phase shifter at each element of the array and are discussed in many standard texts, so they will not be considered further here.

Scanning arrays have special problems in regard to polarization, because they require elements that have, for example, circular polarization not only on axis, but throughout some given scanning sector. This problem is apparent when one considers the principal E- and H-planes of a crossed dipole, as in Figure 5.25(a). In the plane $\phi = 0$, the far field has a component in the θ direction due to the dipole with its axis along the x-axis, and an orthogonal component due to the dipole that lies along the y-axis. The component with θ polarization has a doughnut-shaped pattern with zeros at $\theta = \pm \pi/2$. The orthogonal polarized field is due to the dipole with axis along the y-axis, and is constant for all θ in the plane $\phi = 0$. Thus, even if the dipoles are excited for circularly polarized radiation on axis $\theta = 0$, that polarization will be linear (horizontal) at $\theta = \pm \pi/2$ and will vary from circular at $\theta = 0$ to linear as θ is increased from zero. To produce patterns that are approximately circular over wide scan angles requires equalizing the constituent E- and H-plane patterns.

The combination of complementary dipole and slot antennas can, with proper phase excitation, produce circularly polarized radiation over a wide angular region. An example is the work of Cox and Rupp [142], who developed the dipole-slot element shown in Figure 5.25(f) for use in an array.

A very practical method of radiating circular polarization is to place a polarizing grid in front of a linearly polarized array. A variety of grids have been developed for that purpose, and the most commonly used are mentioned here. In Figure 5.26(a), the classic quarter-wave plate polarizer [143,144] is shown as consisting of a number of closely spaced plates aligned normal to the incident wave, but at 45 deg relative to the polarization of the incident electric field. This 45-deg orientation is used for all the other polarizers discussed as well. Since the incident field can be decomposed into components perpendicular and parallel to the plates, the polarizer action is achieved because the component of incident field perpendicular to the plates passes unaltered, but the component parallel to the plates propagates through the structure by means of a parallel plate mode with phase velocity greater than light. By choosing the length W so that the parallel component is advanced 90 deg relative to the perpendicular component, the radiated field is made a circularly polarized wave.

The Lerner polarizer (Figure 5.26(b)) is composed of resonant grids and arranged in panels mounted normal to the array aperture at 45 deg from the plane of the linearly polarized electric field. The net result of the polarizer is to delay one component of polarization by 90 electrical degrees of phase relative to the other polarization. In the upper part of Figure 5.26(b) is shown a row of metallic strips, which serves to explain the operation. The strips act as a shunt inductance for the E-field along the strips and a shunt capacitance for the E-field perpendicular to the strips. As the strip width is reduced, the inductance dominates, so that in the limit the wire is invisible to orthogonal polarization and inductive to parallel polarization. The Lerner polarizer uses combinations of wires and solid metallic rectangles. The rectangles are capacitive to both polarized components, with the amount of capacitance adjusted by varying the dimensions. The wires are inductive to the parallel E-field. For E-fields along the wire, the wires and rectangles combine to form a parallel resonant circuit. At higher and lower frequencies, the sheet is capacitive and inductive, respectively. The polarizer is made up of two or more

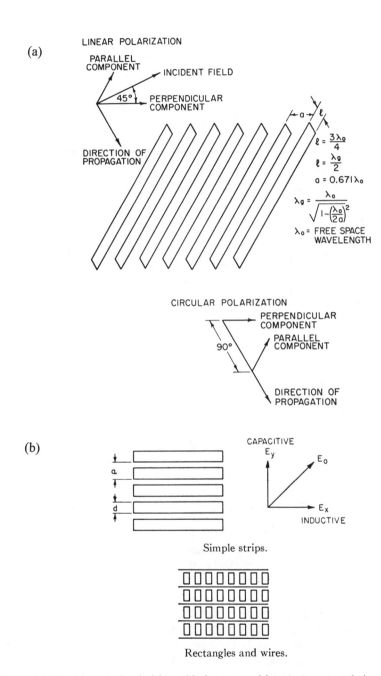

Figure 5.26 Polarizers and polarizing grids for arrays: (a) quarter-wave polarizer (after [143]); (b) Lerner polarizer (after [145]), simple strips and rectangle and wire grids.

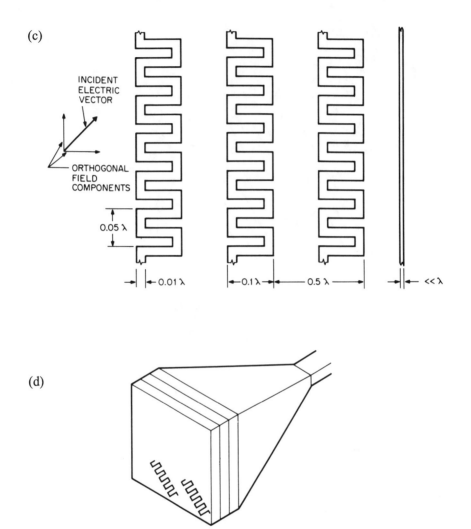

Figure 5.26 (cont.) Polarizers and polarizing grids for arrays: (c) meander line polarizer (after [146]);
(d) element rotation for circular polarized array (after [150]).

such sheets, spaced so that their reflections cancel, while at the same time matching impedance and producing the required 90-deg phase shift at one or several frequencies. Polarizers of this type have been shown to operate over 20% bandwidth, with as little as 0.5 dB insertion loss and good axial ratios.

Meander line polarizers [146,147], shown in Figure 5.26(c), employ printed metallic meander line patterns on dielectric substrates to provide different reactive loadings to orthogonal components of the electric field. The line pattern is again oriented at 45 deg relative to the incident field polarization, and presents inductive loading to the field component parallel to the pattern and capacitive loading to that component normal to the pattern. Design details for such polarizers are given in [146–148].

A new and interesting approach, presented in the work of Teshirogi et al. [149], is to produce circular polarization using groups of linearly polarized elements. The arrangement produces wide-band circular polarization using a technique of sequential rotations and phase shifts to each element. The basic property of this method is apparent even if the elements are linearly polarized. In this case, by sequentially rotating the elements and therefore sequentially rotating their radiated polarization, and advancing the element phase so that a polarization tilt of Φ is associated with a phase advance of Φ, one produces a circularly polarized wave on boresight. Depending on how the elements are combined, one can also obtain a cancellation of reflected signals. The result is wide-band, matched circular polarization radiated from an array of elliptically or even linearly polarized elements. In an extension of this technique shown in Figure 5.26(d), Hall [150] feeds sequentially rotated four-element subarrays with sequentially rotated feeds, so that there are two scales of sequential rotation. This technique is shown to decrease the sidelobes due to feed radiation.

Although the explanation above used a linearly polarized element as basis, the use of subarrays with linearly polarized and phased elements, repeated throughout the array, leads to grating lobe-like sidelobes. The technique is more advantageous when applied to elliptically polarized or nearly circularly polarized radiators. In this case, it improves circularity bandwidth and ellipticity ratio without creating high sidelobes.

5.2 TRANSMISSION LINES FOR PHASED ARRAY FEEDS

The transmission line network of choice for a particular array is determined by the frequency of operation, the array power requirements, and compatibility with array fabrication technology. At UHF frequencies and below, coaxial line networks are often the preferred medium; for frequencies between the upper UHF (about 200 MHz) and 10 GHz, waveguides are the lowest loss, highest power transmission line, and so many ground-based arrays in this frequency range are built with waveguide technology. In this frequency range, strip line feed networks are often less expensive than waveguides (though more lossy), and microstrip networks find application to arrays fabricated by monolithic techniques. At frequencies above 20 GHz, microstrip technology has a special role as the medium for integrating solid-state devices in hybrid or monolithic microwave integrated circuits and so has a unique place as the basis of subarrays on dielectric or semiconductor substrates.

Since there are numerous texts that give detailed equations and graphs of the basic characteristics of most of the important transmission lines, the following section will include only a qualitative description and comparison of array applications. For more details of each of the transmission lines mentioned here, the reader is referred to several modern texts [151–153].

5.2.1 Coaxial Transmission Line

The transverse electric mode (TEM) coaxial transmission line (Figure 5.27) has been the medium of choice for high-power dipole or monopole arrays at frequencies below 1 GHz. Like waveguide, it is self-shielded, and there are a large number of standard devices and design data available for component engineering. Assembly is costly, and corporate feed design is not compatible with computer-aided manufacture. In addition, coaxial line networks are heavy and bulky. However, when high precision, broadband operation coupled with excellent physical stability and high-power operation are required, coaxial line is the main competition to waveguide transmission line. At frequencies below about 1 GHz, waveguide is often unacceptably large, and the small cross section of coaxial line makes it the medium of choice for many applications.

Coaxial line supports a dominant TEM of propagation and so is dispersionless when operated at frequencies below the cutoff of its higher order modes. In this mode, its characteristic impedance is given by

$$Z_0 = \frac{60}{(\epsilon^{1/2})} \ln(b/a) \tag{5.48}$$

The propagation constant is $k/(\epsilon^{1/2})$.

Coaxial lines are ideal feeds for monopole antennas over ground screens and can couple efficiently to waveguide or microstrip lines by means of probe feeds. To feed dipole antennas, however, requires a balun designed to transform the unbalanced coaxial line fields to a balanced two-wire line. A number of coaxial line balun circuits are discussed in [154], and the split tube balun shown in Figure 5.7(a) is one example of a wide variety of circuits that perform impedance transformation and balance.

5.2.2 Rectangular, Cylindrical, and Ridged Waveguides

Although some very-high-power antennas have used waveguides at UHF frequencies, waveguides (Figure 5.27) are mainly useful above 1 GHz. At these frequencies, waveguides have less loss and is only slightly bulkier but can be lighter weight than

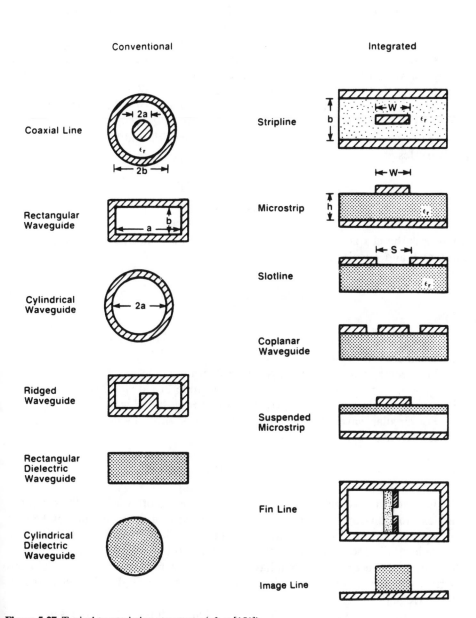

Figure 5.27 Typical transmission structures (after [151]).

coaxial lines if a thin-walled guide is used. Computer-aided manufacturing has also reduced the cost of waveguide corporate feed networks. Space-fed architectures and lenses have been particularly successful using waveguide technology.

Waveguides are excellent high-power transmission media and have sufficient bandwidth for most array applications. They are closed, shielded structures and so provide good isolation for active microwave circuitry. Waveguide characteristic impedance is ambiguous, but for all waveguides that are "cylindrical" (that is, have a constant cross section), one can define a *wave impedance* as the ratio of transverse electric (TE) to magnetic field for the lowest order TE, and transverse magnetic (TM) modes as

$$Z_{TE} = \eta/[1 - (\lambda/\lambda_c)^2]^{1/2}$$
$$Z_{TM} = \eta[1 - (\lambda/\lambda_c)^2]^{1/2}$$

(5.49)

where η is the characteristic impedance of free space, and the guide propagation constant β and wavelength λ_G are defined as

$$\beta = \frac{2\pi}{\lambda_G} = \frac{2\pi}{\lambda}[1 - (\lambda/\lambda_c)^2]^{1/2}$$

(5.50)

Here the cutoff wavelength λ_c corresponds to the lowest frequency that propagates the fundamental (lowest order) mode.

For rectangular waveguide, the lowest order mode is the TE_{10} mode with cutoff given by $\lambda_c = 2a$, where a is the waveguide broad wall inside dimension. The waveguide electric field is chosen normal to the broad wall.

For circular waveguide of radius a, the lowest order mode is the TE_{11} mode with cutoff wavelength

$$\lambda_c = 3.41a$$

(5.51)

For ridged waveguide (see Figure 5.27), the cutoff wavelength is a more complex function of the geometry. The early results that were obtained by Hopfer [80] were based on equivalent circuit techniques. These results are tabulated in two studies [155,156]—one for a single- and the other for double-ridged guide. There are more recent results that have used finite element methods [157,158] and integral equation methods [159] in order to obtain the modal characteristics of ridge waveguides.

The ridged waveguide owes its wide bandwidth to the large separation between the cutoff wavelength for the lowest order mode (TE_{01}) and the next higher mode. This frequency spread increases as shown in [80] with increasing ridge height.

5.2.3 Planar Transmission Lines

The following listing of transmission media includes most of the important transmission lines that are constructed from printed circuit boards. One partial exception is that strip line is most often made from machined sheets of metal, but has also been assembled from circuit boards. This assembly results in simplified fabrication and, in most cases, a compatibility with integrated circuit technology key to low-cost circuit development.

Strip Line

Strip line is a relatively high-power TEM transmission medium that is very compatible with computer-aided design and manufacture. Strip line technology is also well developed, and one can achieve precision design of power dividers and corporate feed networks. Air strip line is particularly favored for low-sidelobe array feeds. Since the center conductor is not exposed, strip line is not used for integration of monolithic solid-state devices. Many standard texts give the relevant parameters [151–153,160].

Microstrip Transmission Line

The main advantage of microstrip line (Figure 5.27) is that the upper surface of the substrate and the strip conductor metallization is exposed. This allows network definition and fabrication by lithographic processes, which has led to monolithic fabrication of circuits and devices at extremely high frequencies (EHF). The basic characteristics of microstrip transmission line are documented in [152].

Calculations using the quasi-TEM approximation have yielded characteristic impedance and effective dielectric constants for the evaluation of propagation velocities. Bahl [151] provides a convenient tabulation of these approximate results. More accurate full-wave numerical solutions are also regularly used for detailed circuit design. Microstrip circuits become lossy at millimeter waves because of radiation from bends, transitions, and other discontinuities. Bahl [151] shows that radiation Q is proportional to $1/(fh)^2$, where f is frequency and h is the substrate thickness. Reducing the radiation loss thus requires decreasing substrate thickness (as one would expect from the scaling of the problem), but this tends to increase the conductor dissipative loss.

With increasing dielectric constant, the relative strip width W/h is decreased for a given characteristic impedance. This can lead to very thin linewidth W for high dielectrics at high frequencies where the substrate is also very thin. The typical impedance values vary from 20Ω to 150Ω, depending on the substrate and dimensions.

Coplanar Waveguide and Grounded Coplanar Waveguide

Coplanar waveguide (Figure 5.27) is particularly suited for the integration of devices on semiconductor substrates like gallium arsenide (GaAs) because of the ease of obtaining dc ground potential for devices using the metallized top surface. The characteristic impedance is given by Wen [161] and ranges from roughly 50Ω to several hundred, depending on substrate. Reference [151] gives approximate results for characteristic impedance and effective dielectric constant and shows curves of attenuation constant for a typical substrate and geometry.

The grounded coplanar waveguide is often used as a microstrip patch feed as shown in Figure 5.21(c), where it is important to move the feed point into the body of the patch. The impedance of this line is given in [162]. The grounded coplanar waveguide is not shown in Figure 5.27, but consists of the coplanar waveguide with a metallic ground screen on the other side of the substrate.

Coplanar Strips Transmission Line

The complementary geometry to coplanar waveguide, this configuration also allows for integration of passive and active circuitry on a single layer. This geometry is not shown in Figure 5.30, but is represented by the coplanar waveguide sketch if planar metallized surfaces are omitted and surfaces not metallized are metallized. It consists of a substrate without ground screen, supporting two parallel conducting strips. The characteristic impedance, effective dielectric constant, and some loss data for coplanar strips are given in [152,153].

Slot Line

Slot line (Figure 5.27) is useful for a number of circuit applications that include integrating devices and transmission lines. It consists of a dielectric substrate with a ground screen having a single longitudinal slot. It has also been used to excite microstrip patch antennas [163]. Mariani [164] and Garg and Gupta [164] give approximate expressions for characteristic impedance and wavelength, and Janaswami and Schaubert [166,167] present numerical results for characteristic impedance and propagation characteristics.

Remarks on Planar Transmission Line

Although most of the references cited here and in the survey texts give the results of quasi-static solutions for characteristic impedance and propagation constant (through static circuit parameters), full-wave solutions have now been carried out for all of the planar lines discussed in this section.

Table 5.2 [152] shows a qualitative comparison of some of the planar transmission lines discussed above. Microstrip line has been used extensively for monolithic integrated circuits in silicon and GaAs, while all four line types have proven useful for hybrid integrated circuit design. A major distinction cited by Bahl [151] is that coplanar lines offer ease in mounting components in series and shunt configurations, while microstrip lines are convenient only for series mounting and slot lines primarily for shunt mounting of components.

A number of other planar transmission lines have proven useful and are described in the literature. These include coupled microstrip lines, suspended and inverted microstrip lines, and a number of "trapped" transmission lines that are basically waveguide-enclosed versions of the planar lines mounted in waveguides that are beyond cutoff.

5.2.4 Other Useful Transmission Lines

Fin Line

Fin transmission lines (Figure 5.27) are metallic fins printed on a dielectric substrate and mounted across a rectangular waveguide. The waveguide size is chosen to be below cutoff to suppress conventional waveguide modes, and so the fin line is

Table 5.2
Qualitative Comparison of Various MIC Lines

Characteristic	Microstrip	Slotline	Coplanar Waveguide	Coplanar Strips
Effective dielectric constant*	~6.5	~4.5	~5	~5
Power-handling capability	High	Low	Medium	Medium
Radiation loss	Low	High	Medium	Medium
Unloaded Q	High	Low	Medium	Low[†]/High[‡]
Dispersion	Small	Large	Medium	Medium
Mounting of components:				
In shunt configuration	Difficult	Easy	Easy	Easy
In series configuration	Easy	Difficult	Easy	Easy
Technological difficulties	Ceramic holes Edge plating	Double side etching		
Elliptically polarized magnetic field configuration	Not available	Available	Available	Available
Enclosure dimensions	Small	Large	Large	Large

Source: [152].
*$\epsilon_r = 10$, $h = 0.025$ inch.
[†]At lower impedances.
[‡]At higher impedances.

basically a captured slot line. Fin line is a wide-band transmission system and is effective throughout the millimeter-wave frequency range. It is compatible with wide-band flared antennas and has also proven a viable medium for the design of components through 100 GHz. Several texts [168,169] give the characteristic impedance and propagation constants of a variety of fin line structures.

Dielectric Image Line

Dielectric image line (Figure 5.27) is frequently used as a wave-guiding medium for millimeter-wave circuits because of its low-loss properties. In addition to the basic image line shown in the figure, there are various trapped and inverted versions that have also found use. Although these lines are generally useful for signal distribution as leaky wave line source radiators (when loaded with periodic discontinuities) and as waveguide feeds of quasi-optical radiating systems, they have not yet found use in scanning arrays.

5.3 ARRAY ARCHITECTURE

5.3.1 Array Manifold

Array cost continues to limit the use of arrays in systems. If cost were no consideration, there would seldom be any need to use other than waveguide- or dipole-type elements. It is primarily this issue that continues to require more innovations and creativity of the array designer, for the solution lies not in mass-produced dipoles or waveguides, but in developed techniques that assemble the array in relatively larger sections and that may incorporate elements or groups of elements, controls, and devices, all in the same fabrication step and all assembled by automatic processes. There is a need for special array architectures, specific ways of collecting, assembling, and mounting array elements, and special types of array feeds to be compatible with various ways of grouping elements.

Particular architectures seem to be appropriate to specific frequency ranges and array geometry requirements (size and depth). Figure 5.28 shows the two basic array constructs and introduces the terms *brick* and *tile* constructs as coined by Kinzel et al. [170]. Figure 5.28(a) shows an array of printed circuit dipoles in a brick arrangement. The brick construct uses the depth dimension to provide the functions that are accomplished in the multiple layers of the array with tile construction. Thus, each brick may contain a row or column power divider, phase shifters, amplifiers, and other devices in addition to maintenance features and cooling. The brick may be produced by monolithic integrated circuit technology, and so be fully compatible with low-cost fabrication. Elements for the brick construction are also quite reasonable, since horizontal dipoles, flared-notch elements,

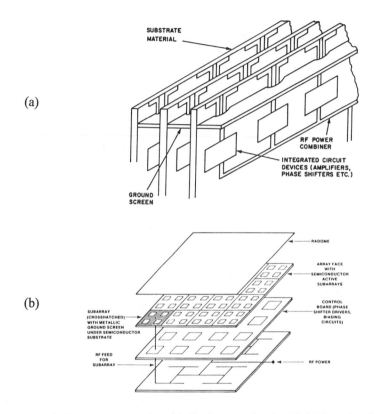

Figure 5.28 Basic array construction: (a) dipole array showing "brick" construction; (b) microstrip patch array showing "tile" construction.

and a variety of others can be integrated into this geometry. These elements generally have broader bandwidth than microstrip patches and this may be a major advantage in some system applications.

In the limiting case, a brick may be a single module and construction is reduced to assembling the array face one element at a time. This has been the established practice for most radar arrays at frequencies through 10 GHz. In this case, the array element modules, which consist of an element and a phase shifter (and perhaps the phase shifter driver circuitry) are inserted into a manifold that provides RF power and phase shifter control. The modules can also include active devices, amplifiers, and switches, and so may be complete transmit-receive front ends. In this way, the transmitter and receiver chain is a part of the array face, and this needs to be accounted for in thermal and mechanical design. The RF power division is accomplished in the manifold, as is logic signal distribution and cooling. This assembly technique is efficient and relatively easy to maintain, though not inexpensive to produce. It seems clear that for frequencies up to K-band (roughly 15 GHz), this type of assembly may always be the most practical because of element size and separation. It now seems that at some time in the future, this architecture may not be practical nor have the lowest cost at EHF and millimeter-wave frequencies, and so may be replaced by brick construction with a multiplicity of elements in each brick or by the tile construction described below. The reason that frequency enters into this selection is that semiconductor substrate size is limited. As frequency is increased, it becomes possible to place more devices and elements on the same chip. At these frequencies, the use of multiple-element brick and tile construction becomes practical.

Figure 5.28(b) shows an architecture that Kinzel et al. [170] call *tile construction*, and that many have called *monolithic array construction*. It appears that the term *tile* is more appropriate because these tiles often have a multiplicity of layers (are not monolithic), and because other architectures seem to be as compatible with monolithic integrated circuit technology and so as equally deserving of that identification. The primary antenna elements used in this type of assembly are the microstrip patch radiator or microstrip dipole, fed by microstrip transmission line, although various other planar transmission lines have also been used to feed microstrip and other planar antennas.

Whether tile or brick construction is used, there is still a significant architectural issue that addresses how the proper array weights are applied to elements at the array face. The array face itself is often organized into subarrays of rows, columns, or areas (Figure 5.29) with each subarray fed separately.

The terms *brick* and *tile* relate to the way the array is assembled, not the organization of the aperture. One could assemble an array of column subarrays using the tile construct if the planar RF power dividers addressed columns of the array, or one could assemble an area subarray by inserting the subarray as a brick from behind the aperture. In terms of the quality of the array radiation pattern, the column subarray organization is usually preferred to area subarrays, because the power distribution network for each row can be made with the proper taper for sidelobe reduction in the plane of the row or column axis. Sidelobe control in the orthogonal plane is provided by a separate power divider. The fabrication of a network to excite the column subarray can be accomplished using power dividers below or in the plane of the aperture, but for most applications, where space permits, the brick fabrication is preferred because it provides more room for phase shifters, power dividers, and other components.

Area subarrays are useful primarily when the array is to be uniformly illuminated, or at least when the area subarrays themselves can have uniform illuminations. To achieve low sidelobes with area subarrays, the subarray amplitude taper would need to be different with each subarray, and that is a costly constraint. When the sidelobe requirements are not too severe, the subarray size can be chosen to use equal amplitude subarrays and to use as amplitude distribution a series of

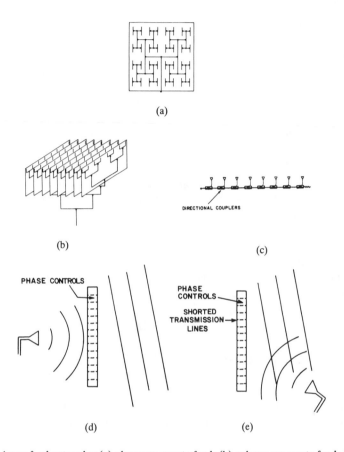

(a)

(b)

(c)

DIRECTIONAL COUPLERS

PHASE CONTROLS

PHASE CONTROLS

SHORTED
TRANSMISSION
LINES

(d)

(e)

Figure 5.29 Array feed networks: (a) planar corporate feed; (b) volume corporate feed; (c) series (line source) feed; (d) lens-fed array; (e) reflectarray.

quantized steps. If the subarrays are the same size, then the periodic amplitude error causes well-defined grating lobes to appear as shown in Chapter 7. These lobes provide the ultimate limit to the sidelobe level.

5.3.2 Array Feed Networks

There are three fundamental array feed configurations for phased arrays: the corporate-fed array, the space-fed array, and the reflectarray.

Corporate-Fed Arrays

Corporate-fed arrays (Figure 5.29(b)) can be assembled with precision power dividers and allow precise control of amplitude and phase illumination across the array as commensurate with low-sidelobe patterns. Some corporate feed networks are also compact and can be made with minimal depth, as in arrays with tile construction. Among the applicable transmission media for corporate feeds, waveguide and coaxial line corporate feeds are expensive, bulky, and difficult to produce, but offer excellent precision and high-power handling capability. Strip line power dividers are applicable at frequencies above 1 GHz, can be extremely precise, and can be fabricated with computer-aided manufacture techniques. These three constrained feed media are still more expensive to produce than corporate feeds of microstrip or other planar transmission lines, but offer better sidelobe control and lower loss, especially at higher frequencies.

The loss in corporate feed networks arises from line dissipation and radiation and reflection from junctions. Whether using brick or tile construction, an N-element square array with elements spaced d apart in both planes has a total transmission line length ℓ in series with every element of at least

$$\ell = (N^{1/2} - 1)d \tag{5.52}$$

In the case of brick construction, the actual length is more than this by the length of the vertical lines connecting the various levels of power division (see Figure 5.29(b)).

One can also show that either type of architecture has a total of $\log_2(N)$ power dividers in series with each element, or, in terms of base 10 logarithms:

$$\text{Number of power dividers} = 3.32 \log_{10}(N) \tag{5.53}$$

An array of only 64 elements needs six power dividers in any series path. Thus, even if the power divider loss is only a few tenths of a decibel, the loss in these components is significant. An illustration of how this can have an impact on array architecture is given in [171].

Series-fed line arrays, as shown in Figure 5.29(c), and previously as waveguide and microstrip line sources (Figures 5.22, 5.23, and 5.24), play a significant role as discussed in Section 5.1, and certainly offer a low-cost alternative whenever system bandwidth and beam squint constraints will allow.

Space-Fed Arrays

A number of very successful space-fed arrays have been built. The space-fed geometry of Figure 5.29(d) provides a convenient way to obtain low-sidelobe patterns with low loss. Space-fed arrays are far less expensive to construct than corporate-fed arrays, and, in addition, offer several advantages in simplicity that are not shared by corporate-fed arrays. They have a natural separation between the array face and the feed that simplifies maintenance if the array face is constructed in the brick configuration. In this case, individual subarrays can be removed for repair. The disadvantages of space-fed arrays are that they suffer spillover loss, reflection loss, do not offer as much pattern control for low-sidelobe radiation, and are bulky. Nevertheless, for many applications the advantages of space-fed arrays far outweigh these disadvantages. Space feeds may also be the only realistic way to illuminate the very large arrays that have been proposed for space-based radar and communications. Moreover, with the use of carefully controlled feed illuminations, space-fed arrays can have excellent sidelobe characteristics.

A unique space feed system that has significant flexibility for pattern control (low-sidelobe formation, nulling) uses a Fourier transform feed to illuminate the lens. Typical transform feeds are Butler matrices or focusing lenses. The design of such array feeds is discussed in Chapter 8.

Reflectarrays

Reflectarrays [172] offer the advantage of simplicity because array controls can be hidden behind the array ground screen. Figure 5.29(e) shows the geometry of a reflectarray and indicates that in this case the array aperture acts as both a receiving and radiating surface. There have not been as many reflectarrays as space-fed lens arrays because the pattern characteristics are not as easy to control as they are space-fed lens arrays. With the reflectarray, the energy scattered from the front face mismatch radiates in the forward direction, and unless suppressed, contributes directly to the sidelobe level. In addition, for reflectarrays it is necessary for the phase shifters to be reciprocal devices (whereas in the lens array they can be nonreciprocal, in which case they are switched between transmit and receive modes).

REFERENCES

[1] Ludwig, A. C., "The Definition of Cross Polarization," *IEEE Trans.*, Vol. AP-21, Jan. 1973, pp. 116–119.
[2] Balanis, C. A., *Antenna Theory, Analysis and Design*, New York: Harper and Row, 1982, p. 338.
[3] Tai, C. T., "Dipoles and Monopoles," Ch. 4 in *Antenna Engineering Handbook*, R. C. Johnson and H. Jasik, eds., New York: McGraw-Hill, 1984.
[4] King, R. W. P., *The Theory of Linear Antennas*, Harvard University Press, 1956.
[5] King, R. W. P., S. S. Sandler, and R. B. Mack, *Arrays of Cylindrical Dipoles*, Cambridge University Press, 1968.

[6] King, R. W. P., and C. W. Harrison, *Antennas and Waves: A Modern Approach*, Cambridge, MA: The MIT Press, 1969.

[7] Brown, G. H., and O. M. Woodward, Jr., "Experimentally Determined Radiation Characteristics of Conical and Triangular Antennas," *RCA Review*, Vol. 13, No. 4, Dec. 1952, p. 425.

[8] Balanis, op. cit. pp. 304–306.

[9] Elliott, R. S., *Antenna Theory and Design*, Englewood Cliffs, NJ: Prentice-Hall, 1951, pp. 301–302.

[10] Fenn, A. J., "Theory and Experimental Study of Monopole Phased Array Antennas," *IEEE Trans.*, Vol. AP-33, No. 10, Oct. 1985, pp. 118–126.

[11] Weeks, W. L., *Antenna Engineering*, New York: McGraw-Hill, 1968.

[12] Balanis, op. cit. pp. 145.

[13] Bowman, D. F., "Impedance Matching and Broadbanding," Ch. 43 in *Antenna Engineering Handbook*, R. C. Johnson and H. Jasik, eds., New York: McGraw-Hill, 1984.

[14] Balanis, C. A., op. cit., pp. 365–368.

[15] Shuhao, "The Balun Family," *Microwave Journal*, Sept. 1987, pp. 227–229.

[16] Wilkinson, W. C., "A Class of Printed Circuit Antennas," *IEEE AP-S Int. Symp. Dig.*, 1974, pp. 270–273.

[17] Wheeler, H. A., "Transmission Line Properties of Parallel Strips Separated by a Dielectric Sheet," *IEEE PGMTT Trans.*, Vol. MTT-13, No. 3, March 1965, pp. 172–185.

[18] Duncan, J. W., and V. P. Minerva, "100:1 Bandwidth Balun Transformer," *Proc. IRE*, Feb. 1960, pp. 156–164.

[19] Edward, B., and D. Rees, "A Broadband Printed Dipole With Integrated Balun," *Microwave Journal*, Vol. 30, May 1987, pp. 339–344.

[20] Tang, R., and R. N. Burns, "Phased Arrays," Ch. 20 in *Antenna Engineering Handbook*, R. C. Johnson and H. Jasik, eds., New York: McGraw-Hill, 1984.

[21] Proudfoot, P. M., "A Wide-Band Printed Circuit Dipole," RADC-TR-88-121, Rome Air Development Center In-House Report, May 1988.

[22] Roberts, W. K., "A New Wideband Balun," *Proc. IRE*, Vol. 45, Dec. 1957, pp. 1628–1631.

[23] Bawer, R., and J. J. Wolfe, "A Printed Circuit Balun for Use With Spiral Antennas," *IEEE Trans.*, Vol. MTT/8, May 1960, pp. 319–325.

[24] Axelrod, A., and D. Lipman, "Novel Planar Balun Feeds Octave-Bandwidth Dipole," *Microwaves and RF News*, Aug. 1986, pp. 91–92.

[25] Cloete, J. H., "Exact Design of the Marchand Balun," *Microwave Journal*, May 1980, pp. 99–110.

[26] Marchand, N., "Transmission-Line Conversion," *Electronics*, Vol. 17, Dec. 1944, pp. 142–145.

[27] Mayer, E. D., and A. Hessel, "Feed Region Modes in Dipole Phased Arrays," *IEEE Trans.*, Vol. AP-30, Jan. 1982, pp. 66–75.

[28] King, R. W. P., and T. T. Wu, "The Cylindrical Antenna With Arbitrary Driving Point," *IEEE Trans.*, Vol. AP-13, Sept. 1965, pp. 710–718.

[29] Bock, E. L., J. A. Nelson, and A. Dorne, "Sleeve Antennas," Ch. 5 in *Very High Frequency Techniques*, pp. 119–137.

[30] King, R. W. P., "Asymmetric Driven Antennas and the Sleeve Dipole," *Proc. IRE*, Oct. 1950, pp. 1154–1164.

[31] Wong, J. L., and H. E. King, "An Experimental Study of a Balun-Fed Open Sleeve Dipole in Front of a Metallic Reflector, *IEEE Trans.*, Vol. AP-20, March 1972, p. 201.

[32] Poggio, A. J., and P. E. Mayes, "Pattern Bandwidth Optimization of the Sleeve Monopole Antenna," *IEEE Trans.*, Vol. AP-14, Sept. 1966, pp. 643–645.

[33] Wunsch, A. D., "Fourier Series Treatment of the Sleeve Monopole Antenna," *IEE Proc.*, Vol. 135, Pt. H, No. 4, Aug. 1988, pp. 217–225.

[34] Balanis, op. cit. p. 333.

[35] Schelkunov, Ch. 11 in *Electromagnetic Waves*, New York: Van Nostrand, 1943.

[36] Brown, G. H. and O. M. Woodward, Jr., op. cit.

[37] Bailey, M. C., "Broad-Band Half-Wave Dipole," *IEEE Trans.*, Vol. AP-32, No. 4, April 1984, p. 412.

[38] Mushiake, Y., "An Exact Step-Up Ratio Chart of a Folded Antenna," *IRE Trans.*, Vol. AP-3, No. 4, Oct. 1954, p. 163.

[39] Thiele, G. A., E. P. Ekelman, Jr., and L. W. Henderson, "On the Accuracy of the Transmission Line Model for Folded Dipole," *IEEE Trans.*, Vol. AP-28, No. 5, Sept. 1980, pp. 700–703.

[40] Hansen, R. C., "Folded and T-Match Dipole Transformation Ratio," *IEEE Trans.*, Vol. AP-30, June 1982.

[41] Herper, J. C., A. Hessel, and B. Tomasic, "Element Pattern of an Axial Dipole in a Cylindrical Phased Array, Part 2: Element Design and Experiments," *IEEE Trans.*, Vol. AP-33, March 1985, pp. 273–278.

[42] Lampe, R. L., "Design Formulas for an Asymptotic Coplanar Strip Folded Dipole," *IEEE Trans.*, Vol. AP-33, Sept. 1985, pp. 1023–1031.

[43] Hilberg, W., "From Approximations to Exact Expressions for Characteristic Impedances," *IEEE Trans.*, Vol. MTT-17, May 1969, pp. 255–265.

[44] Proudfoot, P. M., "A Printed Circuit Folded Dipole with Integrated Balun," RADC-TR-89-237, Oct. 1989.

[45] James, J. R., and G. J. Wilson, "Microstrip Antennas and Arrays, Part 1: Fundamental Action and Limitations," *IEE Journal, Microwaves, Optics and Acoustics*, Vol. 11, Sept. 1977, pp. 165–174.

[46] Mise, M., "Characteristics of Microstrip Antennas," *Inst. Electron. Commun. Engs.*, Japan, Papers Tech. Group Antennas Propagat., 1976–77, Series No. 89, 1976, pp. 19–24.

[47] Oltman, H. G., "Electromagnetically Coupled Dipole Antenna Element," *8th European Microwave Conf.*, Paris, 1978, Vol. 9.

[48] Oltman, H. G., and D. Z. Huebner, "Electromagnetically Coupled Microstrip Dipoles," *IEEE Trans. Ant. and Propagat.*, Vol. AP-29, Jan. 1981, pp. 151–157.

[49] Katehi, P. B., and N. G. Alexopoulos, "On the Modeling of Electromagnetically Coupled Microstrip Antennas—The Printed Strip Dipole," *IEEE Trans.*, Vol. AP-32, No. 11, Nov. 1984, pp. 1179–86.

[50] Rana, I. E., and N. G. Alexopoulos, "Current Distribution and Impedance of Printed Dipoles," *IEEE Trans.*, Vol. AP-29, Jan. 1981, pp. 99–105.

[51] Pozar, D. M., "Analysis of Finite Phased Arrays of Printed Dipoles," *IEEE Trans. Ant. and Propagat.*, Vol. AP-33, Oct. 1985, pp. 1045–1053.

[52] Katehi, P. B., N. G. Alexopoulos, and I. Y. Hsia, "A Bandwidth Enhancement Method for Microstrip Antennas," *IEEE Trans.*, Vol. AP-35, No. 1, Jan. 1987, pp. 5–12.

[53] Kraus, J. D., *Antennas*, McGraw-Hill Book Company, New York, second edition, 1988, pp. 692–694.

[54] Kerr, J. L., "Short Axial Length Broad Band Horns," *IEEE Trans.*, Vol. AP-21, Sept. 1973, pp. 710–714.

[55] Lewis, L. R., M. Fassett, and J. Hunt, "A Broadband Stripline Array Element," *IEEE AP-S Symp. Dig.*, Atlanta, GA, June 1974, pp. 335–337.

[56] Monser, G. J., G. S. Hardle, and J. R. Ergadt, "Closely Spaced Orthogonal Dipole Array," U.S. Patent #3,836,976, 17 Sept. 1974.

[57] Monser, G. J., "Performance Characteristics of Notch Array Elements Over a 6/1 Frequency Band," *1987 Antenna Applications Symp.*, University of Illinois, 1987.

[58] Gibson, P. J., "The Vivaldi Aerial," *9th European Microwave Conf.*, Brighton, U.K., 1979, pp. 101–105.

[59] Prasad, S. N., and S. Mahapatra, "A New MIC Slot-Line Aerial," *IEEE Trans.*, Vol. AP-31, No. 3, May 1983. See also "A Novel MIC Slot-Line Antenna," *9th European Microwave Conf.*, 1979, pp. 120–124.

[60] Yngvesson, K. S., D. H. Schaubert, T. L. Korzeniowski, E. L. Kollberg, T. Thungren, and J. F. Johansson, "Endfire Tapered Slot Antennas on Dielectric Substrates," *IEEE Trans. Ant. and Propagat.*, Vol. AP-33, No. 12, pp. 1392–1400.

[61] Franz, K. M., and P. E. Mayes, "Broadband Feeds for Vivaldi Antennas," *Antenna Applications Symp.*, Univ. of Illinois, 1987.

[62] Povinelli, M. J., "Wideband Dual Polarized Apertures Utilizing Closely Spaced Printed Circuit Flared Slot Antenna Elements for Active Transmit and Receive Array Demonstration," *1989 Antenna Applications Symp.*, Sept. 1989.

[63] Klopfenstein, R. W., "A Transmission Line Taper of Improved Design," *Proc. IRE*, Vol. 44, No. 1, Jan. 1956, pp. 31–35.

[64] Povinelli, M. J., "A Planar Broad-Band Microstrip Slot Antenna," *IEEE Trans.*, Vol. AP-35, No. 8, Aug. 1987, pp. 968–972.

[65] Balanis, op. cit., pp. 496–501.

[66] Long, S. A., "Experimental Study of the Input Impedance of Cavity Backed Slot Antennas," *IEEE Trans.*, Vol. AP-35, No. 8, Jan. 1975, pp. 1–7.

[67] Mailloux, R. J., "On the Use of Metallized Cavities in Printed Slot Arrays With Dielectric Substrates," *IEEE Trans.*, Vol. AP-35, No. 5, May 1987, pp. 477–487.

[68] Yoshimura, Y., "A Microstrip Slot Antenna," *IEEE Trans. Microwave Theory Tech.*, Vol. MTT-20, Nov. 1972, pp. 760–762.

[69] Collier, M., "Microstrip Antenna Array for 12-GHz TV," *Microwave Journal*, Vol. 20, Sept. 1977, pp. 67–71.

[70] Dorne, A., and D. Latarus, Ch. 7 in *Very High Frequency Techniques*, Radio Research Labs. Staff, New York: McGraw-Hill, 1947.

[71] Ragan, G. L., Section 6-12 in *Microwave Transmission Circuits*, New York: McGraw-Hill, 1948.

[72] Newman, E. H., and G. A. Thiele, "Some Important Properties in the Designs of T-Bar Fed Slot," *Antennas, IEEE Trans.*, Vol. AP-23, No. 1, Jan. 1975, pp. 97–100.

[73] Arkind, K. D., and R. L. Powers, "Printed Circuit Antenna for Wide Bandwidth Requirements," *IEEE AP-S Symp. Dig.*, Los Angeles, CA, 1981, pp. 359–362.

[74] Bahl, I. J., and P. Bhartia, *Microstrip Antennas*, Dedham, MA: Artech House, 1980.

[75] Nakaoka, K., K. Itoh, and T. Matsumoto, "Microstrip Line Array Antenna and Its Application," *Int. Symp. on Antennas and Propagation*, Japan, 1978, pp. 61–64.

[76] Amitay, N., V. Galindo, and C. P. Wu, *Theory and Analysis of Phased Array Antennas*, New York: Wiley Interscience, 1972.

[77] Wheeler, H. A., "A Systematic Approach to the Design of a Radiator Element for a Phased Array Antenna," *Proc. IEEE*, Vol. 56, 1968, pp. 1940–1951.

[78] McGill, E. G., and H. A. Wheeler, "Wide Angle Impedance Matching of a Planar Array Antenna by a Dielectric Sheet," *IEEE Trans.*, Vol. AP-14, 1966, pp. 49–53.

[79] Lewis, L. R., L. J. Kaplan, and J. D. Hanfling, "Synthesis of a Waveguide Phased Array Element," *IEEE Trans.*, Vol. AP-22, 1974, pp. 536–540.

[80] Hopfer, S., "The Design of Ridged Waveguides," *IRE Trans.*, Vol. MTT-3, No. 5, Oct. 1955, pp. 20–29.

[81] Harvey, A. F., Ch. 1 in *Microwave Engineering*, London, New York: Academic Press, 1963, pp. 21–26.

[82] Itoh, T., "Waveguides and Resonators," Ch. 30 in *Reference Data for Engineers, Radio, Computer, and Communications*, 7th edition, E. C. Jordan, ed., Indianapolis, IN, H. W. Sams and Co., 1986.

[83] Chen, M. H., and G. N. Tsandoulas, "Bandwidth Properties of Quadruple Ridged Circular and Square Waveguide Radiators," *IEEE Trans. AP-S Int. Symp. Dig.*, 1973, pp. 391–394.

[84] Montgomery, J. P., "Ridged Waveguide Phased Array Elements," *IEEE Trans. Antenna Propagat.*, Vol. AP-24, No. 1, Jan. 1976, pp. 46–53.

[85] Chen, C. C., "Quadruple Ridge-Loaded Circular Waveguide Phased Arrays," *IEEE Trans. Microwave Theory Tech.*, Vol. MTT-22, May 1974, pp. 481–483.

[86] Wang, S. S., and A. Hessel, "Aperture Performance of a Double-Ridge Rectangular Waveguide in a Phased Array," *IEEE Trans. Ant. and Propagat.*, Vol. AP-26, March 1978, pp. 204–214.

[87] Munson, R. E., "Conformal Microstrip Antennas and Microstrip Phased Arrays," *IEEE Trans. Ant. and Propagat.*, Vol. AP-22, Jan. 1974, pp. 74–78.

[88] Howell, J. Q., "Microstrip Antennas," *IEEE Trans.*, Vol. AP-23, 1975, pp. 90–93.

[89] Carver, K. R., and J. W. Mink, "Microstrip Antenna Technology," *IEEE Trans.*, Vol. AP-29, Jan. 1981, pp. 2–24.

[90] Bahl, I. J., and P. Bhartia, op. cit., pp. 48–55.

[91] Mailloux, R. J., J. F. McIlvenna, and N. P. Kernweis, "Microstrip Array Technology," *IEEE Trans.*, Vol. AP-29, No. 1, Jan. 1981, pp. 25–37.

[92] James, J. R., P. S. Hall, and C. Wood, *Microstrip Antenna Theory and Design*, London: Peter Peregrinus, 1981.

[93] Lo, Y. T., "Microstrip Antennas," Ch. 10 in *Antenna Handbook, Theory, Applications and Design*, Y. T. Lo and S. W. Lee, eds., New York: Van Nostrand Reinhold, 1988.

[94] McGrath, D. T., F. A. Mullinix, and K. D. Huck, "Fortran Subroutines for Design of Printed Circuit Antennas," RADC-TM-86-08, 1986.

[95] Munson, R. E., "Microstrip Antennas," Ch. 7 in *Antenna Engineering Handbook*, R. C. Johnson and H. Jasik, eds., New York: McGraw-Hill, 1984.

[96] Harrington, R. F., *Time Harmonic Electromagnetic Fields*, New York: McGraw-Hill, 1961, p. 183.

[97] Schaubert, D. H., F. G. Farrer, A. Sindoris, and S. T. Hayes, "Microstrip Antennas With Frequency Agility and Polarization Diversity," *IEEE Trans.*, Vol. AP-29, Jan. 1981, pp. 118–123.

[98] Hammerstadt, E. O., "Equations for Microstrip Circuit Design," *Proc. 5th European Microwave Conf.*, Hamburg, Sept. 1975, pp. 268–272.

[99] Derneryd, A. G., and A. G. Lind, "Extended Analysis of Rectangular Microstrip Antennas," *IEEE Trans.*, Vol. AP-27, 1979.

[100] Vandensand, J., H. Pues, and A. Van de Capelle, "Calculation of the Bandwidth of Microstrip Resonator Antennas," *Proc. 9th European Microwave Conf.*, 1979, pp. 116–119.

[101] Lo, Y. T., D. Solomon, and W. F. Richards, "Theory and Experiments on Microstrip Antennas," *IEEE Trans. Ant. and Propagat.*, Vol. AP-27, March 1979, pp. 137–145.

[102] Aksun, M. I., S. L. Chuang, and Y. T. Lo, "On Slot Coupled Antennas and Their Applications to Circularly Polarized Operation, Theory and Experiment," *IEEE Trans.*, Vol. AP-38, No. 8, Aug. 1990, pp. 1224–1230.

[103] Pozar, D. M., "Input Impedance and Mutual Coupling of Rectangular Microstrip Antennas," *IEEE Trans. Ant. and Propagat.*, Vol. AP-30, Nov. 1982, pp. 1191–1196.

[104] Katehi, P. B., and N. G. Alexopoulis, "On the Modeling of Electromagnetic Coupled Microstrip Antennas—The Printed Strip Dipole," *IEEE Trans.*, Vol. AP-32, No. 11, Nov. 1984, pp. 1179–1185.

[105] Pozar, D. M., "Microstrip Antenna Aperture Coupled to a Microstrip Line," *Elect. Letters*, Vol. 21, No. 2, Jan. 1985, pp. 49–50.

[106] Herd, J. S., "Full Wave Analysis of Proximity Coupled Rectangular Microstrip Antenna Arrays," *Electromagnetics*, Vol. 11, Jan. 1991, pp. 21–46.

[107] Chew, W. C., and Q. Liu, "Resonance Frequency of a Rectangular Microstrip Patch," *Proc. 1987 Antenna Applications Symp.*, U. of Illinois.

[108] Sanford, G., and L. Klein, "Increasing the Beamwidth of a Microstrip Radiating Element," *Int. Symp. Dig. of Ant. and Propagat. Soc.*, Univ. of Washington, June 1979, pp. 126–129.

[109] Hansen, R. C., "Cross Polarization of Microstrip Patch Antennas," *IEEE Trans.*, Vol. AP-35, No. 6, June 1987, pp. 731–732.

[110] James, J. R., and P. S. Hall, eds., *Handbook of Microstrip Antennas*, Peter Peregrinus, on behalf of the Institute of Electrical Engineers, London, 1989.

[111] Long, S. S., and M. D. Walton, "A Dual-Frequency Stacked Circular Disk Antenna," *IEEE Trans.*, Vol. AP-27, No. 2, March 1979, pp. 270–273.

[112] Richards, W. F., S. E. Davidson, and S. A. Long, "Dual Band Reactively Loaded Microstrip Antennas," *IEEE Trans.*, Vol. AP-33, May 1985, pp. 556–561.

[113] Paschen, D. A., "Practical Examples of Integral Broadband Matching of Microstrip Elements," *Proc. 1986 Antenna Applications Symp.*, University of Illinois, Sept. 1986.

[114] Sabban, A., "A New Broadband Stacked Two Layer Microstrip Antenna," *IEEE AP-S Int. Symp. Dig.*, 1983, pp. 63–66.

[115] Yasuo, S., N. Miyano, and T. Nd Chiba, "Expanding the Bandwidth of a Microstrip Antenna," *IEEE AP-S Int. Symp. Dig.*, 1983, pp. 366–369.

[116] Hall, P. S., C. Wood, and C. Garrett, "Wide Bandwidth Microstrip Antennas for Circuit Integration," *Elect. Letters*, Vol. 15, 1979, pp. 458–460.

[117] Collings, R., U.S. Patent 3680136, July 1972.

[118] Byron, E. V., "A New Flush Mounted Antenna Element for Phased Array Application," *Proc. Phased Array Antennas Symp.*, 1970, pp. 187–192, reprinted in *Phased Array Antennas*, A. A. Oliner and G. H. Knittel, eds., Dedham, MA: Artech House, 1972.

[119] Bauer, R. L., and J. J. Schuss, "Axial Ratio of Balanced and Unbalanced Fed Circularly Polarized Patch Radiator," *IEEE Ant. and Propagat. Society Symp.*, 1987.

[120] Hanfling, J. D., and J. J. Schuss, "Experimented Results Illustrating Performance Limitations and Design Tradeoffs in Probe Fed Microstrip-Patch Element Phased Arrays," *IEEE Ant. and Propagat. Society Symp.*, 1986.

[121] Schuss, J. J., and J. D Hanfling, "Observation of Scan Blindness Due to Surface Wave Resonance in an Array of Printed Circuit Patch Radiators," *IEEE AP-S Int. Symp. Dig.*, 1987.

[122] Schuss, J. J., and J. D Hanfling, "Nonreciprocity and Scan Blindness in Phased Arrays Using Balanced-Fed Radiators," *IEEE Trans.*, Vol. AP-35, No. 2, Feb. 1987, pp. 134–138.

[123] Greiser, J., "Coplanar Stripline Antenna," *Microwave Journal*, Vol. 19, 1976, pp. 47–49.

[124] Conte, R., J. Toth, T. Dowling, and J. Weis, "The Wire Grid Microstrip Antenna," *IEEE Trans.*, Vol. AP-29, No. 1, Jan. 1981, pp. 158–167.

[125] Kraus, J. D., *Antennas*, 2nd edition, McGraw-Hill, p. 492.

[126] Elliott, R. S., "The Design of Waveguide-Fed Slot Arrays," Ch. 12 in *Antenna Handbook, Theory Applications and Design*, Van Nostrand Reinhold, 1968.

[127] Compton, R. T., Jr., and R. E. Collin, "Slot Antennas," Ch. 14 in *Antenna Theory, Part 1*, R. J. Collin and F. J. Zucker, eds., New York: McGraw-Hill, 1969.

[128] Yee, H. Y., "Slot-Antenna Arrays," Ch. 9 in *Antenna Engineering Handbook*, R. C. Johnson and H. Jasik, eds., New York: McGraw-Hill, 1961, 1984.

[129] Rama Rao, B., "94 GHz Slotted Waveguide Array Fabricated by Photolithographic Techniques," *Elect. Letters*, Vol. 20, No. 4, 16 Feb. 1984, pp. 155, 156.

[130] Yee, H. Y., "Impedance of a Narrow Longitudinal Shunt Slot in a Slotted Waveguide Array," *IEEE Trans.*, Vol. AP-22, 1974, pp. 589–592.

[131] Oliner, A. A., "The Impedance Properties of Narrow Radiating Slots in the Broadface of Rectangular Waveguide," *IEEE Trans.*, Vol. AP-5, 1957, pp. 12–20.

[132] Gruenberg, H., "Second Order Beams of Slotted Waveguide Arrays," *Canadian J. of Physics*, Vol. 31, Jan. 1953, pp. 55–69.

[133] Kurtz, L. A., and J. S. Yee, "Second Order Beams of Two-Dimensional Slot Arrays," *IEEE Trans.*, Vol. AP-5, Oct. 1957, pp. 356–362.

[134] Derneryd, A., "Butterfly Lobes in Slotted Waveguide Antennas," *IEEE Ant. Society Int. Symp. Dig.*, 15 June 1987.

[135] Green, J., H. Shnitkin, and P. J. Bertalan, "Asymmetric Ridge Waveguide Radiating Element for a Scanned Planar Array," *IEEE Trans.*, Vol. AP-38, No. 8, Aug. 1990, pp. 1161–1165.

[136] Compton, R. T., Jr., and R. E. Collin, "Slot Antennas," op. cit., pp. 587–590.

[137] James, J. R., and G. J. Wilson, "Microstrip Antennas and Arrays: Part 1—Fundamental Action and Limitations, Part 2—New Design Techniques," *IEEE J. MOA*, Sept. 1977, pp. 165–181.

[138] Offutt, W. B., and L. K. DeSize, "Methods of Polarization Synthesis," Ch. 23 in *Antenna Engineering Handbook*, R. C. Johnson and H. Jasik, eds., New York: McGraw-Hill, 1984.

[139] Kerr, J. L., "Microstrip Polarization Techniques," *Proc. 1978 Ant. Applications Symp.*, University of Illinois, Sept. 1978.

[140] Weinschel, H. D., "A Circularly Polarized Microstrip Antenna," *Dig. Int. Symp. Ant. and Propagat.*, June 1975, pp. 177–180.

[141] Schaubert, D. H., F. G. Farrer, A. Sindoris, and S. T. Hayes, op. cit.

[142] Cox, R. M., and W. E. Rupp, "Circularly Polarized Phased Array Antenna Element," *IEEE Trans.*, Vol. AP-18, Nov. 1970, pp. 804–807.

[143] Ruze, J., "Wide Angle Metal Plate Optics," *IRE Proc.*, Vol. 38, No. 1, Jan. 1950, pp. 53–59.

[144] Ramsay, J. F., "Circular Polarization for CW Radar," Marconi Wireless Telegraph Co., 1952, *Proc. Conf. on Centimeter Areals for Marine Navigational Radar*, 15–16 June, 1950, London.

[145] Lerner, D. S., "A Wave Polarizer Converter for Circular Polarization," *IEEE Trans.*, Vol. AP-13, No. 1, Jan. 1965, pp. 3–7.

[146] Young, L., L. A. Robinson, and C. A. Hacking, "Meander Line Polarizer," *IEEE Trans.*, Vol. AP-21, May 173, pp. 376–378.

[147] Terret, C., J. R. Levrel, and K. Mahdjoubi, "Susceptance Computation of a Meander-Line Polarizer Layer," *IEEE Trans.*, Vol. AP-32, No. 9, Sept. 1984, pp. 1007–1011.

[148] Johnson, R. C., and H. Jasik, Ch. 3 in *Antenna Engineering Handbook*, New York: McGraw-Hill, 1984.

[149] Teshirogi, T., N. Tanaka, and W. Chujo, "Wideband Circularly Polarized Array Antenna with Sequential Rotations and Phase Shift of Elements," *Proc. Int. Symp. on Ant. and Propagat.*, 1985, pp. 117–120.

[150] Hall, P. S, "Feed Radiation Effects in Sequentially Rotated Microstrip Patch Arrays," *Elect. Letters*, Vol. 23, No. 17, 13 Aug. 1987, pp. 877–878.

[151] Bahl, I. J., "Transmission Lines," Ch. 1 in *Handbook of Microwave and Optical Components, Vol. 1, Microwave Passive and Antenna Components*, K. Chang, ed., Wiley Interscience, 1989.

[152] Gupta, K. L., R. Garg, and L. J. Bahl, *Microstrip Lines and Slot-Lines*, Dedham, MA: Artech House, 1979.

[153] Itoh, T., "Transmission Lines," Ch. 29 in *Reference Data for Engineers: Radio, Computer, and Communications*, 7th edition, E. C. Jordan, ed., Indianapolis, IN: H. W. Sams and Co., 1986.

[154] Bowman, D. F., "Impedance Matching and Broadbanding," Ch. 43 in *Antenna Engineering Handbook*, R. C. Johnson and H. Jasik, eds., New York: McGraw-Hill, 1984.

[155] Tai, C. T., op. cit., Ch. 5.

[156] Itoh, T., "Waveguides and Resonators," op. cit.

[157] Sylvester, P., "High Order Finite-Element Waveguide Analysis," *IEEE Trans.*, Vol. MTT-17, Aug. 1969, p. 651.

[158] Chen, M. H., and G. N. Tsandoulas, "Modal Characteristics of Quadruple Ridged Circular and Square Waveguides," *IEEE Trans., Microwave Theory Tech.*, Vol. MTT-22, Aug. 1974, pp. 801–804.

[159] Montgomery, J. P., "On the Complete Eigenvalue Solution of Ridged Waveguide," *IEEE Trans. Microwave Theory Tech.*, Vol. MIT-19, June 1971, pp. 547–555.

[160] Howe, H., *Stripline Circuit Design*, Dedham, MA: Artech House, 1974.

[161] Wen, C. P., "Coplanar Waveguide: A Surface Strip Transmission Line Suitable for Nonreciprocal Gyromagnetic Device," *IEEE Trans.*, Vol. MTT-17, No. 12, Dec. 1969, pp. 1087–1090.

[162] Rowe, D. A., and B. Y. Lao, "Numerical Analysis of Shielded Coplanar Waveguide," *IEEE Trans.*, Vol. MTT-31, Nov. 1983, pp. 911–915.

[163] Tanner, D. R., and Y. T. Lo, "Analysis of a Rectangular Microstrip Antenna Excited by a Slot Line," presented at National Radio Sciences Meeting (URSI), June 1988.

[164] Mariani, E. A., C. P. Heinzman, J. P. Agrios, and S. B. Cohn, "Slotline Characteristics," *IEEE Trans.*, Vol. MTT-17, 1969, pp. 1091–1096.

[165] Garg, R., and K. C. Gupta, "Expressions for Wavelength and Impedance of Slot Line," *IEEE Trans.*, Vol. MTT-24, Aug. 1976, p. 532.

[166] Janaswamy, R., and D. H. Schaubert, "Dispersion Characteristics for Wide Slotlines on Low Permittivity Substrates," *IEEE Trans. Microwave Theory Tech.*, Vol. MTT-33, No. 8, 1985, pp. 723–726.

[167] Janaswamy, R., and D. H. Schaubert, "Characteristic Impedance of a Wide Slotline on Low Permittivity Substrates," *IEEE Trans. Microwave Theory Tech.*, Vol. MTT-34, No. 8, 1985, pp. 900–902.

[168] Meier, P. J., "Integrated Finline Millimeter Components," *IEEE Trans.*, Vol. MTT-22, Dec. 1974, pp. 1209–1210.

[169] Knorr, J. B., and P. M. Shayda, "Millimeter Wave Fin-Line Characteristics," *IEEE Trans.*, Vol. MTT-28, No. 7, July 1980, pp. 737–743.

[170] Kinzel, J., B. J. Edward, and D. E. Rees, "V-Band Space-Based Phased Arrays," *Microwave J.*, Vol. 30, No. 1, Jan. 1987, pp. 89–102.

[171] Mailloux, R. J., "Phased Array Architecture," *IEEE Proc.*, Vol. 80, No. 1, Jan. 1992, pp. 163–172.

[172] Berry, D. J., R. G. Malech, and W. A. Kennedy, "The Reflectarray Antenna," *IEEE Trans.*, Vol. AP-11, 1963, pp. 646–651.

Chapter 6
Summary of Element Pattern and Mutual Impedance Effects

6.1 MUTUAL IMPEDANCE EFFECTS

Although a detailed consideration of array mutual coupling is beyond the scope of this book, this chapter introduces the subject by presenting and describing simple examples of the analysis, coupled with a number of figures that show the current state of research in this important area.

Throughout this text, the equations for pattern analysis and synthesis are given for arrays of radiators with known currents or aperture fields. Implicit in that formulation are three assumptions: that the current or fields are proportional to applied excitations, that the distribution of current or aperture field is the same for each radiator, and that the distribution does not change as the array is scanned. A primary challenge to modern array theory is that, in general, none of these statements is true. In a finite array, all of the currents and fields differ from element to element in magnitude, phase, and distribution, and these differences vary as a function of frequency and array scan angle. This complex dependence on geometry, frequency, and scan angle results from the mutual interaction among all of the elements of the array. It can be evaluated by writing the radiated field from all elements as generalized integrals that include current and charge distributions over the surface of the radiating antennas and nearby diffracting bodies, and using these to require satisfaction of boundary conditions at each radiator and each diffracting body. This procedure usually results in a multiplicity of simultaneous integral or integrodifferential equations, and has not been solved exactly except for infinite waveguide arrays over an infinite ground screen and several other special (idealized) cases, some of which are included in this chapter.

Although the analysis of isolated elements remains useful for predicting the gross parameters of an element (like polarization, general pattern shape, and resonant frequency), the elements generally behave very differently in an array than when isolated. The array behavior is dominated by the mutual coupling of the

324

various elements. Figure 6.1 shows the element pattern of an isolated dipole and one in an array with several different spacings. The data are for a finite (7-by-9-element) array over a ground plane. The curve for an isolated element is dashed. The pattern is most like that of the isolated element for close element spacing, a fact which results from the presence of grating lobes in the scan plane for larger separations. For element spacing d_x ($0.5 \leq d_x/\lambda \leq 1.0$), there are scan angles θ_0 such that

$$|\sin \theta_0| > (\lambda/d_x - 1) \tag{6.1}$$

for which the grating lobe radiates into real space ($\sin \theta_{-1} > -1$). As $|\theta_0|$ is increased, this lobe can take a growing share of the power. For larger d_x, the grating lobe onset occurs progressively closer to broadside, and the resulting element pattern then falls off rapidly to accommodate power distributed to the grating lobe.

Because of the complexity of a mutual coupling phenomenon, it is not possible to simply list tabular data or show generic curves that apply to all elements. There are, however, similarities between arrays of different elements that reveal the array grating dimensions and orientation to be far more important than the elements

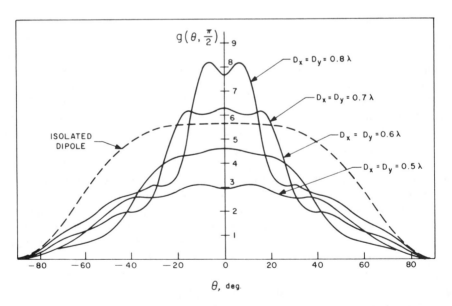

Figure 6.1 *H*-plane element gain functions for a center element of a 7-by-9-element dipole array ($\lambda/2$ dipoles, $\lambda/4$ above ground. Element spacings denoted D_x and D_y (after Diamond [1,2]). Note: dashed curve is for isolated dipole over ground.

themselves. Such grating-related phenomena have been the subject of several studies [3,4] dealing with the radiation of continuous current sheets. It sometimes happens that subtle changes to an array element produce major changes to the radiation characteristics and can even introduce the catastrophic pattern "blindness" that will be described later. For this reason, the examples in this chapter are intended to be illustrative of the sort that one might expect; for any array design to be complete, it is necessary to perform the detailed evaluation of its mutual coupling (or element pattern) performance, or to measure these parameters in the actual array.

Although they do not include mutual coupling, the synthesis and analysis presented elsewhere in this text are still valid because the current or aperture distributions remain very similar for all elements of the array, even though mutual interaction may alter the relative amplitudes and phases between various elements. This is true primarily because the elements are small and usually resonant. Thus, in an array of dipoles, the first-order result of mutual coupling is to alter the impedance of each of the array elements. The shape of the current distribution on each element is nearly the same as that on any other element of the array. In this case, the standard synthesis procedures specify the required currents. The mutual coupling analysis is used to solve for the input voltages that produce these currents.

6.2 FINITE ARRAYS, INFINITE ARRAYS, AND THE POISSON SUMMATION FORMULA

6.2.1 Formulation and Results for Finite Arrays

Figure 6.2 shows a finite one- or two-dimensional array of wire elements. In the simplest approximation of mutual coupling, one assumes that all dipoles have exactly the same current distribution $f(z)$ so that the current on the nth element is

$$i_n(z) = I_n f(z) \tag{6.2}$$

One can express the complete mutual coupling relationship as an impedance matrix relating the wire currents to the applied voltages.

$$V_m = \sum Z_{mn} I_n \tag{6.3}$$

where the V_m are the applied voltages and the I_n are the complex amplitudes of the current distribution, as noted above. In this case, the behavior of the coupled circuit is clear. If the Z_{mn} matrix were diagonal, there would be no interdependence (coupling) between the various array elements. In fact, however, each current I_n

(a)

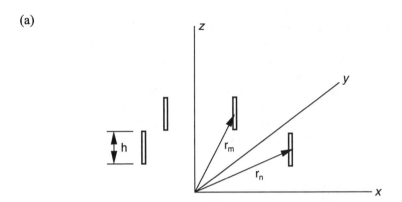

(b)

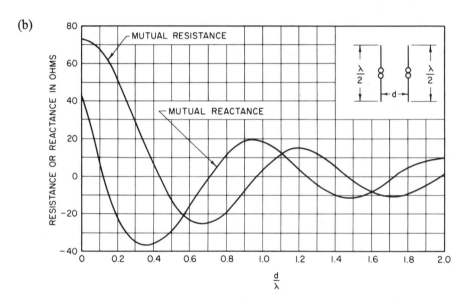

Figure 6.2 Mutual coupling of dipole antennas: (a) geometry of array of dipoles with centers at $r_n = \hat{x}x_n + \hat{y}y_n$; (b) mutual impedance Z_{12} between two dipoles (after Tai [5])

is excited not only by its applied voltage V_n but also by coupling from all the other currents I_m, and the network is a generalized N-port impedance network. The summation in this case is over each of the elements of the array.

Coefficients Z_{mn} in the mutual impedance matrix have the form of integrals over the free-space scalar Green's function kernel

$$G(r_n, r_m) = (1/4\pi)\{\exp[-jk\,|\mathbf{r_n} - \mathbf{r'_m}|]/|\mathbf{r_n} - \mathbf{r'_m}|\} \tag{6.4}$$

where the distance

$$|\mathbf{r_n} - \mathbf{r'_m}| = [(x_n - x'_m)^2 + (y_n - y'_m)^2 + (z_n - z'_m)^2]^{1/2}$$

as indicated in Figure 6.2, with the primed coordinates indicating the domain of integration and the unprimed indicating the observation point. When the elements are close together, the integrals are quite complex and must be evaluated numerically. However, when the elements are far apart, the integrals can be approximated using standard methods. Figure 6.2 also shows the mutual impedance parameters Z_{mn} for elements in a two-element array, as computed by Tai [5].

The example below is illustrative of the procedure used in formulating the array interaction problem. In the general procedure, one obtains the fields radiated by the unknown current (or aperture field) on each antenna by expanding the unknown distribution in a series of basis functions. Next, one requires that the appropriate boundary conditions are satisfied on all surfaces. The basic expressions for the radiated fields are usually obtained in terms of integrals over expressions that involve some Green's function operator acting on the unknown current or field. In general, the Green's function is a 9-term dyad.

Most of the published element and array studies are based on the use of the free-space Green's functions, although the emergence of microstrip patch and dipole arrays has required solutions based on the method of moments using a spectral (transform) Green's function, and these analytical studies have yielded useful design data for such arrays. Some of the examples cited later utilize the spectral formulation.

A simple illustration of mutual coupling is the analysis of the interaction of two dipole antennas with their axes both parallel to the z-axis and with centers in the plane $z = 0$. The radiated field due to either dipole centered at location (x'_1, y'_1) or (x'_2, y'_2) is given by the potential function

$$A_z^{(1,2)}(x, y) = \frac{\mu}{4\pi} \int_{h/2}^{h/2} i^{(1,2)}(z') \frac{e^{-jkR^{(1,2)}}}{R^{(1,2)}} \, dz' \qquad (6.5)$$

where

$$R^{(1,2)} = [(x - x'_{(1,2)})^2 + (y - y'_{(1,2)})^2 + (z - z'_{(1,2)})^2]^{1/2}$$

This expression assumes that the current $i^{(1,2)}$ is centered at the axis of dipole 1 or 2 and not distributed across each dipole cross section. This *filamentary current* approximation is commonly used and gives an accurate field representation, even at the surface of the dipole. The radiated electric field is given by

$$E^s = -j\omega A_z - \frac{j}{\omega\mu\epsilon} \nabla(\nabla \cdot A_z) \tag{6.6}$$

Both dipoles have impressed sources (V_1 and V_2), which we will assume are *delta function* sources, which means that the potential gradient or electric fields of the sources is singular, and the potential is a step function of position. More generalized sources are described in [6,7].

Since the boundary condition at the surface of the dipoles is that the tangential electric field be zero, the scattered tangential electric field at each dipole is therefore required to be the negative of the incident field, or

$$E_z^s = -E_z^{inc} \tag{6.7}$$

at $-h/2 \le z \le h/2$. At each element, from (6.6) above,

$$\left(\frac{d^2}{dz^2} + k^2\right)A_z^2 = -j\omega\epsilon E_z^{inc} \tag{6.8}$$

The incident field E_z^{inc} consists of a delta function source $V_{1,2}\delta(z')$ added to the radiated field incident from the other dipole, and one can write both equations in the following form:

$$\left(\frac{d^2}{dz^2} + k^2\right)\{A_z^1(x_n, y_n, z) + A_z^2(x_n, y_n, z)\} = -j\omega\epsilon\delta(z)V_n \tag{6.9}$$

for x and y on each dipole surface ($n = 1$ or 2). This pair of integrodifferential equations must be solved simultaneously for the currents $I_1(z)$ and $I_2(z)$. The equations above are in the form known as Pocklington's equation [8], but a number of authors have chosen to solve the integrated form due to Hallén [9,10].

Simple and useful solutions have been obtained using a single *basis function* for the currents $[i_1(z) = I_1 f(z), i_2(z) = I_2 f(z)]$, which might be sinusoidal, or other basis functions as in the previous example of an N-port coupled network. With this substitution, the integrals can be performed (numerically) and the equations satisfied at one point on each antenna (a procedure called *point matching*). The resulting simultaneous algebraic equations are solved for I_1 and I_2.

An alternative to point matching is to require that the equations be satisfied to an average sense by multiplying the equations by a weighting function and integrating this weighted average along each antenna element. If the weighting function has the same form as the basis function, this procedure is known as Galerkin's method [11] and possesses stationary characteristics that improve its accuracy.

The procedure outlined above is general and is extended to the case of any array by including all of the elements of a large array in the simultaneous equations and inverting the set to obtain the solution for all currents.

In the general case of an array of dipoles (oriented with axes in the z-direction) and the locations (x_n, y_n, z), the same equation is written

$$\left(\frac{d^2}{dz^2} + k^2\right)\left\{\sum_m \int i_m(z')\, G(\mathbf{r_n}, \mathbf{r'_m})\, dz'\right\} = -j\omega\epsilon\delta(z)\, V_n \qquad 1 \leq n \leq N \qquad (6.10)$$

where

$$G(r_n, r'_m) = \frac{e^{-jk|\mathbf{r_n} - \mathbf{r'_m}|}}{4\pi\, |\mathbf{r_n} - \mathbf{r'_m}|}$$

This integrodifferential equation can be solved approximately by the point matching or Galerkin's method and, if a single basis function is used to represent the current in each element, results in N equations in the N unknown values of coefficients , (using $i_n(z) = I_n f(z)$).

More accurate solutions than those obtained with a single basis function can e obtained using higher order expansions of the current and some variation of he method of moments to obtain a matrix solution. Thorough treatments of the method of moments as applied to antenna and scattering problems are found in he texts by Harrington [11], Balanis [6], and Stutzman and Thiele [7]. If the current is expanded in p-basis functions for each antenna, the resulting matrix formulation will consist of $p \times N$ simultaneous equations.

The texts [6,7,11] illustrate expanding the current in basis functions. On any particular nth antenna, the current $i_n(z)$ is written as the sum

$$i_n(z) = \sum_{p=1}^{P} I_{p,n} f_p(z) \qquad (6.11)$$

The proper choice of basis functions $f_p(z)$ depends on the kind and size of the antenna element, but both entire domain basis functions (which span the entire element) and subdomain basis functions (piecewise continuous over the element) have been used successfully. The writings of King [10] are notable examples of using only two or three selected entire domain current expansions and point-matching techniques for wire elements. Similarly, entire domain Fourier series expansions have been found very practical for a variety of waveguide problems, where the waveguide modal functions serve as the basis functions [12,13]. It is, however, the use of various piecewise continuous subdomain basis functions that have led to generalized flexible software for the solution of a wide variety of antenna and array problems. The disadvantage of subdomain basis functions is that about 7 to 10 basis functions are required per wavelength, and that can lead to large numbers of equations. An array of 20 elements might require the solution of 150 to 200 simultaneous equations, and an array of 200 elements requires that over 1,000 simultaneous equations be solved. For this reason, entire domain basis functions with very few basis functions per element are often used for larger arrays.

Pocklington's equation has been written in other forms more suitable for computation. Examples are given in the literature [14,15].

Figure 6.3 shows the results of Wu [16] who analyzed finite parallel plane arrays. The figure shows element patterns of a 15-element array and clearly indicates the asymmetry expected of edge elements. Infinite-array data are included for comparison, and are noted here in one example that the element pattern is always zero at the horizon for an infinite periodic array, but finite for the finite array.

For large arrays, most of the central elements are very far from the array edges, see the same embedded impedance, and behave very similarly to each other. In such arrays, edge effects are less important than they are for small arrays, and it is often convenient to analyze the associated infinite-array structure, in which

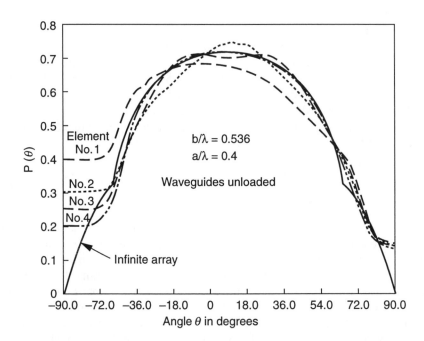

Figure 6.3 Element patterns in a 12-element parallel plane array. Infinite array data included for comparison (after C. P. Wu [16]). Element spacing, $b/\lambda = .5636$, aperture $a/\lambda = 0.4$.

all elements have exactly the same impedance and all voltages and currents differ from one another by only a complex constant. Certain details of infinite-array analysis are outlined in the next section.

6.2.2. Formulation and Results for Infinite Arrays

The result of studies of the infinite array is to obtain accurate predictions of element impedance as a function of scan and element patterns for an element embedded in the infinite array. These results then serve as a good approximation for elements away from the edge of the large finite array.

Infinite-array theory can be formulated from a mode-matching or integral equation approach. The integral equation formulation is based on a Green's function that can be derived from an infinite set of free-space Green's functions by a simple transformation, the Poisson summation formula [17], as described below.

Equation (6.6) gives the general expression for the field of a finite linear array of any size. In this section, the Green's function G is the free-space Green's function, and the currents are a direct response to the applied source. However, in the case of an infinite array with periodic progressive sources, the currents are all related by a complex constant, and it is convenient to incorporate that progressive phase into the Green's function itself. For a one-dimensional infinite dipole array with elements centered at $x = md_x$ as in Figure 6.4(a), one expresses the relationship between the current at location (x, y, z) in the element at $m = 0$ and all others as

$$i(x + md_x, y, z) = i(x, y, z)e^{-jkmd_x \sin \theta_0} \tag{6.12}$$

for a beam at θ_0.

With this simplification, all the elements satisfy the same integral equation, and the solution of the mutual coupling problem is much simplified. Equation (6.10) for an infinite one-dimensional array, assuming the current centered along the y-axis of each element and written at the nth element, now has the form

$$\left(\frac{d^2}{dy^2} + k^2\right)\left\{\sum_{m=-\infty}^{\infty} \int i(0, y', z')e^{-jk(md_x u_0)} G(\mathbf{r_n}, \mathbf{r'_m}) dy'\right\} = -j\omega\epsilon\delta(y)e^{-jknd_x u_0} \tag{6.13}$$

for $-\infty \leq n \leq \infty$, where

$$G(\mathbf{r_n}, \mathbf{r'_m}) = \frac{e^{-jk|\mathbf{r_n} - \mathbf{r'_m}|}}{4\pi|\mathbf{r_n} - \mathbf{r'_m}|}$$

and

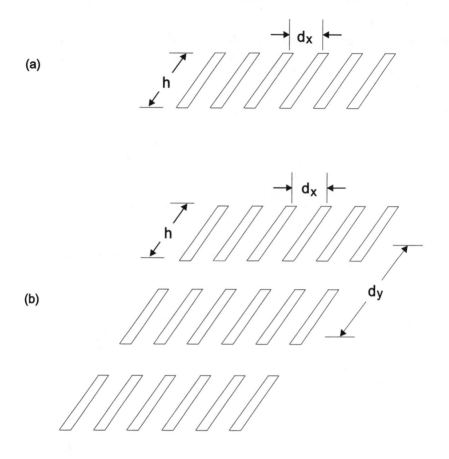

Figure 6.4 Infinite array geometries: (a) one-dimensional array; (b) two-dimensional array on a rectangular grid.

$$|\mathbf{r_n} - \mathbf{r'_m}| = [(x + nd_x - x' - md_x)^2 + (y - y')^2 + (z - z')^2]^{1/2}$$

Since all equations differ only by the complex constant, only the equation at $m = 0$ is needed.

For a two-dimensional array (Figure 6.4(b)) with elements centered at locations $(x, y) = (md_x, nd_y)$ scanning a beam to (u_0, v_0), the expression is

$$i(x + md_x, y + nd_y, z) = i(x, y, z)e^{-jk(md_x u_0 + nd_y v_0)} \tag{6.14}$$

and the integrodifferential equation takes the form

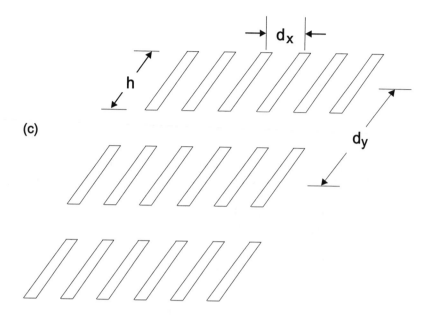

Figure 6.4 (cont.) Infinite array geometries: (c) two-dimensional array on triangular grid.

$$\left(\frac{d^2}{dy^2} + k^2\right)\left\{\sum_n \sum_m \int i(x,y,z)e^{-jk(md_xu_0 + nd_yv_0)} G(\mathbf{r}_{ij}, \mathbf{r}'_{mn})\,dy'\right\}$$

$$= -j\omega\epsilon\delta(y)e^{-jk(id_xu_0 + jd_yv_0)} \quad (6.15)$$

for $-\infty \leq i \leq \infty$ and $-\infty \leq j \leq \infty$, and where

$$\left|\mathbf{r}_{ij} - \mathbf{r}'_{mn}\right| = [(x_i - x'_m)^2 + (y_j - y'_n)^2 + (z - z')^2]^{1/2}$$

and $x'm = md_x + x'$, and $y'n = nd_y + y'$.

It is often convenient to change the form of the summations in the expressions for mutual coupling. This is done using the Poisson summation formula [17], which is stated

$$\sum_{-\infty}^{\infty} f(\alpha n) = \frac{1}{\alpha}\sum_{p=-\infty}^{\infty} F\left(\frac{2p\pi}{\alpha}\right) \quad (6.16)$$

where

$$f(t) = \frac{1}{2\pi} \int_{-\infty}^{\infty} e^{+j\omega t} F(\omega)\, d\omega$$

and the Fourier transform $F(\omega)$ is

$$F(\omega) = \int_{-\infty}^{\infty} e^{-j\omega t} f(t)\, dt \tag{6.17}$$

The advantage of this transformation is to transform a slowly convergent series in $f(t)$ into a new series in its transform $F(\omega)$, which can be much more rapidly convergent.

The Poisson summation formula is useful for linear and two-dimensional arrays. For a linear array, the two forms of the summation are written

$$\sum_{m=-\infty}^{\infty} \frac{e^{-jk[(x-x'_m)^2+(y-y')^2+(z-z')^2]^{1/2}}}{[(x-x'_m)^2+(y-y')^2+(z-z')^2]^{1/2}}\, e^{-jku_0 m d_x}$$

$$= -\frac{j\pi}{d_x} \sum_{p=-\infty}^{\infty} e^{-jku_p(x-x')} H_0^2\{K_p[(y-y')^2+(z-z')^2]^{1/2}\} \tag{6.18}$$

where $x'_m = x' + md_x$, and $K_p = [1 - u_p^2]^{1/2}$, and $u_p = u_0 + p\lambda/d_x$. H_0^2 is the Hankel function of the second kind as associated with outward traveling waves in y- and z-directions. The result of this manipulation is to transform the infinite summation of free-space $\exp(-jkR)/R$ type Green's functions into an infinite set or *discrete spectrum* of waves, each with the Hankel function dependence in the transverse direction.

With this form, the integral in (6.13) written at locations $(x_n = x + nd_x, y, z)$ is modified by replacing

$$\sum_{m=-\infty}^{\infty} 4\pi G(r_n, r_m) e^{-jkmu_0 d_x} \tag{6.19}$$

with the above, using $x = x_n$.

In the two-dimensional array with currents given as in (6.14) above, the summations that appear in the mutual impedance expressions or in terms of the integral equations are transformed as

$$\sum_{m=-\infty}^{\infty} \sum_{n=-\infty}^{\infty} \frac{e^{-jk[(x-x'_m)^2+(y-y'_n)^2+(z-z')^2]^{1/2}}}{[(x-x'_m)^2+(y-y'_n)^2+(z-z')^2]^{1/2}}\, e^{-jk(u_0 m d_x + v_0 n d_y)}$$

$$= -\frac{j2\pi}{d_x d_y} \sum_{p=-\infty}^{\infty} \sum_{q=-\infty}^{\infty} \frac{e^{-jk[u_p(x-x')+v_q(y-y')]-jK_{pq}|z-z'|}}{K_{pq}} \tag{6.20}$$

where $K_{pq} = k[1 - u_p^2 - v_q^2]^{1/2}$, and $x'_m = md_x + x'$, $y'_n = nd_y + y'$, and again $u_p = u_0 + p\lambda/d_x$, $v_q = v_0 + q\lambda/d_y$.

In the integral equation (6.14), this expression replaces

$$\sum_m \sum_n 4\pi G(r_{ij}, r'_{mn}) e^{-jk(md_x u_0 + nd_y v_0)} \tag{6.21}$$

for the equation written at $x_i = x + id_x$, $y_j = y + jd_y$.

The u_p and v_q are the grating lobe locations as indicated above, and so this series is often referred to as the *grating lobe series*. We will also refer to it as the *spatial harmonic series* or *Floquet series*, both terms from periodic structure theory. This expression illustrates some of the convenient properties of this most useful transformation, since the complicated square root function is replaced by much simpler exponential terms representing all the plane waves corresponding to points (u_p, v_q) on the grating lobe lattice, some propagating and some evanescent. The propagating grating lobes, those within the unit circle, are the only ones that represent true radiation and are used to compute far-field radiated power.

Note that the transformations above bring the formulation from one that is called an *element-by-element* formulation to a periodic structure *wave-type* formulation. Although one can always formulate the analysis using the element-by-element approach and then transform to the periodic form, as indicated in previous paragraphs, it is often more convenient to formulate the entire set of boundary conditions and even to derive the integral equations from the periodic structure point of view at the outset. Alternatively, one can employ a mode-matching approach that does not explicitly require the Green's function or the solution of an integral equation, but proceeds to solve the differential equations directly.

To employ either of these two alternative approaches that exploit the periodic nature of the fields, one requires at the outset that all fields repeat periodically across the array and that they have the form below for the two-dimensional case.

$$A(x, y, z) = B(x, y, z) \exp\{-jk(md_x u_0 + nd_y v_0)\} \tag{6.22}$$

where, once the exponential dependence has been removed, $B(x, y, z)$ is fully periodic in x and y. Therefore, $B(x, y, z)$ can be written

$$B(x, y, z) = \sum_{-\infty}^{\infty} \sum_{-\infty}^{\infty} b_{pq} g(z) e^{-j2\pi(px/d_x + qy/d_y)} \tag{6.23}$$

and

$$A(x, y, z) = \sum_{-\infty}^{\infty} \sum_{-\infty}^{\infty} b_{pq} g(z) e^{-jk(u_p x + v_q y)} \tag{6.24}$$

The z dependence $g(z)$ above must satisfy the Helmholtz equation in the region above the array. If that region is free space, the dependence is

$$g(z) = \exp[-jK_{pq}|z|] \tag{6.25}$$

and one can clearly note the same dependence as obtained from the transformed free-space element Green's function. To obtain the Green's function equivalent to (6.5), one could require the periodic form $A(x, y, z)$ to satisfy the inhomogeneous Helmholtz equation for a periodic infinitesimal source, and follow the usual procedure of integrating over the source discontinuity. The use of Green's theorem allows the potential function then to be cast in terms of this Green's function (similar to (6.5)). Expressing the boundary conditions then results in an integral equation equivalent to 10.

As an alternative to solving an integral equation, one can use the periodic structure perspective to solve the differential equations directly. One need only satisfy the boundary conditions in one periodic cell of the array to automatically satisfy all boundary conditions across the array. The mode-matching approach does not explicitly require the solution for a Green's function, but instead matches boundary conditions that include the source field. Examples of this approach can be found in the work of Diamond [18], Amitay et al. [19], and others, where the technique has found much utility for solving infinite waveguide array boundary problems.

With the fields written in the infinite-array form, one can write the most general form using the TE and transverse magnetic (TM) modes for a periodic structure for $\mathbf{E}$ and $\mathbf{H}$ (in free space) as

$$\mathbf{E} = \sum_{p=-\infty}^{\infty} \sum_{q=-\infty}^{\infty} \mathbf{E_0}(p, q)e^{-jk[xu_p+yv_q]-jK_{pq}|z|}$$

$$\mathbf{H} = \sum_{p=-\infty}^{\infty} \sum_{q=-\infty}^{\infty} \mathbf{H_0}(p, q)e^{-jk[xu_p+yv_q]-jK_{pq}|z|} \tag{6.26}$$

where $\mathbf{H_0}(p, q)$ and $\mathbf{E_0}(p, q)$ are constants evaluated for the particular geometry, and

$$\mathbf{E_0}(p, q) = \left[\frac{Z_0H_z}{u_p^2 + v_q^2}\right][-\hat{\mathbf{x}}v_q + \hat{\mathbf{y}}u_p]$$

$$+ \left[\frac{E_z\cos\theta_{pg}}{(u_p^2 + v_q^2)}\right][-\hat{\mathbf{x}}u_p - \hat{\mathbf{y}}v_q] + \hat{\mathbf{z}}E_z$$

$$\mathbf{H_0}(p, q) = \left[\frac{H_z\cos\theta_{pq}}{u_p^2 + v_q^2}\right][-\hat{\mathbf{x}}u_p - \hat{\mathbf{y}}v_q] + \hat{\mathbf{z}}H_z$$

$$+ \left[\frac{E_z}{Z_0(u_p^2 + v_q^2)}\right][\hat{\mathbf{x}}v_q - \hat{\mathbf{y}}u_p] \tag{6.27}$$

This field representation is used to match boundary conditions within one cell of the periodic structure, and hence for the entire array.

Although this procedure has obvious similarities to the Green's Function approach in the character of the fields, the procedure for solving the resulting equations can be quite different. In the above, the b_{pq} ((6.23) and (6.24)) are the terms used in expanding the aperture fields (or dipole currents). Each term corresponds to a term of the grating lobe series, and one solves for these term by term. To include, for example, 200 grating lobe terms requires solution of a matrix equation of 200 unknowns. Alternatively, with the use of the Green's function formulation, one uses as unknowns the aperture modal coefficients and evaluates the sum over all grating lobes for each term of an n-term basis function expansion of the current $i(z)$. The size of the matrix is the number of terms N in the expansion of $i(z)$.

Another advantage of either infinite-array formulation is the immediate identification of propagating and nonpropagating grating lobes that allows one to write the normalized power transmitted through the network over the space of a single periodic cell:

$$P = 1/2 \, \text{Re} \int \mathbf{S} \cdot d\mathbf{a} \qquad (6.28)$$

where the periodic cell area $d\mathbf{a}$ is normal to the array (for a rectangular lattice, the total cell area is $d_x d_y$).

When the Poynting vector $\mathbf{S} = \mathbf{E} \times \mathbf{H}^*$ is integrated over a unit cell, the orthogonality of the various spatial harmonics makes the power integral a simple two-dimensional summation over the individual spatial harmonic powers. For the p, qth mode, the net radiated peak power density is

$$\mathbf{S}_{pq} = \left[\frac{1}{(u_p^2 + v_q^2)^2} \right] \left\{ Z_0 H_z H_z^* + \frac{E_z E_z^*}{Z_0} \right\} [\hat{\mathbf{x}} u_p + \hat{\mathbf{y}} v_q + \hat{\mathbf{z}} \cos \theta_{pq}] = S_{pq}^0 \, \hat{\mathbf{r}}_{pq} \qquad (6.29)$$

where

$$\hat{\mathbf{r}}_{pq} = \hat{\mathbf{x}} u_p + \hat{\mathbf{y}} v_q + \hat{\mathbf{z}} \cos \theta_{pq}$$

is a unit vector in the direction of propagation of the p, qth grating lobe and is directed radially.

If all unwanted grating lobes are suppressed by restricting the element spacing, then the entire radiated power is in the $\mathbf{S}_{00}$ mode. All higher order grating lobes have imaginary Poynting vectors and so do not contribute real power. The real power radiated through one cell normal to the array is given by the integral above and is (for a rectangular lattice)

$$P = 1/2 \int \mathbf{S_{00}} - \hat{\mathbf{z}} \, dxdy = \frac{d_x d_y S_{00}}{2} \cos \theta_0 = P_0 \cos \theta_0 \qquad (6.30)$$

where the integration has been performed over the array lattice unit cell.

The term P_0 is the total power radiated through the unit cell, although directed at the angle (θ_0, ϕ). This expression says that the net power radiating out from the array surface is the product of the total power times the projection factor $\cos \theta_0$.

The input power to that cell is computed as the incident line power, so the normalized power transmitted through the cell is equal to the incident power less the reflected power, or

$$\frac{P_0}{P_{in}} = 1 - |\Gamma|^2 \qquad (6.31)$$

Usually the reflection coefficient Γ is evaluated directly with the solution of the integral equation, and the above expression from conservation of energy is automatically satisfied. In fact, Amitay and Galindo [20] point out that one cannot use conservation of energy to test the degree of convergence of method of moments solutions, which are satisfied term by term. However, conservation of energy does serve as a test of software errors and numerical accuracy.

As indicated above, the projection of the array unit cell in the θ direction introduces the factor $\cos \theta$, so that the effective element pattern radiated from a two-dimensional infinite array at the scan angle θ_0 is given by

$$f(\theta, \phi) = (1 - |\Gamma(\theta, \phi)|^2) \cos \theta_0 \qquad (6.32)$$

where Γ is the infinite array reflection coefficient when no grating lobe radiates. Γ also depends on the scan angles θ and ϕ. The perfectly matched infinite array thus has a $\cos \theta$ element pattern and the array gain must fall off at least as fast as $\cos \theta$ if the array is matched at broadside ($\Gamma(0, \phi) = 0$). With careful array design, one can approximate the $\cos \theta_0$ dependence out to 60 deg and beyond in one plane. More typical gain falloff varies like $\cos \theta_0^{3/2}$ or $\cos \theta_0^2$, depending on the plane of scan and element design.

If the infinite array solution is based on the Floquet modes (periodic structure) type of formulation, then there is a minimum number of terms of the infinite series that need to be included before the series is truncated. Typically, for a two-dimensional array it is customary to include in the spatial harmonic summations all terms corresponding to $\pm m$, $\pm n$, each with magnitude of at least 10 (corresponding to $21 \times 21 = 441$ terms). More terms are often required, depending on geometric considerations.

Unless a very large number of terms is used, simply using more terms in the spatial harmonic summations is sometimes not adequate to ensure convergence. In some cases, the convergence is optimized when the ratio of spatial harmonic terms to (modal) basis function terms is set to some fixed number dictated by geometrical considerations. This phenomenon is called *relative convergence* [21,22]. Most often, however, the number of spatial harmonics is increased until absolute convergence is ensured. This may take a very large number of terms, and convergence acceleration techniques are often used to reduce computation time. These techniques range from the use of Kummer's transformation [23], in which one adds and subtracts an asymptotic approximation that can be summed in closed form, or acceleration techniques based on the Poisson summation formula or other transformations [24–27]. These methods significantly improve convergence, but at the cost of added complexity that may also be significant.

Infinite-array theory gives an excellent approximation of the impedance behavior of central elements in large arrays and is often more appropriate than dealing with the severe difficulties of inverting large matrices in the element-by-element analysis for these cases. Figure 6.5(a,b), due to Steyskal (28), shows a comparison of the infinite-array calculation with a finite-array calculation for E- and H-plane scans of a triangular grid array of circular waveguides. The figure shows the complex active reflection coefficient Γ (here written R) for the two calculations. The magnitude of Γ is shown solid and the phase dashed. It is obvious that infinite-array theory (long dashes) gives a good approximation of the average reflection coefficient for this central element. Figure 6.5(b) also shows array parameters for the scan angle $u > 1$, which indicates that the array is scanned "beyond endfire" or fed with a phase progression more rapid than that of endfire. The beam in this case radiates like a trapped surface wave, which radiates at endfire, but can have tailored characteristics related to supergain. In the case of Figure 6.5, Steyskal was attempting to move the radiated beam closer to endfire.

6.3 ARRAY BLINDNESS AND SURFACE WAVES

The phenomenon of array blindness is a condition that results from array mutual coupling and can bring about essentially complete cancellation of the antenna-radiated beam at certain scan angles. This result is accompanied by near-unity reflection coefficient [29] at most of the central elements of the array. From the element pattern point of view, it is seen as a zero in the array element pattern. Figures 6.6 and 6.7 show element pattern and data due to Farrell and Kuhn [13] for an array of waveguides on a triangular grid. Figure 6.6 shows experimental H-plane scan data (solid line) for a finite array of 95 waveguide elements compared with modal expansion data (dashed line) and that computed with a single-mode theory (dotted line). The experiment and infinite-array theory show good corre-

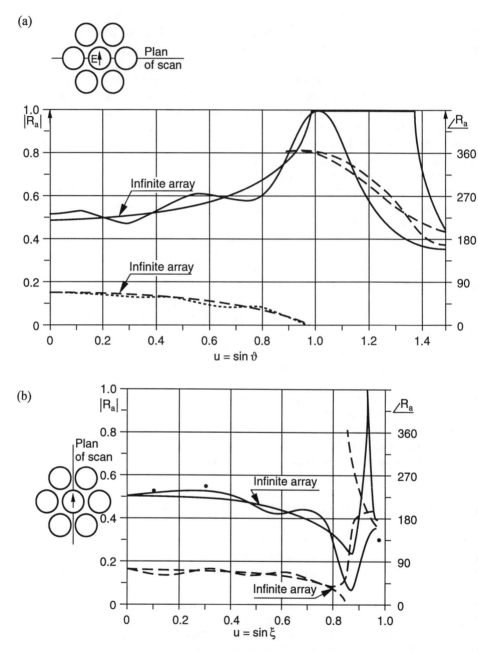

Figure 6.5 Active reflection coefficient of central element of a 127-element array. Infinite array data shown for comparison. Solid lines indicate magnitude, dashed lines indicate phase. Measured values indicated by dots (after Steyskal [28]). (a) *H*-plane scan; (b) *E*-plane scan.

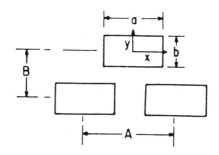

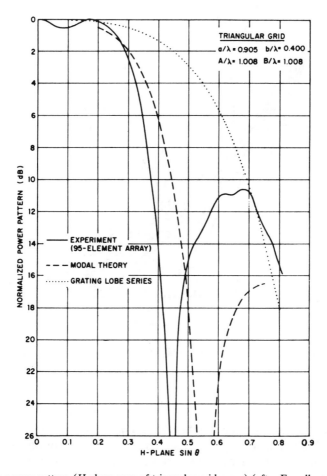

Figure 6.6 Array power pattern (*H*-plane scan of triangular grid array) (after Farrell and Kuhn [13]).

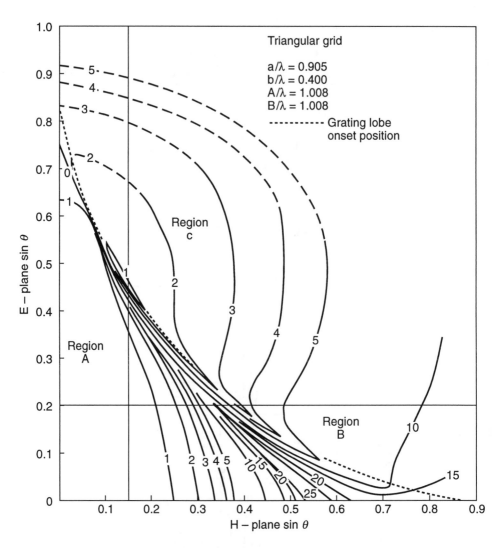

Figure 6.7 Contour map (in dB) of scanned peak of power pattern (after Farrell and Kuhn [13]).

lation, while the single-mode theory (called *grating lobe series* in the figure) does not exhibit the blindness. Figure 6.7 shows a contour plot of power radiated in all real space for the same array, indicating the observed blindness occurring in all scan planes.

The blindness phenomenon is associated with a kind of surface wave on the array and is often associated with higher order odd modes on the radiating element or with some other mode of cancellation. In the data of Farrel and Kuhn, the lowest order odd mode was shown to be responsible for the blindness. Blindness can also be associated with a true surface wave that is supported by the structure itself (like a dielectric slab loaded array). In addition, the existence of an array blind spot will usually (but not always) occur at array spacings less than a wavelength, and at a scan angle less than that at which a grating lobe enters real space. Each of these relationships will be clarified in the following paragraphs.

Before proceeding, the term *surface wave* should be defined. Certain dielectric structures, like the slab shown in Figure 6.8(a) or the periodic corrugated surface shown in Figure 6.8(c,d) support lossless wave propagation along their axes with velocities of propagation less than light. For example, the dielectric slab over a ground screen of Figure 6.8(a) supports TM waves and TE waves, with propagation constant β given by the solution of the transcendental equations below for TE wave:

$$k_1 \cos k_1 d + jk_2 \sin k_1 d = 0 \tag{6.33a}$$

for TM wave:

$$\epsilon_r k_2 \cos k_1 d + jk_1 \sin k_1 d = 0 \tag{6.33b}$$

where $k_1^2 = \epsilon_r k_0^2 - \beta^2$, and $k_2^2 = k_0^2 - \beta^2$, and $\beta^2 = k_x^2 + k_y^2$.

As shown in Figure 6.8(b) for a low permeability substrate ($\epsilon_r = 2.55$), these waves are slower than light ($\beta > k_0$). The TM_0 wave exists for all dielectric thicknesses and dielectric constants (does not cut off). This wave will later be shown to be a significant detriment to the design of microstrip antennas. The wave is a surface wave because the propagation constant in the z-direction is given by

$$k_z = [k_0^2 - \beta^2]^{1/2} = -j[\beta^2 - k_0^2]^{1/2} \tag{6.34}$$

and is purely imaginary, thus leading to the z-dependence

$$\exp[-jk_z z] = \exp[-|k_z|z] \tag{6.35}$$

This real exponential decay means that there is no power propagated in the z-direction, and the wave is bound closely to the dielectric layer surface as it travels unattenuated down the slab in the x-direction. The term *surface wave* relates to this behavior.

Periodic structures, like those in Figures 6.8(c,d), with spatial period less than one-half wavelength, can also support slow traveling waves. The existence of such

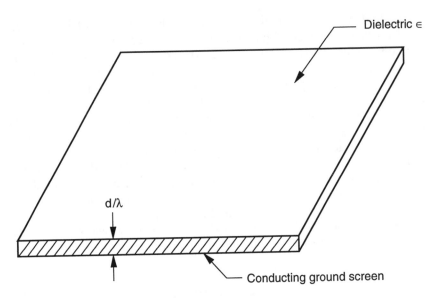

Figure 6.8(a) Structures supporting "surface waves": dielectric slab structure.

lossless *normal modes* propagating on passive metallic gratings has been known to be possible for many years. For such structures, the relevant phase velocities correspond to slow wave propagation (phase velocity is less than C), or in the language of array theory, imaginary space ($u = \sin \theta > 1$). With this choice of spacing ($dx < \lambda/2$), none of the waves in the grating lobe spectrum has its direction cosine in real space ($|u| \leq 1$) and there is no loss due to radiation. This type of passive metal grating thus supports lossless transmission, and has been implemented as corrugated structures, monopole arrays, and many other periodic structures. This type of wave solution is often termed a *surface wave* because of its similarity to that supported by the various dielectric structures described earlier. In fact, the wavelike solution supported by these periodic structure open waveguides is actually a spectrum of surface waves (or grating lobes), with one wave of the spectrum having a propagation constant similar to that of the surface waveguide, and all others very tightly bound to the structure (larger wave numbers and stronger exponential dependence). This propagating nonradiating mode of operation is utilized in Yagi-Uda arrays by terminating the array and letting the wave spectrum radiate endfire.

Array blindness results when the array geometry with short-circuited input ports would support a normal mode (lossless nonradiating propagation) along the structure at some given scan angle. At the angle of array blindness with the array

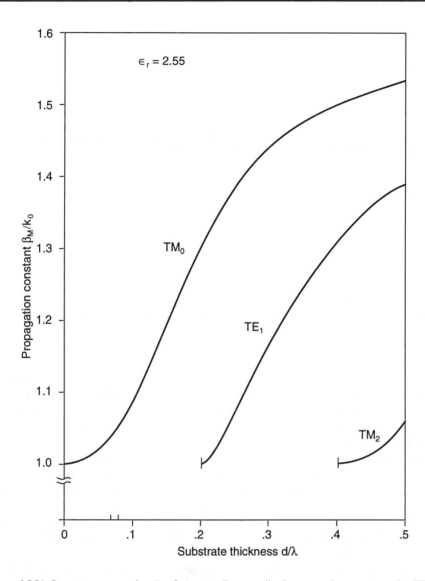

Figure 6.8(b) Structures supporting "surface waves": normalized propagation constants for TE and TM waves (after Pozar and Schaubert [39]).

excited at all input ports, the input impedance at all ports is identically zero, with the structure supporting a nonradiating lossless mode. Mathematically, this is analogous to a resonant L-C circuit. In the L-C circuit case, at resonance, the input current is unbounded ($I = V/Z_{in}$) because of the zero in the impedance ($Z_{in} = 0$)

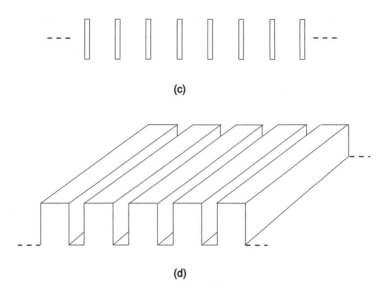

(c)

(d)

Figure 6.8 (cont.) Structures supporting "surface waves": (c) monopole array supporting "surface wave (spectrum)"; (d) shorted corrugated structure.

or the pole in admittance ($Y_{in} = \infty$). The resonance can be defined as the condition at which a nonzero current is supported with no input signal. The resonant frequency is the solution of the eigenvalue problem, and is that frequency of undamped oscillation of the circuit with input terminals shorted (by the zero resistance path of the ideal voltage source).

The array, however, is more complex than the L-C circuit because it has a distributed set of input ports, with signals applied to each port. In a manner entirely analogous to the resonant circuit, if there is a propagating nonradiating solution that would satisfy the boundary conditions of the shorted array structure, then applying a set of signals with that phase progression would result in zero input impedance at all input ports. If the input impedance is zero, one can place short circuits at the terminals without changing the solution.

Extension of the above logic to the array case for scan angles in real space and with spacing greater than one-half wavelength is not obvious. For such spacings the inter-grating lobe separation is less than 2 in u-space, and there is always at least one beam in real space. Thus, at least one array beam should radiate, and the combined network should have loss. If one were to look at the equivalent shorted array, one would argue that it cannot support a normal mode solution because the radiation would preclude a lossless solution. Yet the blindness phenomenon is caused by the existence of a normal mode solution that exists precisely because it allows no radiation.

Part of the answer to this intuitive dilemma came from the study of Farrell and Kuhn [12,13], who provided an essential key to understanding blindness and performed rigorous analysis of a waveguide array with a blind spot. They were the first to observe that waveguide higher order modes play a dominant role in achieving the cancellation necessary for a null. The null occurs when radiation contributions from the lowest order symmetric and antisymmetric modes cancel to produce the element pattern zero. They also showed that the null is accompanied by a zero in input conductance, as distinguished from the infinite susceptance obtained at the grating lobe point using a one-mode analysis. Diamond [30] and later Borgiotti [31] confirmed all of these findings for waveguide arrays.

Oliner and Malech [2] suggested what is now generally accepted as true: the blind spot is associated with the normal mode solution of an equivalent, reactively loaded passive array, and the condition of a complete null on the real array occurs when the elements are phased to satisfy the boundary conditions for the equivalent passive array. Knittel et al. [32] developed this theory and showed that in the vicinity of the null the solution corresponds to a leaky wave of the passive structure, but that surface-wave-like fields exist immediately at the null. This is consistent with the results of an analysis made earlier by Wu and Galindo [33], who demonstrated that the only radiating (fast) wave of the periodic structure spatial harmonic spectrum is identically zero at a null (because of cancellation by the odd mode), and that for this reason a normal mode can exist even for a structure with a period greater than one-half wavelength.

Along with these contributions to the understanding of the physics of a phased array blindness, other authors have shown that both waveguide aperture and lattice dimensions are critical in determining the likelihood of a blind spot. Ehlenburger et al. [29] proved that reducing either of these dimensions moves the position of a null further out in u-space, and that certain higher order waveguide modes can cause nulls in certain scan planes. Their analysis explains, for example, that for pure H-plane scan with a rectangular grid, there is no null, but for a triangular grid, the TE_{20} mode can cause a null in the same plane. Finally, to avoid nulls, Ehlenburger et al. list the choices of waveguide sizes for rectangular apertures on several size grids. Figure 6.9 from Ehlenberger et al. [29] shows the displacement of a null by reducing the waveguide dimension while maintaining constant lattice spacing.

The critical role that array lattice dimensions play in the occurrence of blindness has been exploited in order to predict its onset. Byron and Frank [34] described a procedure for combining simulator measurements and an approximate mathematical model to predict array blind spots, and Knittel [35] used the $k - \beta$ diagram to reveal a direct relation between the blindness effect and the cutoff conditions of the next higher waveguide mode and lattice mode (grating lobe).

Figures 6.10 and 6.11 from Knittel [35] show the locus on a $k - \beta$ diagram of the blind spot for the array with grating studied by Farrell and Kuhn, but with

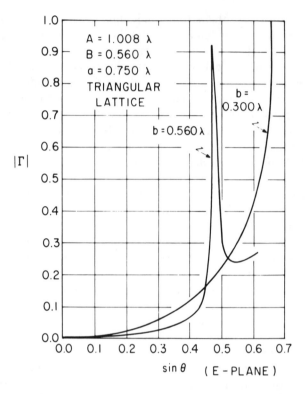

Figure 6.9 Blindness location versus waveguide aperture size with constant lattice spacing (after [29])

frequency varied over a wide range. The locus of array blindness is shown solid on both curves. It is significant that the locus never crosses any of these higher order mode loci, because crossing the TE$_{20}$ mode cutoff would allow energy to leak back into the waveguides, and crossing the grating lobe cutoff line would allow energy to radiate by means of a grating lobe. In neither case could the passive equivalent array sustain an unattenuated normal mode. Figure 6.11 also shows that if the waveguide size is reduced and no changes are made to the periodic grid dimensions, then the blindness is moved to wider angles.

These two figures were included to demonstrate the power of this graphical technique for predicting the onset of blindness difficulties. In all cases shown by Knittel, the blindness locus remained nearly asymptotic to the waveguide or grating lobe loci, whichever occurred at lower frequency. The implication for design is obviously that the null can be avoided by choosing dimensions sufficiently smaller than those for the cutoff conditions.

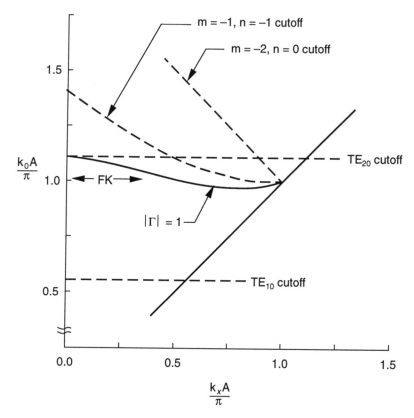

Figure 6.10 $k - \beta$ diagram showing array null locus for the triangular grid array of Farrell and Kuhn, with $B/A = 1$; $a/A = 0.898$; $b/A = 0.397$ (after Knittel [35]).

Dipole arrays with thin wire elements do not appear to have blindness [36]. However, when they are driven by real feed lines, there may be a mode of feed line radiation that can produce a cancellation effect that results in blindness. An example is the published study of Mayer and Hessel [37].

It appears that nearly any embellishment one might add to the array face can also be the cause (at some frequency, at some angle) of blindness [38]. The author investigated the use of metallic fences on the array surface in an attempt to reduce or alter mutual coupling, and so to improve wide-angle element match. In fact, fences did add another dimension to optimize match and so did improve wide-angle impedance match. However, Figure 6.12 shows that some choices of dimen-

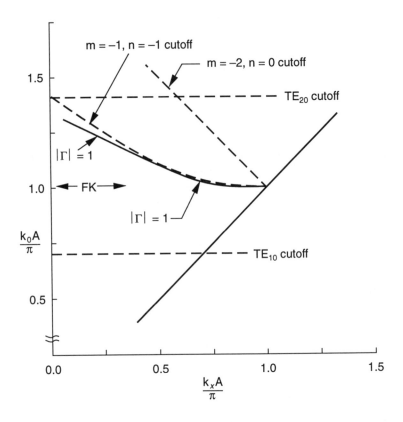

Figure 6.11 $k - \beta$ diagram showing array null locus of the triangular grid array of Farrell and Kuhn (see Fig. 10) but with dimension a/A reduced to 0.709 (after Knittel [35]).

sions do produce the deep infinite-array element pattern nulls that indicate an array blindness.

Although dielectric layers have been used for many years to improve scan match (see Section 6.5), they too can be the source of array blindness. This phenomenon is not new and was observed in the early work of Wu and Galindo [33]. More recently, blindness has been observed to occur in microstrip patch arrays or microstrip dipoles when the combination of dielectric constant and substrate thick-

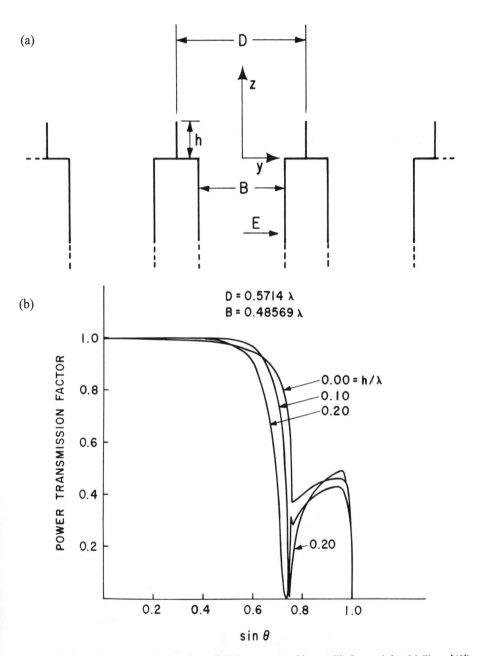

Figure 6.12 Radiating characteristics of parallel plane arrays with metallic fences (after Mailloux [48]). Note: power transmission factor is $1 - |\Gamma|^2$. (a) Geometry; (b) array power transmission factor for various fence heights.

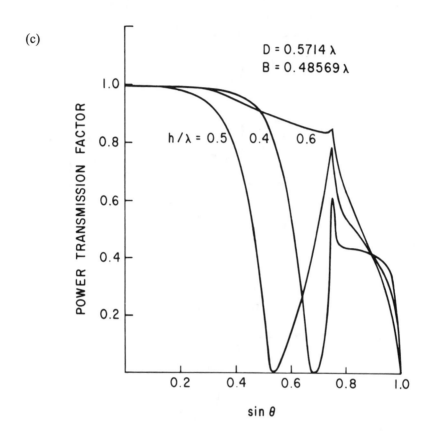

(c)

$D = 0.5714\,\lambda$
$B = 0.48569\,\lambda$

POWER TRANSMISSION FACTOR

$h/\lambda = 0.5$ 0.4   0.6

sin θ

Figure 6.12 (cont.) Radiating characteristics of parallel plane arrays with metallic fences (after Mailloux [48]). Note: power transmission factor is $1 - |\Gamma|^2$. (c) Array power transmission factor for various fence heights.

ness is such as to support a tightly bound surface wave, one with a phase velocity that is sufficiently slow so that it couples to an array grating lobe. A particular case is illustrated by arrays of microstrip patches etched on dielectric substrates. Pozar and Schaubert [39] have correlated the TM surface wave propagation constant as given by the previous transcendental equation (6.33) with the observed blindness angle. The dielectric layer itself supports a surface wave, and although the boundary conditions are perturbed by the array patch or dipole structure, the location of the blindness is often predicted very accurately by the surface wave propagation constant. The mechanism for coupling into the surface wave is depicted in Figure 6.13(a) [39], where several solid circles and dashed circles of larger radius are shown. The central solid circle defines real space bounded by $u^2 + v^2 = 1$, and

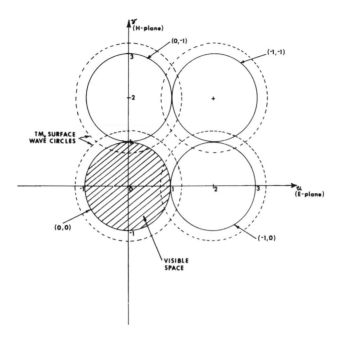

Figure 6.13(a) Array radiation and grating lobe loci compared with slow "surface wave" loci (after Pozar and Schaubert [39]): surface wave circle diagram.

the other solid circles define the corresponding regions that bound the array grating lobes and so have centers at $u = p\lambda/d_x$, $v = q\lambda/d_y$. The spacings in both dimensions are chosen to be equal to $\lambda/2$, so the grating lobe circles touch the central circle. The dashed circle, centered at the origin, is the locus of possible surface wave normalized wave numbers k_S/k_0 for the TM_{01} surface wave, which is the only one not cut off for very thin dielectrics. In this case, it is important to understand that it is only the circle itself that is the locus of allowed surface waves, not the whole region enclosed within the circle. The circle radius k_S/k_0 is greater than unity, and so the locus of the grating lobe can intersect the surface wave circle. When this happens, the impedance seen by the microstrip feed is a short circuit, and again there is array blindness. This logic is therefore the same as had been previously understood for arrays without dielectric layers that have blindness. The locus of scan angles for which intersections with the surface wave circle (dashed with center $(0, 0)$) occur are also circles. These circles (dashed) have centers at the points $(p\lambda/d_x, q\lambda/d_y)$, and where these intersect the scan space is where the interference can occur. A necessary logical extension to this argument is that the "perturbed" surface wave also has a spectrum of allowed grating lobes due to its periodicity. Since these

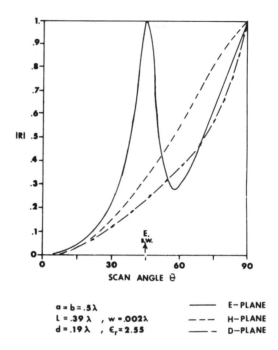

-a = b = .5λ
L = .39 λ , w = .002λ
d = .19 λ , ϵ$_r$ = 2.55

——— E – PLANE
– – – H – PLANE
—– – D – PLANE

Figure 6.13(b) Array radiation and grating lobe loci compared with slow "surface wave" loci (after Pozar and Schaubert [39]): E, H, and diagonal plane scan data showing blindness in E-plane.

fall exactly at the same lattice points as the forced array excitation, there is complete cancellation at the main lobe radiation angle in the central circle.

Pozar and Schaubert [39] point out that polarization plays a major role in whether the surface wave is excited. In the case of printed dipoles, the lowest order TM surface wave is not excited for H-plane scan because of polarization mismatch. Figure 6.13(b) shows the data of Pozar and Schaubert for a printed dipole array in E- and H-plane scans. A clear blindness is evident in the E-plane, but none in the H-plane. Figure 6.13(c) shows the computed reflection coefficient of the array, with the locus of the surface wave circles indicated to confirm that the blindness occurs at the angles predicted by the surface wave theory.

Other types of printed circuit arrays can be subject to blindness due to coupling with modes within the substrate. For example, Figures 6.14 and 6.15 [40] show printed circuit slot arrays with and without conducting cavity walls. In this example, the case without backing cavities is seen in Figure 6.14 to have very strong blindness due to coupling with a wave spectrum that propagates in the parallel plane region

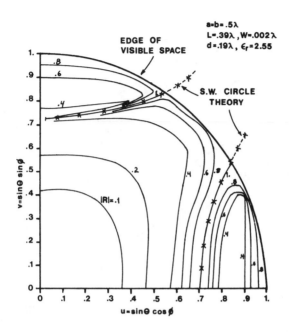

Figure 6.13(c) Array radiation and grating lobe loci compared with slow "surface wave" loci (after Pozar and Schaubert [39]): reflection coefficient contour plot showing loci of the surface wave for unloaded dielectric surface.

and has the same periodicity as the forced excitation. An array with cavity-backed slots, shown in Figure 6.15, has no blindness for any scan angle if the spacing λ/2 is maintained.

6.4 IMPEDANCE AND ELEMENT PATTERNS IN WELL-BEHAVED INFINITE SCANNING ARRAYS

By the proper choice of array element design and lattice, one can avoid the array blindness phenomenon and obtain a satisfactory, "well-behaved" array aperture. This section lists a number of the cases for which there are analytical/numerical solutions for the infinite array radiation characteristics and describes the impedance and element pattern behavior that is observed. Before proceeding further, it is important to understand that the one most significant choice toward ensuring well-behaved array scan match is to keep element spacing small.

Arrays of slots or dipoles with element spacings of one-half wavelength or less are free of blindness or grating lobes and have generally well-behaved scanning

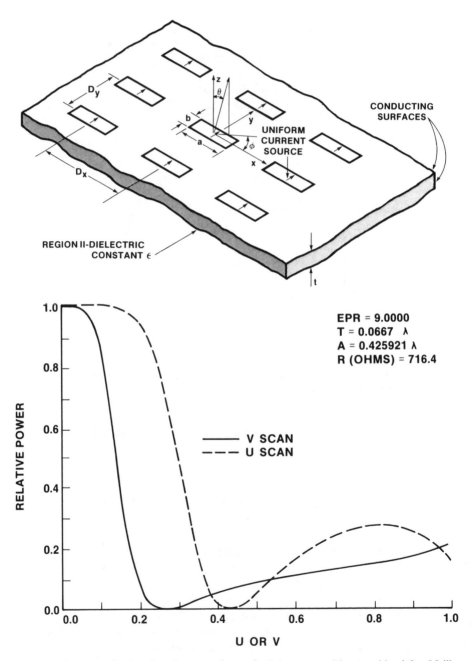

Figure 6.14 Geometry (top) and performance (bottom) of slot arrays without cavities (after Mailloux [40]).

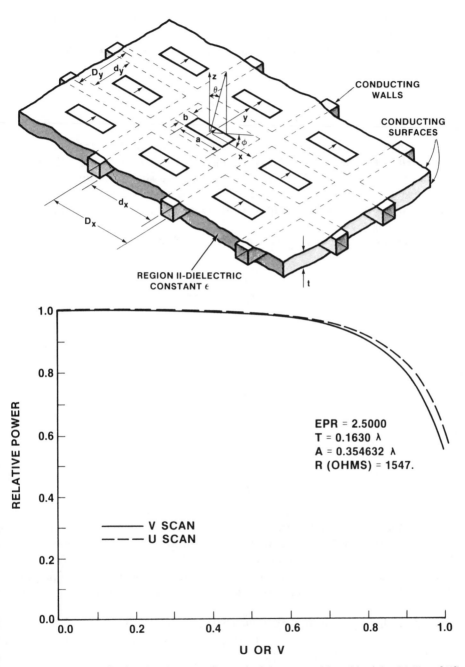

Figure 6.15 Geometry (top) and performance (bottom) of slot arrays with cavities (after Mailloux [40]).

characteristics. Infinite arrays of such elements exhibit particularly simple scan behavior, which has been well documented. Two-dimensional infinite arrays of infinitesimal slots in a ground screen have input conductance of the form [2]:

$$G_a = \frac{N}{d_x d_y} (\epsilon/\mu)^{1/2} \frac{(1 - u^2)}{\cos \theta} \qquad (6.36)$$

In this expression, the factor N is a constant of proportionality, and d_x and d_y are the interelement spacings. In the two principal planes $\phi = 0$ and $\pi/2$, the conductance varies like $\cos \theta$ and $(\cos \theta)^{-1}$, respectively. Therefore, for a two-dimensional array, there is no one scan angle θ at which one can match the array for all azimuth angles ϕ, unless one introduces some kind of matching that also depends on the scan angle.

The array susceptance is given by Oliner and Malech as

$$B_a = -\frac{N}{d_x d_y} (\epsilon/\mu)^{1/2} \sum_m{}' \sum_n{}' \Phi_{mn} \qquad (6.37)$$

where

$$\Phi_{mn} = \frac{u_m^2 - 1}{\{u_m^2 + v_n^2 - 1\}^{1/2}}$$

and again

$$u_m = u_0 + \frac{m\lambda}{d_x} \qquad v_n = v_0 + \frac{n\lambda}{d_y}$$

In (6.37), the primes indicate that the sums exclude the propagating modes. This simple expression, written in terms of the nonpropagating grating lobe direction cosines, is an accurate representation of the scan behavior for arrays of short slots. Oliner and Malech point out that for spacings that allow propagation of no grating lobes, the term Φ_{mn} is always positive, so short-slot arrays are always inductive for all scan angles.

The radiating patterns and scan characteristics of infinite arrays of short dipoles with no ground screen are readily related to those of slots. In this case, the array impedance is given by [2]

$$Z_a = 1/2 \left(\frac{N}{d_x d_y}\right) [Z_0 + jX] \qquad (6.38)$$

where

$$Z_0 = (\mu/\epsilon)^{1/2} \left(\frac{1 - u_0^2}{\cos \theta}\right)$$

and

$$X = -1/2 \, (\mu/\epsilon)^{1/2} \sum' \sum' \Phi_{mn}$$

Here it is seen that the impedance varies like $\cos \theta$ for $\phi = 0$ and like $1/\cos \theta$ for $\phi = \pi/2$, which is the inverse of the variation for the slot impedance case. In fact, as pointed out by Oliner and Malech, if the factor N were the same, this expression shows that both the active resistance and reactance of the short dipole array are just half of the admittance and susceptance of the slot array. The factor of two is introduced because the slot array radiates into only a half space, but the dipole array radiates into a full space.

The addition of a ground plane implies that the dipoles are mounted over the screen by some height h.

In this case, the input impedance is given by [2] as

$$Z_a = R_a + jX_a \qquad (6.39)$$

where

$$R_a = [N/(d_x d_y)] Z_0 \sin^2 (kh \cos \theta)$$

and

$$X_a = [N/(d_x d_y)] X + 1/2 \, [N/(d_x d_y)] Z_0 \sin (2kh \cos \theta)$$

and X and Z_0 are given as in the previous equations.

The expressions $\sin^2 (kh \cos \theta)$ and $\sin (2kh \cos \theta)$ account for the dipole height above the ground screen. A typical choice of the height h is $\lambda/4$, in which case the signal is maximum at the zenith ($\theta = 0$).

The expressions above assume not only that the array elements are infinitesimal, but that the spacings are small enough that no grating lobes alter the conductance. At spacings greater than one-half wavelength, the presence of grating lobes at large scan angles significantly alters the array impedance variation and element patterns. These effects can be very complex, as compared to the behavior of closely spaced arrays indicated above. Figures 6.16 and 6.17 show the impedance and reactance of an infinite array of dipoles with spacing $d_x = d_y = 0.6$.

For an array with relatively large spacings, the E- and H-plane behavior of dipole arrays are significantly different from each other, with the H-plane depen-

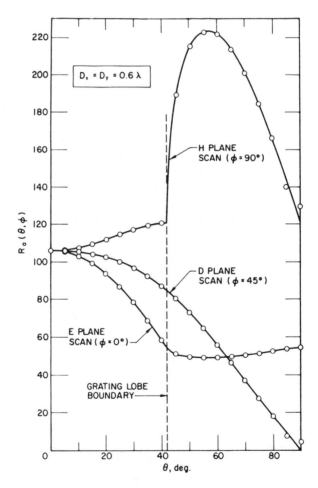

Figure 6.16 Dipole array scanning data for three different planes of scan. Dipoles are $\lambda/4$ above ground, array spacing 0.6λ (after Diamond [1,2]).

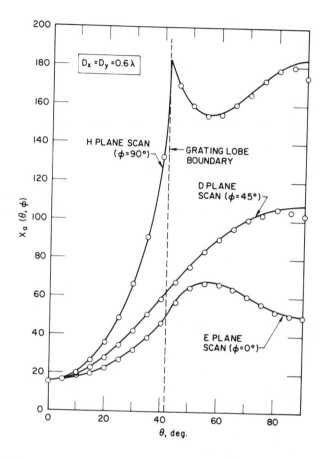

Figure 6.17 Variation in active reactance as a function of scan angle for three different planes of scan and for array of $\lambda/2$ dipoles mounted $\lambda/4$ over a ground screen. Solid lines from infinite array theory. Circles from element-by-element calculation for 65 × 149-element array (after Diamond [1,2]).

dence exhibiting a significant discontinuity in both impedance and reactance at the scan angle of the grating lobe entrance. Figures 6.16 and 6.17 show these reactance parameters as computed by Oliner and Malech [2] using infinite array theory. The circles shown on these figures are from earlier results (circles) obtained by Diamond, who used one-term representations of the dipole current in the element-by-element formulation for a 65 × 149 element array. Solutions including higher order mode representations of currents are available in the literature, but for $\lambda/2$ dipoles a single term is sufficient.

Infinite-array element patterns are also strongly dependent on the element spacing. Figure 6.1 [2] shows the element pattern corresponding to various d/λ, including the value 0.6, where the element pattern is exhibiting significant narrowing. The patterns that use smaller element spacings are broader and follow a cos θ dependence more closely.

Rigorous infinite-array solutions, some of which are referenced in the previous section, have been obtained for a large number of array types. Figure 6.18 shows a few of the basic waveguide array configurations for which infinite-array solutions have been published. Included in the figure are flush-mounted arrays of rectangular and circular waveguide elements, ridge-loaded elements [41], protruding dielectric (TEM solution) [42] and dual-frequency elements [43]. Among other published solutions are numerous interlaced multiple frequency configurations [43–46], iris loading [47], and fence [48] or corrugated plate [49] loadings for impedance matching, as well as very-wide-band strip line configurations [50].

Recent studies of finite and infinite microstrip arrays have included circular and rectangular patch arrays excited by probe feeds, or electromagnetically coupled to aperture feeds or microstrip patches beneath other substrates, and all of the above with superstrate dielectric layers. Figure 6.19, due to Aberle and Pozar [51], shows the scan characteristics of a probe-fed infinite array of circular microstrip patches. In these data, the authors compare the results of using idealized feed modes with those obtained by including the feed as part of the boundary value problem. In the idealized feed model, the probe-feed is considered a short electric line source, the moment method is used to solve for patch current, and the input impedance is found as the reaction of the patch electric field and the probe current. This model is accurate for patch arrays with substrate thickness up to about 0.02λ. The improved feed model represented in the figure treats the probe as part of the boundary value problem, solving for the zero tangential electric field on both probe and patch. Simulator measurements show significantly improved estimation of array reflection coefficients using this more complex probe model. The data of Figure 6.19 shows E-, H-, and diagonal-plane scans for infinite arrays of circular patches fed by a single probe. The authors also treated arrays fed by balanced two-probe feeds, with a 180-deg phase shift between the probes (Collings radiators).

Figure 6.20 shows several results published by Schuss [52] using an analytical procedure that placed special emphasis on evaluating probe currents accurately [53]. Figure 6.20(a) shows the balanced feed patch radiator geometry as fed by 180-deg reactive baluns. The use of reactive baluns can lead to an element pattern blindness at one frequency within the operating band, as shown in Figure 6.20(b). This example shows the E-plane embedded element pattern of a patch radiator with $d_x = 0.508\lambda_0$, $d_y = 0.5\lambda_0$, $a_x = a_y = 0.363\lambda_0$, $h = 0.057\lambda_0$, probe separation $0.531a_x$, and relative dielectric constant 2.2. This array element is excited by a matching network to optimize impedance match throughout the scan sector over all frequencies. The matching network, which consists of quarter-wave transmission

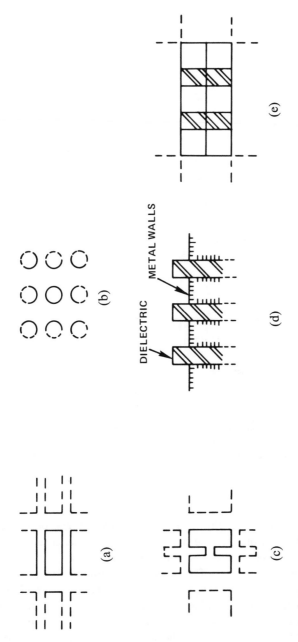

Figure 6.18 Infinite-array geometries. (a) Rectangular waveguide array; (b) circular waveguide array; (c) ridge-loaded waveguide array; (d) Protruding dielectric waveguide array; (e) Dual-frequency array.

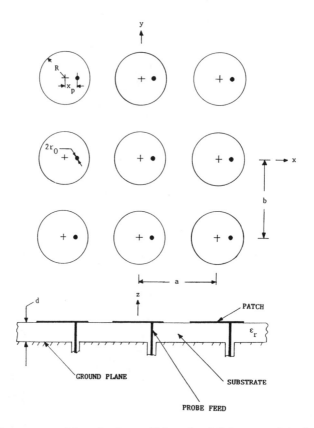

Figure 6.19(a) The geometry of the reflection coefficient of an infinite array of circular patch elements fed by single probes as computed with idealized source and improved source models (ϵ_r = 2.55, $d = 0.06\lambda_0$, $R = 0.166\lambda_0$, $x_p = 0.083\lambda_0$, $r = 0.0004\lambda_0$, $a = 0.51\lambda_0$ (after Aberle and Pozar [51]).

line matching sections and open-circuit stub transmission lines, produces a doubly tuned frequency response that can be precisely designed because of the accuracy of the analytical model. The data shown are taken at $1.05\lambda_0$ and compare theory (solid) versus element pattern data for two experimental arrays of different size. The array element pattern follows a cos θ dependence out to about 50-deg. Beyond that point, one can observe a blindness due to a surface wave resonance. This blindness was shown to be related to the use of a reactive power divider. The blindness shown in the figure is absent if the same radiators were excited by a four-port (Wilkinson) power divider. At lower frequencies, the array follows a cos θ pattern to much wider angles.

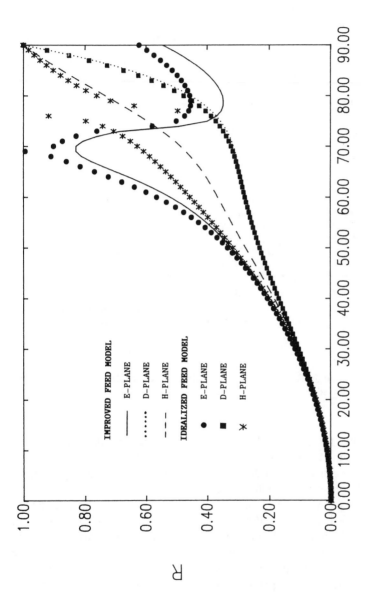

Figure 6.19(b) Reflection coefficient of an infinite array of circular patch elements fed by single probes as computed with idealized source and improved source models ($\epsilon_r = 2.55$, $d = 0.06\lambda_0$, $R = 0.166\lambda_0$, $x_p = 0.083\lambda_0$, $r = 0.0004\lambda_0$, $a = 0.5\lambda_0$ (after Aberle and Pozar [51]): reflection coefficient magnitude.

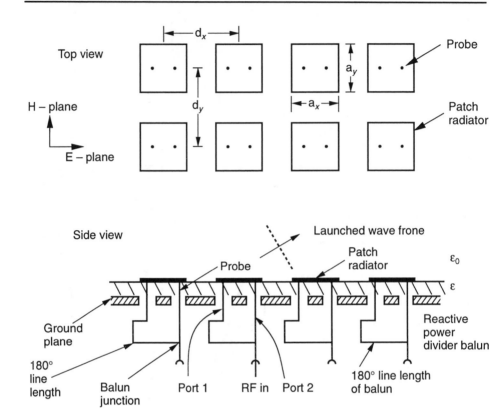

Figure 6.20(a) Characteristics of dual-probe fed microstrip arrays of rectangular patches (after Schuss [52]): balanced fed patch radiator geometry.

Figure 6.21 shows the data of Herd [54,55], who analyzed infinite arrays of electromagnetically coupled patches. The electromagnetically coupled geometry is of particular interest because the bandwidth of these antennas can be significantly broader than that of conventional patch antennas.

Figure 6.21(a) shows three of the geometries investigated by Herd. The wide variety of electromagnetically coupled geometries allows for control of additional degrees of freedom to optimize scan performance throughout the array scan sector. Figure 6.21(b) gives the geometric parameters and shows a Smith chart plot of the element impedance for an infinite array at broadside. The element shown has over 18% bandwidth at broadside, due primarily to the double-tuned behavior, as evidenced in the Smith chart looped characteristic. Other array designs provided less bandwidth but better scan characteristics. An improved design provided 10% bandwidth over a 50-deg half-angle cone of scan with less than 2:1 VSWR.

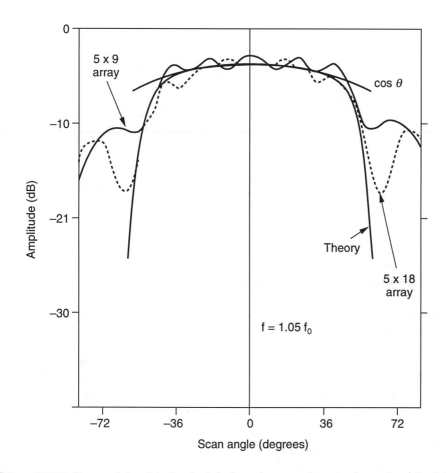

Figure 6.20(b) Characteristics of dual-probe fed microstrip arrays of rectangular patches (after Schuss [52]): measured and predicted *E*-plane embedded element pattern of radiator at 1.05f_0.

6.5 WIDE-ANGLE IMPEDANCE MATCHING

A comprehensive survey by Knittel [56] lists many of the approaches used to impedance match dipole and waveguide array antennas throughout their frequency and scan range. Specifically, Knittel categorizes all wide-angle impedance (WAIM) techniques as *transmission line region* techniques and *free-space region* techniques that alter the wide-angle match. This categorization is still convenient, and we will follow it here.

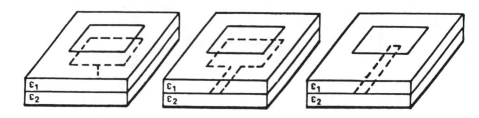

Figure 6.21(a) Impedance data for scanned electromagnetically coupled patch antennas (after Herd [54,55]): geometries investigated.

6.5.1 Transmission Line Region Techniques

Since conventional single-mode matching can produce an impedance match at one frequency and one scan angle, but cannot compensate for impedance variations with scan, the transmission line region techniques that accomplish varying degrees of wide-angle match do so with passive circuits either by controlling higher order modes in the aperture, or by separate interconnections between the various elements, or by using an active tuning circuit to improve impedance match. The transmission line schemes identified by Knittel are given below.

The Use of Connecting Circuits Between the Various Transmission Line Feeds to Partially Cancel the Variation of Array Admittance With Scan

Early studies of using interconnecting circuits between elements of an array show that substantial improvement in scanned match can be obtained [57–60]. The technique remains difficult to implement in a practical array, and so has not been used in practice, although in fact there has been much debate as to whether perfect scan match is attainable. It should be noted that in Chapter 8 techniques are described for interconnecting feed lines that produce specially shaped element or subarray patterns within the array. These techniques differ from the above and are not for mutual coupling alteration. They can be considered as simply combining element patterns to form more directive subarray patterns.

CASE 2B:

h1 = .159 cm

h2 = .318 cm

X1 = Y1 = 1.87 cm

X2 = Y2 = 1.95 cm

ε_1 = 2.33−j.002

ε_2 = 2.33−j.002

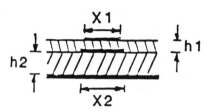

measured ●   predicted △

(2:1 B.W. = 18.7%)   (2:1 B.W. = 19.2%)

Figure 6.21(b) Impedance data for scanned electromagnetically coupled patch antennas (after Herd [54,55]): electromagnetically coupled wideband microstrip array (broadside data).

Dielectric Loaded and Multimode Waveguides

The use of higher order transmission line modes (whether propagating or nonpropagating) can alter the wide-angle impedance properties of waveguide arrays. Waveguides with single-mode transmission lines with dielectric plugs have been investigated by Wu and Galindo [33] and Amitay and Galindo [20]. In this case, the higher order modes are cut off, but in the work of Tang and Wong [61], propagating higher order modes are allowed, and they play a role in achieving wide-angle matching of the array elements.

Electrically Tunable Matching

Active tuning of the array elements can be used to match the array at all scan angles. In 1970, Knittel's assessment [56] was that adding the additional devices to implement this tuning method would probably be unwarranted because of the increased cost and possibly decreased reliability. Indeed, this assessment is still true with regard to fielded arrays, but there has been recent interest in these techniques, since they offer the advantage of replacing the ferrite circulators now commonly used with monolithic microwave integrated circuit (MMIC) transmit-receive (TR) modules, and thus potentially reduce cost, weight, and size of these devices. The tuning circuitry can be implemented with monolithic technology to reduce cost and improve reliability, and so may yet find a place in practical systems.

Lossy Matching

A final matching technique discussed by Oliner and Malech [2] is to use a circulator and load to improve the match for transmit arrays. Although not truly an impedance matching technique, this means is commonly used in wide-band matching in (single element) communications systems, especially when generator mismatch becomes a serious problem. In the array application, it has also been used with solid-state TR modules to provide some isolation between the transmit and receive ports, but not specifically for the purpose of impedance matching.

6.5.2 Free-Space Region WAIM Techniques

The five free-space techniques outlined by Knittel in 1970 are all still viable and have found application in practical systems. Clearly, the most important of these is the reduction of element spacing, which was mentioned earlier in this chapter in reference to Oliner and Knittel's work on blindness elimination. The other techniques include the use of dielectric slabs of thin dielectric sheets parallel to the array aperture, and the addition of some kind of periodic loading structure to alter the scan matching conditions at the array face. The two loading methods surveyed by Knittel were metallic baffles and periodic loading of the array ground screen.

Reduced Element Spacing

Reduced element spacing was first shown to produce significantly less impedance variation with scan than more widely spaced elements. However, it was the Knittel et al. studies [32,35] that best illustrated the reduced impedance variation and elimination of blindness difficulties. The results of these studies, shown in Section 6.3, have led to standard design guidelines that maintain element spacing at a few percent less than that required to exclude grating lobes from real space at the extreme scan angles. Although these guidelines are adequate for most waveguide and dipole arrays, one must always obtain theoretical or experimental verification of element scan properties before committing to a final design. Recent investigations have shown [62] that by using very close spacing and numerical optimization, one can develop array apertures with very little impedance variation due to scan.

Baffles Between Array Elements

Investigations of various periodic loadings have shown that it is possible to improve array match with these techniques. One of the earliest geometries proposed was the use of baffles, metallic fences separating rows of elements, as shown in Figure 6.22(a). These studies demonstrated some improvement in *E*-plane scan of dipole arrays due to baffles mounted parallel to the array *H*-plane. Other studies of baffles were published by Mailloux [48], previously shown in Section 6.3 and Figure 6.12, as were some very complex baffle-corrugation arrangements [49], as shown in Figure 6.22(b). In general, these techniques were shown to improve the scan properties of particular arrays and allow scanning to wider angles.

Loaded Ground Planes

A loading technique introduced by Hessel and Knittel [64] consists of alternating feed waveguides and shorted parasitic waveguides used as microwave "chokes" to interrupt ground screen current. The technique showed some wide-angle scan improvement in the *E*-plane and permitted scanning slightly beyond the point at which the *E*-plane grating lobe enters at endfire, while leaving the *H*-plane scan characteristics unaltered.

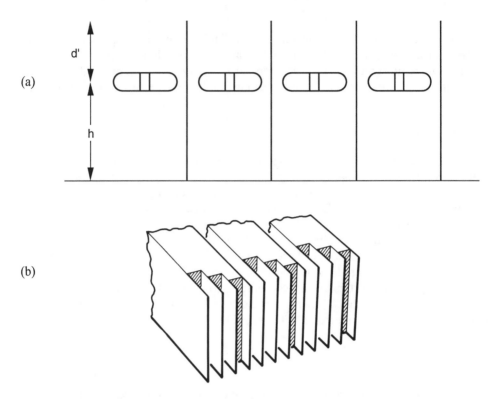

Figure 6.22 Wide-angle impedance matching using several geometries. (a) Baffles in the *H*-plane of a
dipole array (after Edelberg and Oliner [63]); (b) matching with corrugated plates (after
Duport [49]).

Dielectric Slabs Used as Cover (Radome) Layers

The use of dielectric (and now more exotic anisotropic materials) as array radomes
has been studied for waveguide, dipole and microstrip patch arrays. These elements
play a role in impedance matching and have been discussed in other sections.

Dielectric WAIM Sheets

One of the more practical means of scan matching waveguide arrays was proposed
by Magill and Wheeler [65]. The technique incorporates a dielectric sheet in front
of the array (Figure 6.23(a)) to remove some of the susceptance variation as the
array is scanned. This dielectric WAIM sheet is placed at a location in front of the
array where the referenced array reflection is most nearly a pure susceptance, and
at this point the susceptance of the WAIM sheet is used to reduce the overall

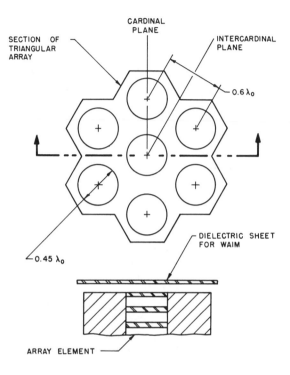

Figure 6.23(a) Dielectric WAIM sheets for scan matching (after McGill and Wheeler [65]): array geometry.

impedance variation. Figure 6.23(a,b) show a section of waveguide array geometry and its reflection coefficient portrayed on a Smith chart. Three scan points are shown for reference. These are at 56-deg in the *E*- and *H*-planes, and 29 deg in the intercardinal plane (this case to simulate a near broadside data point). Since the thin dielectric layer presents a pure susceptance, the dielectric in this example is placed a distance in front of the array where the scanned reflection coefficient lies as close as possible to the unity conductance circle. Magill and Wheeler [65] point out that this may not necessarily result in the best scan match, and so in some cases it may be preferable to match near some other conductance circle, and match the conductance using some means internal to the transmission line. For simplicity, the example proceeds assuming a grouping near the unity conductance circle.

The susceptance of a thin dielectric layer at broadside $B(0)$ is given approximately by:

$$\frac{B(0)}{G_0} = (\epsilon_R - 1)\frac{2\pi t}{\lambda_0} \tag{6.40}$$

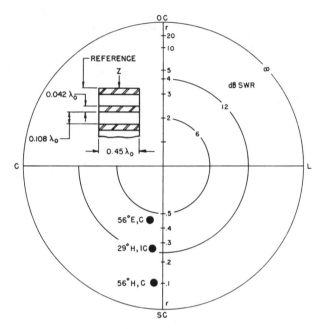

MEASUREMENTS PERFORMED AT λ_0, TEST WAVELENGTH.
DATA OBSERVED IN WAVEGUIDE SIMULATOR.

Figure 6.23(b) Dielectric WAIM sheets for scan matching (after McGill and Wheeler [65]): measured reflection of array element before impedance matching.

where t is the dielectric thickness, $B(0)$ is the broadside susceptance, G_0 is the free-space conductance, and ϵ_r is the relative dielectric constant.

The thin dielectric layer has the approximate scan dependence given for E- and H-plane scan planes.

$$H\text{-plane} \quad \frac{B(\theta)}{B(0)} = \frac{1}{\cos\theta} \tag{6.41}$$

$$E\text{-plane} \quad \frac{B(\theta)}{B(0)} = \cos\theta - \frac{\sin^2\theta}{k\cos\theta} \tag{6.42}$$

Figure 6.24 illustrates the wide-angle impedance matching procedure. Figure 6.24(a) repeats the array Smith chart data of Figure 6.23(b). Figure 6.24(b) shows that same data referenced forward (into space) in front of the array to an area near the unity conductance circle. The relative position of the points is changed

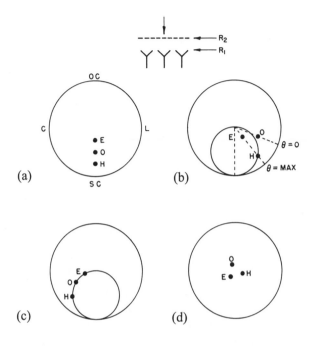

Figure 6.24 Calculation procedure for matching with thin high-k dielectric sheet (after McGill and Wheeler [65]). (a) Array at R_1; (b) array at R_2; (c) dielectric WAIM sheet; (d) array with sheet at R_2.

because the propagation constant $k_z = k \cos \theta$ (or the wavelength $\lambda_z = \lambda_0/\cos \theta$) is used for the Smith chart manipulations, so the relative shift in reference plane location is $t/\lambda_z = t \cos \theta/\lambda_0$ and is greater for the broadside case than the wide-angle cases. Since the dielectric slab reflection coefficient falls along the unit conductance curve (Figure 6.24(c)), the sum of its susceptance added to that of the array combines to produce scan-matched behavior approximately like that of 6.24(d).

Recent attempts at wide-angle impedance matching, especially for arrays of printed circuit elements, has primarily been accomplished by using reduced spacing to keep the grating lobe out of real space, and then by merely optimizing element dimensions and matching at broadside or some other chosen scan angle. Dielectric sheets are often used as a cover (radome) to protect the array face, and so the sheet dimensions, spacing, and dielectric constant are parameters included in the optimization. Electromagnetically coupled patches and multilayer patches are usually scan matched using the available degrees of freedom within the patch and cover geometry.

6.6 MUTUAL COUPLING PHENOMENA FOR NONPLANAR SURFACES

The electromagnetics of array mutual coupling is complex, even for planar arrays, but substantially more so for nonplanar arrays. In the planar case for an array without a dielectric substrate, the coupling is written in terms of the free-space potential functions (Section 2.1.1). However, if the array is nonplanar, there are relatively few configurations that can even be written in terms of closed form (or series) Green's functions. Arrays on cylinders can be rigorously formulated in terms of Bessel functions [66], and solutions are available for elements and arrays on spheres [67] and cones [68]. In each of these cases, there are problems of convergence that arise in obtaining far-field element patterns and describing element mutual coupling. These are treated in some detail in the text [66] by Borgiotti. For elements on a cylinder, the rigorous solutions offer exact analysis of array coupling, but the Bessel function series is very slowly convergent for large cylinders and must be transformed to asymptotic series for faster convergence. The resulting approximate results are not uniformly valid, and are usually written as separate expressions for the cylinder "lit" and "shadow" regions as seen from a far-field source.

Borgiotti applied the method of symmetrical components to obtain element patterns on arrays that were infinite in the axial plane and extended entirely around the cylinder. In this case, one can obtain exact element patterns in the circumfer-

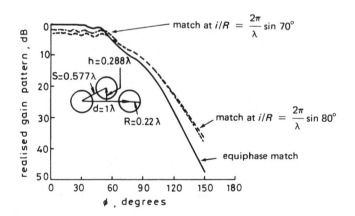

Figure 6.25 Element gain pattern, circumferential plane, and circumferential polarization ($R = 11.61\lambda$). Other dimensions shown (after Borgiotti [66]).

ential and axial planes. Figure 6.25 shows an example in which the array is matched at the equiphase condition and at angles nearer the endfire direction. These data show that one can obtain several decibels of increased gain near endfire by matching at angles nearer to endfire.

Detailed investigations into the phenomenology of element patterns on curved surfaces have revealed the role of creeping wave radiation in determining the array azimuthal element patterns. Creeping waves are known to contribute to the radiation of isolated apertures on conducting cylindrical surfaces. They propagate with nearly the free-space velocity and radiate along the local tangent. On smooth conducting surfaces, their radiation is significant only in the shadow region. On periodic cylindrical arrays with azimuthal spacing greater than $\lambda/2$, however, these play a significant role in the forward direction. This occurs because the creeping wave velocity is fast and radiates near the local tangent, but the waves also excite grating lobes that radiate back into the forward region element pattern. Figure 6.26 indicates this procedure for a cylinder with azimuthal element spacing greater than a half wavelength, and shows the angular locations of these grating lobes for the clockwise and counterclockwise creeping waves. The resulting element power pattern dips are proportional to $(kr_0)^{2/3}$. Figure 6.27 shows H-plane azimuthal element power patterns for an array of dipoles, as shown in Figure 6.27(a), completely covering a cylinder of various radii. Figure 6.27(b) compares the cylindrical array element pattern with one from a planar array with the same grid. The essential differences between the two curves are that the planar azimuth pattern is zero for $\Phi > 90$ deg, while the circular array pattern radiates into this shadow region, and

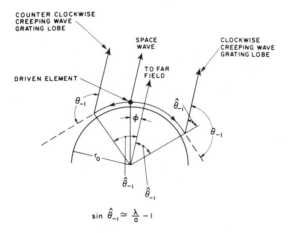

Figure 6.26 Wave contributions in forward (lit) region of active element patterns of circular array on a conducting cylinder (after Hessel [69]).

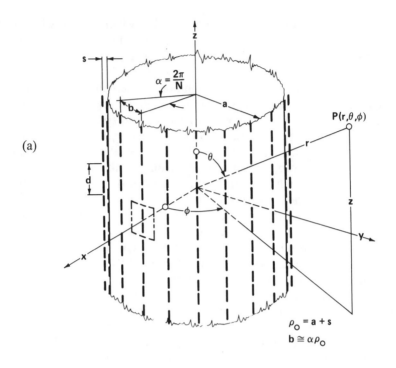

Figure 6.27 Periodic dipole array on a conducting circular cylinder (after Herper et. al. [70]). (a) Geometry.

the circular array has a rippled element pattern while the planar array element pattern is smooth. Both of these effects are caused by creeping waves. Figure 6.27(d) shows a logarithmic plot of the element pattern of cylindrical arrays with the same grid, but on cylinders with different radii. The slope of these curves in the shadow region is steeper for larger radii and is in fact proportional to $(ka \sin \theta)^{1/3}$ for θ, the elevation angle. The ripple shown in these curves near the 180-deg azimuth angle is due to the interaction of clockwise and counterclockwise creeping waves.

The need for more general solutions that apply to arbitrary concave and convex surfaces has been satisfied primarily by high-frequency asymptotic methods. Early studies [71,72] obtained approximate formulas for mutual impedance of slots on cylinders and cones. Studies of radiation from cylinders using high-frequency diffraction methods have led to excellent descriptions of element patterns and arrays. More recent studies of concave arrays by Tomasic and Hessell [73] have presented techniques for the analysis of arrays on generalized concave surfaces. Concave surfaces are often used in array feeds, seldom as radiating arrays them-

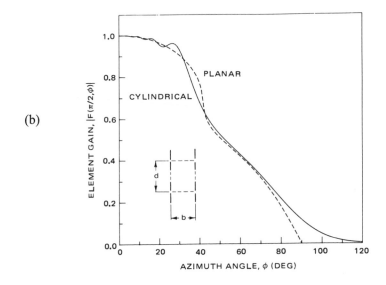

(b)

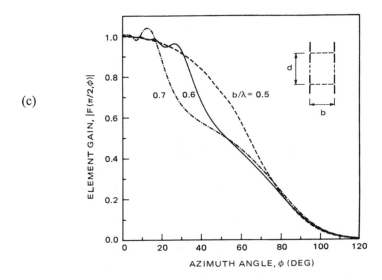

(c)

Figure 6.27 (cont.) Periodic dipole array on a conducting circular cylinder (after Herper et. al. [70]). (b) *H*-plane voltage element gain pattern for cylindrical and reference planar arrays (*b*/λ = 0.6; *d*/λ = 0.7; *ka* = 120); (c) *H*-plane voltage element gain pattern (*d*/λ = 0.7; *ka* = 120) for azimuth spacing *b*/λ = 0.5, 0.6, 0.7.

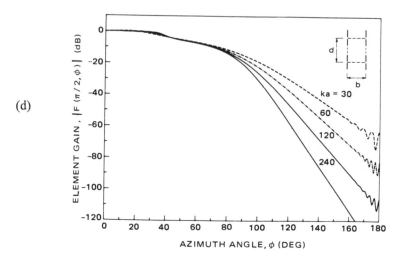

Figure 6.27 (cont.) Periodic dipole array on a conducting circular cylinder (after Herper et. al. [70]). (d) *H*-plane element gain power pattern (dB) ($b/\lambda = 0.6$, $d/\lambda = 0.7$) for cylinder radii *ka* = 30, 60, 120, 240.

selves. The analytical methods introduced in these mutual coupling analyses consisted of assuming the array to be locally periodic and using ray tracing methods to account for the mutual interaction of the elements. Several recent summaries of numerical techniques for solving electromagnetic problems highlight combined geometrical theory of diffraction and moment method solutions [74,75] that have direct and significant relevance to arrays on complex bodies.

6.7 SMALL ARRAYS AND WAVEGUIDE SIMULATORS FOR THE EVALUATION OF PHASED ARRAY SCAN BEHAVIOR

Because of the cost of building phased arrays, it is extremely important to obtain reliable measurements of the designed array scan properties before committing to a final design. The two primary methods of design verification are to build a small array and/or to construct a waveguide simulator.

Tests on a small array usually involve the measurement of radiating element patterns of all or at least certain of the array elements, one at a time, with all other elements terminated in matched loads. This provides a measurement of $[1 - |\Gamma|^2]$ $\cos \theta$ if the array is matched at broadside and so can provide an excellent indication of the array scanning characteristics. Figures 6.3, 6.5, 6.16, 6.17, and 6.20 show the relationship between element patterns in several small arrays and infinite array data. One of the most important factors in the use of small test arrays is to make sure that the array is large enough to indicate the occurrence of a blind spot, which can appear as a small dip in the element pattern if the array is too small.

The *waveguide simulator* simulates the performance of an infinite array using the natural imaging that takes place in a rectangular waveguide. By way of introduction, Figure 6.28(a) depicts a set of planar wavefronts for two waves with polarization in the plane perpendicular to the paper and traveling in directions ± 0 relative to the z-axis. The points of intersection between these waves are chosen to represent places where the electric fields of the two waves are equal and opposite. The locus of these points is a set of vertical lines (shown dashed) along which the net electric field is zero, and along this locus one could pass perfectly conducting metallic sheets (perpendicular to the paper) without disturbing the field. The dashed lines are shown ending at $z = 0$ to indicate that the array of sheets could be terminated at any point or continued on to infinity without changing the field distribution. The dashed lines represent an infinite set of finite-length, parallel-plate waveguides, each supporting waves that travel in the positive z-direction. Within any one of these waveguides, the field distribution is recognized as that corresponding to the TE_{10} mode, while the whole set of waveguides comprises an infinite array receiving waves from the ± 0 direction. Since the only electric field is in the y-direction, one could also pass conducting sheets perpendicular to the y-axis without disturbing the fields, and so instead of a parallel-plate simulator, it is more convenient to use a rectangular waveguide geometry.

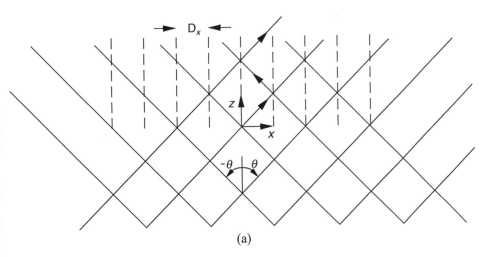

(a)

Figure 6.28 Waveguide simulator geometries. (a) Basic TE_{10} mode simulator.

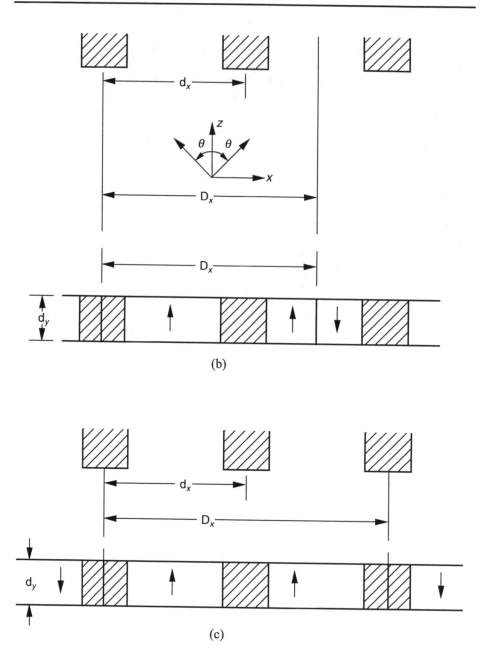

Figure 6.28 (cont.) Waveguide simulator geometries. (b) equivalent infinite array represented by simulator; (c) higher order TE$_{10}$ mode simulator.

Useful array simulators are also based on the simple principle introduced above, with the region representing free space modeled by a single (oversize) waveguide. If the wall locations of that waveguide are chosen to correspond to zero field points in the actual array, then the entire configuration can be simulated by the junction of the two waveguides. Array simulators have evolved from the original work of Brown and Carberry [76] and Hannon and Balfour [77].

The waveguide simulator gives one scan angle data point at each frequency, so one simulator is not adequate to test the wide-scan behavior of a given feed lattice. However, by comparing the results from a theoretical model of the infinite array with the simulator data, one can confirm the scan angle performance as a function of frequency and so uncover any frequency-dependent scan anomalies. Equally important, with a confirmed theoretical model, one can confidently investigate all scan angles. The simulator thus is often used as an adjunct to a theoretical solution.

The simplest simulator is the one shown in Figure 6.28(b), which simulates the H-plane scan of an infinite array of open ended waveguide elements with spacings d_x and d_y. The simulator is fed by the larger waveguide, with inner dimensions D_x and D_y, which propagates the incident fundamental TE_{10} mode traveling in the z-direction. The field in the y-direction is constant, and so the waveguides in Figure 6.28(b) are excited with constant phase and zero-thickness walls. The figure shows the horizontal walls that pertain to both sets of waveguides. For certain incident angles, the imaging of aperture fields by the waveguide wall simulates the remainder of the array, indicated in Figure 6.28(b). The TE_{10} incident fields as represented by the two waves are

$$
\begin{aligned}
E(x) &= \exp[-jk_z z + jkx \sin \theta] + \exp[-jk_z z - jkx \sin \theta] \\
&= 2 \exp[-jk_z z] \cos(kx \sin \theta)
\end{aligned}
\tag{6.43}
$$

Only certain incident angles θ satisfy the waveguide boundary conditions $E(x) = 0$ at $x = \pm D_x/2$, but for these angles the simulator exactly simulates the infinite phased array, whether on transmit or receive. For the case chosen, this angle occurs for $2(D_x/\lambda) \sin \theta = \pm 1$, or

$$
\sin \theta = \pm \frac{\lambda}{2D_x} = \pm \frac{\lambda}{3d_x}
\tag{6.44}
$$

since $D_x = 1.5d_x$ (Figure 6.28(b)).

Depending on the element spacing d_x/λ, the angle of the main beam radiation of the simulated infinite array can vary with frequency. A useful range is between about $\theta = 35$ deg for $d_x = 0.575\lambda$ to about 42 deg for half-wavelength element spacing.

For this type of simulator, the input impedance measured in the oversize feed (simulator) waveguide is the infinite array impedance measured from the free-space side, as seen from the angle θ. The array elements are waveguides, and one is a half-width guide, so the full waveguide element is terminated in a matched load or the generator impedance, and the half-width waveguide is beyond cutoff, so there is no need to terminate it if it is long enough. The simulator can also be used for current-carrying elements (dipoles, patches, etc.), provided proper care is taken to include the image of the feed lines.

Operation of this most fundamental simulator gives the impedance measured from free space. The impedance looking into the array element transmission lines has the same reflection coefficient magnitude, but not necessarily the same phase. For the present case, since there is only one full element in the simulator, one can excite the array element directly and measure its input impedance. In the other simulators discussed in the following paragraphs, which have several complete array elements, one needs to excite all elements with the proper phase relationship in order to measure the array impedance from the transmitter side. In all cases, however, one can excite the simulator from the free-space side using only a single incident mode (two waves) per polarization. Hannon and Balfour [77] give a detailed description of how this impedance "looking in" to the array from free space is then used to obtain the impedance "looking out" from an equivalent circuit and measurements "looking in" with elements terminated in two impedance states: a matched load and a short circuit. These two measurements allow the full determination of the array equivalent circuit.

The above is the widest angle single-mode simulator one can devise. Other simulators can give results nearer broadside, but the broadside angle itself is excluded, since it would require an infinite number of elements. In practice, one must limit the simulator width in order to avoid generating higher order modes in the feed guide.

The simulator of Figure 6.28(d) also uses TE_{10} excitation and again has one-half period (D_x) for the incident two-wave combination. In this case, $2D_x = 8d_x$, and so the associated angle θ is given by

$$\sin \theta = \frac{\lambda}{2D_x} = \frac{\lambda}{8d_x} \tag{6.45}$$

and is about 12.5 deg for $d_x = 0.575\lambda$ or about 14.5 deg for $d_x = \lambda/2$. This angle is very close to broadside, and since one cannot devise a simulator for broadside incidence, this is normally considered adequate for predicting broadside behavior.

E-plane scanning simulators have been built, and again the work of Hannon and Balfour [77] is cited. In order to simulate E-plane scanning, it is required that the higher order TM_{11} mode be used. This mode again can be represented by two plane waves, but in this case the waves are tilted with respect to the waveguide walls.

Several Useful Simulators

This section lists several other simulators, but many more are possible and some are listed in the literature. The following definitions are given relative to Figure 6.29. Cardinal (solid) and intercardinal (dashed) planes of scan are shown on the figure, which is given as a square grid. The planes shown are planes of symmetry for the grid and do not refer to the array polarization.

H-polarization and E-polarization refer here to the polarization in the plane of scan, irrespective of the array symmetries. E-polarization is when the electric field is parallel to the scan plane, and H-polarization is when the magnetic field is parallel to the plane of scan. Here, "H-polarization and an intercardinal scan plane" means that the array scans in an intercardinal plane relative to its square grid, with the electric field normal to the scan plane and the magnetic field in the scan plane.

Figure 6.29(b) shows five simulators designed for circular waveguide arrays. The notation C and IC at the top of the figure indicate cardinal and intercardinal planes of the scan simulated, while the notation H and E refer to the scan plane as indicated above. Although these simulators operate over a range of angles, the angles noted represent typical operating angles. The upper simulator simulates near broadside scan angle for H-polarization and cardinal scan plane. This configuration, like the two discussed earlier, is simulated by the simple TE_{10} mode in the rectangular simulator waveguide.

The remaining four simulators represent relatively wide scan angles. The two shown in the central row are the H-polarization, and are excited by the TE_{10} mode. The one at left was discussed earlier, while the one at right represents a scan in an intercardinal plane, with the electric field vertical (because the scan plane is horizontal in the sketch). The E-polarization cases in the bottom row require the TM_{11} mode incident in a square, oversize waveguide. The simulator at the left of the bottom row simulates E-plane polarization for scanning in a cardinal plane. The array sample is rotated 45 deg relative to the TE_{10} case, because the incident TM wave is composed of two plane waves with their plane of propagation in the square waveguide at 45 deg to the walls of the waveguide. The simulator at right, bottom row simulates intercardinal plane scan for E-polarization, and again the array sample is rotated 45 deg relative to the TE_{10}-fed waveguide above.

A number of other authors have listed useful simulators. Balfour [78] presents a list of one-port simulators for rectangular and triangular grid arrays, and Wheeler [79] surveys a wide variety of simulators and their use in array element design. Gustinsic [80] presented a fundamentally new simulator concept, called a *multielement waveguide simulator*, and showed that a single simulator containing $N \times M$

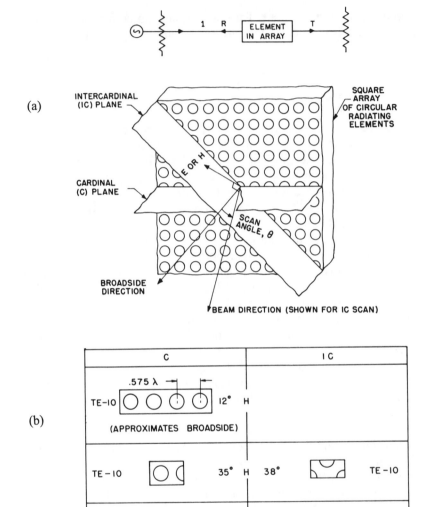

Figure 6.29 Simulator geometries and scan plane definitions (after Hannon and Balfour [77]): (a) cardinal and intercardinal planes; (b) a collection of five simulators for arrays of circular waveguides.

elements can be used to measure the reflection coefficient for an infinite array at $N \times M$ scan angles. The measurement involves the determination of the $N \times M$ transmit coefficients between one element and each of the other elements of the simulator, with the simulator waveguide itself terminated in a matched load. The simulator procedure is too detailed to describe here, but relies on the fact that any mode of the rectangular waveguide simulator can be considered to be composed of four plane waves within the simulator. Like more traditional simulators, which represent scan in only one plane, there is a single angle (θ, Φ) in space that fully describes the direction of the incident plane wave. In the simulators that only represent one plane of scan, the two plane waves emanate from angles (θ, Φ) and (θ, $\Phi + \pi$). In the present case for a two-dimensional array, there are planes of symmetry that represent waves from (θ, Φ) and three other directions: (θ, $\phi + \pi$), (θ, $-\phi$), and (θ, $\pi - \phi$).

Not all θ, ϕ incident angles can be simulated. Gustinsic's technique uses the fact that in the simulator waveguide, only $N \times M$ modes can propagate for simulator size $Nd_x \times Md_y$, with d_x and d_y both equal to $\lambda/2$. If the dimensions were larger, there would be at least one angle at which an extra mode could form the given mode excitation, and this would correspond to a grating one, which would then satisfy the simulator conditions, as it does in space.

Assuming there are no grating lobes, one could then excite a given simulator normal mode (m, n) by applying signals at the array ports that had the symmetry of the (n, m) mode. In each case, a direct measurement of the input impedance gives the infinite-array input impedance at that angle. Since this requires a very complex array element, however, Gustinsic applies signals consecutively to each input port while measuring the signals from all other ports. Superposition of the data exactly simulates the array results at all $N \times M$ angles. Derneryd and Gustinsic [81] further refined the procedure by developing an interpolation scheme to allow computation of active array impedance from multielement simulators.

REFERENCES

[1] Diamond, B. L., "Phased Array Radar Studies," Jan 63 to July 64, (Group 44 Rept. MIT Lincoln Lab., Tech. Rept. No. TR-381).

[2] Oliner A. A., and R. G. Malech, "Mutual Coupling in Infinite Scanning Arrays," Ch. 3 in *Microwave Scanning Antennas*, Vol. 2. R. C. Hansen, ed., New York: Academic Press, 1966.

[3] Wheeler, H. A., "Simple Relations Derived From a Phased-Array Antenna Made of an Infinite Current Sheet," *IEEE Trans.*, Vol. AP-13, No. 4, July 1965, pp. 506–514.

[4] Pozar, D. M., "General Relations for a Phased Array of Printed Antennas Derived From Infinite Current Sheets," *IEEE Trans.*, Vol. AP-33, No. 5, May 1985, pp. 498–503.

[5] Tai, C. T., "Dipoles and Monopoles," Ch. 4 in *Antenna Engineering Handbook*, R. C. Johnson and H. Jasik, eds., McGraw-Hill, 1984.

[6] Balanis, G. *Antenna Theory: Analysis and Design*, New York: Harper and Row, 1982.

[7] Stutzman, W. L., and G. A. Thiele, *Antenna Theory and Design*, New York: John Wiley and Sons, 1966.

[8] Pocklington, H. C., "Electrical Oscillations in Wire," *Cambridge Philosophical Society Proc.*, London, Vol. 9, 1897, pp. 324–332.

[9] Hallén E., "Theoretical Investigations Into the Transmitting and Receiving Qualities of Antennae," *Nova Acta Regiae, Soc. Sci. Upsaliencis*, Ser. IV, 11, No. 4, 1938, pp. 1–44.

[10] King, R. W. P., *The Theory of Linear Antennas*, Cambridge, MA: Harvard University Press, 1956.

[11] Harrington, R. F., *Field Computation by Moment Methods*, New York: Macmillan Co., 1968.

[12] Farrell, G. F., Jr., and D. H. Kuhn, "Mutual Coupling Effects of Triangular Grid Arrays by Modal Analysis," *IEEE Trans.*, Vol. AP-14, 1966, pp. 652–654.

[13] Farrell, G. F., Jr., and D. H. Kuhn, "Mutual Coupling Effects in Infinite Planar Arrays of Rectangular Waveguide Horns," *IEEE Trans.*, Vol. AP-16, 1968, pp. 405–414.

[14] Mittra, R., ed., *Computer Techniques for Electromagnetics*, New York: Pergamon Press, 1973, pp. 7–70.

[15] Richmond, J. H., "Digital Computer Solutions of the Rigorous Equations for Scattering Problems," *Proc. IEEE*, Vol. Aug. 1985, pp. 796–804.

[16] C. P. Wu, "Analysis of Finite Parallel-Plate Waveguide Arrays," *IEEE Trans.*, Vol. AP-18, No. 3, May 1970, pp. 328–334.

[17] Collin, R. E., *Field Theory of Guided Waves*, McGraw-Hill, 1960.

[18] Diamond, B. L., "A Generalized Approach to the Analysis of Infinite Planar Phased Arrays of Apertures," *Proc. IEEE*, Vol. 56, Nov. 1968, pp. 1837–1851.

[19] Amitay, N., C. P. Wu, and V. Galindo, "Methods of Phased Array Analysis," *Phased Array Antennas*, A. Oliner and G. H. Knittel, eds., Artech House, 1972, pp. 68–82.

[20] Amitay, N., and V. Galindo, "On Energy Conservation and the Method of Moments in Scattering Problems," *IEEE Trans.*, Vol. AP-17, Nov. 1969, pp. 722–729.

[21] Lee, S. W., W. R. Jones, and J. J. Campbell, "Convergence of Numerical Solutions of Iris-Type Discontinuity Problems," *IEEE Trans.*, Vol. MTT-19, 1971, pp. 528–536.

[22] Shuley, N. V., "Relative Convergence for Moment-Method Solutions of Integral Equations of the First Kind as Applied to Dichroic Problems," *Elec. Letters*, Vol. 21, No. 2, 31 Jan. 1985, pp. 95–97.

[23] Abramowitz, M., and I. Stegun, eds., *Handbook of Mathematical Functions With Formulas Graphs and Mathematical Tables*, Washington, D.C., U.S. Government Printing Office, June 1964.

[24] Jordan, K. E., G. R. Richter, and P. Sheng, "An Efficient Numerical Evaluation of the Green's Function for the Helmholtz Operator on Periodic Structures," *J. Comp. Phys.*, Vol. 63, 1986, pp. 222–235.

[25] Cohen, E., "Critical Distance for Grating Lobe Series," *IEEE Trans.*, Vol. AP-39, No. 5, May 1991, pp. 677–679.

[26] Jorgenson, R. E., and R. Mittra, "Efficient Calculation of the Free Space Periodic Green's Function," *IEEE Trans.*, Vol. AP-38, May 1990, pp. 633–642.

[27] Richards, W. F., K. McInturff, and P. S. Simon, "An Efficient Technique for Computing the Potential Green's Functions for a Thin, Periodically Excited Parallel-Plate Waveguide Bounded by Electric and Magnetic Walls," *IEEE Trans. Microwave Theory and Tech.*, Vol. MTT-35, No. 3, March 1987, pp. 276–281.

[28] Steyskal, H. J., "Mutual Coupling Analysis of a Finite Planar Waveguide Array," *IEEE Trans.*, Vol. AP-22, No. 4, July 1974, pp. 594–597.

[29] Ehlenberger, A. G., L. Schwartzman, and L. Topper, "Design Criteria for Linearly Polarized Waveguide Arrays," *IEEE Proc.*, Vol. 56, DATE, pp. 1861–1872.

[30] Diamond, B. L., "Resonance Phenomena in Waveguide Arrays," *IEEE G-AP Int. Symp. Dig.*, 1967, pp. 110, 111.

[31] Borgiotti, G. V., "Modal Analysis of Periodic Planar Phased Arrays of Apertures," *IEEE Proc.*, Vol. 56, 1968, pp. 1881–1892.

[32] Knittel, G. H., A. Hessel, and A. A. Oliner, "Element Pattern Nulls in Phased Arrays and Their Relation to Guided Waves," *Proc. IEEE*, Vol. 56, 1968, pp. 1822–1836.

[33] Wu, C. P., and V. Galindo, "Surface Wave Effects on Dielectric Sheathed Phased Arrays of Rectangular Waveguide," *Bell Syst. Tech. J.*, Vol. 47, 1968, pp. 117–142.

[34] Byron, E. V., and J. Frank, "On the Correlation Between Wide-Band Arrays and Waveguide Simulators," *IEEE Trans.*, Vol. AP-16, No. 5, Sept. 1968, pp. 601–603.

[35] Knittel, G. H., "The Choice of Unit Cell Size for a Waveguide Phased Array Element and Its Relation to the Blindness Phenomenon," Boston Chapter AP-S, 1970.

[36] Chang, V. W. H., "Infinite Phased Dipole Array," *Proc. IEEE*, Vol. 56, No. 11, Nov. 1968, pp. 1892–1900.

[37] Mayer, E. D., and A. Hessel, "Feed Region Modes in Dipole Phased Array," *IEEE Trans.*, Vol. AP-30, Jan. 1982, pp. 66–75.

[38] Frazita, R. F., "Surface Wave Behavior of a Phased Array Analyzed by Grating Lobe Series," *IEEE Trans.*, Vol. AP-15, Nov. 1967, pp. 822–824.

[39] Pozar, D. M., and D. H. Schaubert, "Scan Blindness in Infinite Arrays of Printed Dipoles," *IEEE Trans.*, Vol. AP-32, No. 6, June 1984, pp. 602–610.

[40] Mailloux, R. J., "On the Use of Metallic Cavities in Printed Slot Arrays With Dielectric Substrates," *IEEE Trans.*, Vol. AP-35, No. 5, May 1987, pp. 477–487.

[41] Wang, S. S., and A. Hessel, "Aperture Performance of a Double-Ridge Rectangular Waveguide in a Phased Array," *IEEE Trans.*, Vol. AP-26, March 1978, pp. 204–214.

[42] Lewis, L. R., A. Hessel, and G. H. Knittel, "Performance of a Protruding Dielectric Waveguide Element in a Phased Array," *IEEE Trans.*, Vol. AP-20, 1972, pp. 712–722.

[43] Mailloux, R. J., and H. Steyskal, "Analysis of a Dual Frequency Array Technique," *IEEE Trans.*, Vol. AP-27, No. 2, March 1979, pp. 130–134.

[44] Hsiao, J. K., "Analysis of Interleaved Arrays of Waveguide Elements," *IEEE Trans.*, Vol. AP-19, Nov. 1971, pp. 729–735.

[45] Hsiao, J. K., "Computer Aided Impedance Matching of an Interleaved Waveguide Phased Array," *IEEE Trans.*, Vol. AP-20, July 1972, pp. 505–506.

[46] Boyns, J. E., J. H. Provincher, "Experimental Results of a Multifrequency Array Antenna," *IEEE Trans.*, Vol. AP-20, Sept. 1972, pp. 589–595.

[47] Lee, S. W., and W. R. Jones, "On the Suppression of Radiation Nulls and Broadband Impedance Matching of Rectangular Waveguide Phased Arrays," *IEEE Trans.*, Vol. AP-19, Jan. 1971, pp. 41–51.

[48] Mailloux, R. J., "Surface Waves and Anomalous Wave Radiation Nulls in Phased Arrays of TEM Waveguides With Fences," *IEEE Trans.*, Vol. AP-16, No. 1, Jan. 1972, pp. 160–166.

[49] Dufort, E. C., "Design of Corrugated Plates for Phased Array Matching," *IEEE Trans.*, Vol. AP-16, No. 1, Jan. 1968, pp. 37–46.

[50] Lewis, L. R., M. Fassett, and J. Hunt, "A Broadband Stripline Array Element," *IEEE AP-S Int. Symp. Record*, 1974.

[51] Aberle, J. T., and D. M. Pozar, "Analysis of Infinite Arrays of One- and Two-Probe Fed Circular Patches," *IEEE Trans.*, Vol. AP-38, No. 4, April 1990, pp. 421–432.

[52] Schuss, J. J., "Numerical Design of Patch Radiator Arrays," *Electromagnetics*, Vol. 11, Jan. 1991, pp. 47–68.

[53] Liu, C. C., J. Shmoys, and A. Hessel, "E-Plane Performance Tradeoffs in Two-Dimensional Microstrip Patch Element Phased Arrays," *IEEE Trans.*, Vol. AP-30, 1982, pp. 1201–0000.

[54] Herd, J. S., *Scanning Impedance of Proximity Coupled Rectangular Microstrip Antenna Arrays*, Ph.D. Thesis, Univ. of MA, 1989.

[55] Herd, J. S., "Full Wave Analysis of Proximity Coupled Rectangular Microstrip Antenna Arrays," *Electromagnetics*, Jan. 1992.

[56] Knittel, G. H., "Wide Angle Impedance Matching of Phased Array Antennas: A Survey of Theory and Practice," *Phased Array Antennas*, Dedham, MA: Artech House, A. A. Oliner and G. H. Knittel, eds., 1972.

[57] Cook, J. S., and R. G. Pecina, "Compensation of Coupling Between Elements in Array Antennas," *1963 IEEE Ant. and Propagat. Int. Symp. Dig.*, pp. 234–236.

[58] Hannon, P. W., D. S. Lerner, and G. H. Knittel, " Impedance Matching a Phased Array Antenna Over Wide Scan Angles by Connecting Circuits," *IEEE Trans.*, Vol. AP-13, No. 1, Jan. 1965, pp. 28–34.

[59] Amitay, N., "Improvement of Planar Array Match by Compensation Through Contiguous Element Coupling," *IEEE Trans.*, Vol. AP-14, No. 5., Sept. 1966, pp. 580–586.

[60] Hannon, P. W., "Proof That a Phased Array Antenna Can Be Impedance Matched for All Scan Angles," *Radio Science*, Vol. 2, New Series, No. 3, March 1967, pp. 361–369.

[61] Tang, R., and N. W. Wong, "Multimode Phased Array Element for Wide Scan Angle Impedance Matching," *Proc. IEEE* (Special Issue on Electronic Scanning), Vol. 56, No. 11, 1968, pp. 1951–1959.

[62] Munk, B. A., T. W. Kornbau, and R. D. Fulton, "Scan Independent Phased Arrays," *Radio Science*, Vol. 14, No. 6, Nov., Dec. 1979, pp. 978–990.

[63] Edelberg, S., and A. A. Oliner, "Mutual Coupling Effects in Large Antenna Arrays 2: Compensation Effects," *IRE Trans.*, Vol. AP-8, No. 4, July 1960, pp. 360–367.

[64] Hessel, A., and G. H. Knittel, "A Loaded Ground Plane for the Elimination of Blindness in a Phased Array Antenna, *1969 IEEE AP-S Int. Symp. Dig.*, pp. 163–169.

[65] Magill, E. G., and H. A. Wheeler, "Wide Angle Impedance Matching of a Planar Array Antenna by a Dielectric Sheet," *IEEE Trans.*, Vol. AP-14, No. 1, Jan. 1966, pp. 49–53.

[66] Borgiotti, G. V., "Conformal Arrays," Ch. 11 in *The Handbook of Antenna Design*, Vol. 2, Rudge, Milne, Olver, and Knight, eds., Peter Peregrinus, 1983.

[67] Hessel, A., Y. L. Liu, and J. Shmoys, "Mutual Admittance Between Circular Apertures on a Large Conducting Sphere," *Radio Science*, Vol. 14, 1979, pp. 35–42.

[68] Balzano, Q., and T. B. Dowling, "Mutual Coupling Analysis of Arrays of Aperture on Cones," *IEEE Trans.*, Vol. AP-22, Jan. 1974, pp. 92–97.

[69] Hessel, A., "Mutual Coupling Effects in Circular Arrays on Cylindrical Surfaces—Aperture Design Implications and Analysis," *Phased Array Antennas*, A. Oliner and G. H. Knittel, eds., Artech House, 1972, pp. 273–291.

[70] Herper, J. C., A. Hessel, and B. Tomasic, "Element Pattern of an Axial Dipole in a Cylindrical Phased Array—Part I: Theory, Part II: Element Design and Experiments," *IEEE Trans.*, Vol. AP-33, March 1985, pp. 259–278.

[71] Golden, K. E., et al., "Approximation Techniques for the Mutual Admittance of Slot Antennas in Metallic Cones," *IEEE Trans.*, Vol. AP-22, Jan. 1974, pp. 44–48.

[72] Steyskal, H., "Analysis of Circular Waveguide Arrays on Cylinders," *IEEE Trans.*, Vol. AP-25, 1977, pp. 610–616.

[73] Tomasic, B., and A. Hessel, "Periodic Structure Ray Method for Analysis of Coupling Coefficients in Large Concave Arrays—Part I: Theory, Part II: Application," *IEEE Trans.*, Vol. AP-37, Nov. 1989, pp. 377–397.

[74] Pathak, P. H., "High Frequency Techniques for Antenna Analysis," *IEEE Proc.*, Vol. 80, No. 1, Jan. 1992, pp. 44–65.

[75] Thiele, G. A., "Overview of Selected Hybrid Methods in Radiating System Analysis," *IEEE Proc.*, Vol. 80, No. 1, Jan. 1992.

[76] Brown, C. R., and T. F. Carberry, "A Technique to Simulate the Self and Mutual Impedance of an Array," *IEEE Trans.*, Vol. AP-11, May 1963, pp. 377–378.

[77] Hannon, P. W., and M. A. Balfour, "Simulation of a Phased Array Antenna in a Waveguide," *IEEE Trans.*, Vol. AP-13, May 1965, pp. 342–353.

[78] Balfour, M. A., "Active Impedance of a Phased Array Antenna Element Simulated by a Single Element in a Waveguide," *IEEE Trans.*, Vol. AP-15, No. 2, March 1967, pp. 313–314.

[79] Wheeler, H. A., "A Survey of the Simulator Technique for Designing a Radiating Element," *Array Antennas*, A. Oliner and G. H. Knittel, eds., *Proc. 1970 Phased Array Antenna Symp.*, Dedham, MA: Artech House.

[80] Gustinsic, J. J., "The Determination of Active Array Impedance With Multi-Element Simulators," *IEEE Trans.*, Vol. AP-20, No. 5, Sept. 1972, pp. 589–595.

[81] Derneryd, A. G., and J. J. Gustinsic, "The Interpolation of General Active Array Impedance From Multielement Simulators," *IEEE Trans.*, Vol. AP-27, No. 1, Jan. 1979, pp. 68–71.

Chapter 7
Array Error Effects

7.1 INTRODUCTION

A variety of errors, both random and spatially correlated, are introduced across the array by imperfect components and signal distribution networks, and these reduce the precision of the array excitation. An array illumination, designed to produce very low sidelobes without errors, may result in only modest sidelobes in the presence of phase and amplitude errors. If the errors are due to tolerance limits on the individual devices, it is usually possible to ensure that the errors have zero mean, at least at the array center frequency of operation. For example, an error in the power divider network that results in a progressively increasing phase error across the array can often be compensated for by measuring the error and resetting the phase shifters to correct for the power divider error. If, however, the power divider error is due to transmission line length errors, then the phase shifter correction will only compensate at center frequency. More serious yet, if errors are correlated from element to element or across large sections of the array, then the resulting radiation pattern can have large, distinct sidelobes.

Usually, it is the intent of the designer to ensure that all correlated errors are removed, so that all that remains are the residual, uncorrelated phase and amplitude errors limited by the ultimate precision of the components. The remaining errors are treated as random, and the residual (average) sidelobe errors, peak sidelobe expectation, gain degradation, and beam pointing error are estimated by statistical procedures. Results of this type are summarized in the next section.

In addition to random phase and amplitude errors, there are several types of highly correlated errors that are vitally important in array design because they result in high peak sidelobes. Examples treated here include the periodic phase or amplitude errors caused by discrete phase shifters, quantized amplitude tapers across the array, and the frequency-dependent phase errors due to contiguous wide-band subarrays with time delay at the subarray level.

7.2 EFFECTS OF RANDOM AMPLITUDE AND PHASE ERRORS IN PERIODIC ARRAYS

The increased sidelobe level, pointing error, and directivity decrease due to random array errors has been extensively documented in the literature. Early studies of Ruze [1,2], Elliott [3], Allen [4], and others [5,6] obtained average values of these parameters and statistical estimates of the peak sidelobe level at points within the pattern. In addition, more recent work by Hsiao [7,8] and Kaplan [9] have given convenient curves of peak sidelobe probability as a function of array parameters.

In the results to follow, the array is assumed to have an amplitude error δ_n and phase error Φ_n at the nth element. The meaning of the amplitude error δ_n is that the signal at the nth element has amplitude $(1 + \delta_n)A_n$, where A_n is the correct amplitude. The meaning of the phase error Φ_n is that the correct phase to steer a beam to the chosen angle is not the correct excitation, but $\exp(j\Phi_n)$ times the correct excitation. In addition, the array has a number of totally failed (zero amplitude signal) elements randomly located throughout the array. The failed elements are modeled by assuming a fixed probability P that any nth element is operating properly except for amplitude and phase errors, so that the probability of that element being completely failed (having zero amplitude) is $(1 - P)$.

The occurrence of "failed" elements of the type included here is primarily limited to active arrays, where a failed amplifier may have zero output, or to thinned arrays where elements are removed from randomly chosen locations. Thinned arrays are discussed in Chapter 3. The most common kind of discrete failure for passive arrays is for a phase shifter to have a failed bit. This and other kinds of discrete failure are not specifically modeled here, but if these occur randomly they are included in the phase error variance.

The following treatment includes only the most common types of errors. Other errors, such as element position errors and polarization errors, have been treated in the literature [10,11].

Including phase and amplitude errors as indicated above, the far-field array factor is given by

$$F(\theta, \phi) = \sum p(n)A_n(1 + \delta_n) \exp[jk(\hat{\boldsymbol{\rho}} \cdot \mathbf{r} - \hat{\boldsymbol{\rho}}_0 \cdot \mathbf{r}_n')] \exp(j\Phi_n) \qquad (7.1)$$

where

$$\hat{\boldsymbol{\rho}} = \hat{\mathbf{x}}u + \hat{\mathbf{y}}v + \hat{\mathbf{z}} \cos \theta \qquad \hat{\boldsymbol{\rho}}_0 = \hat{\mathbf{x}}u_0 + \hat{\mathbf{y}}v_0 + \hat{\mathbf{z}} \cos \theta_0$$

and

$$\mathbf{r} = \hat{\mathbf{x}}x + \hat{\mathbf{y}}y + \hat{\mathbf{z}}z \qquad \mathbf{r}_n' = \hat{\mathbf{x}}x_n + \hat{\mathbf{y}}y_n$$

In this representation, the factor $p(n)$ accounts for the failed elements by randomly setting $p(n) = 1$ with probability P, and zero with probability $(1 - P)$. The summation Σ is written as one-dimensional for convenience. It can be taken as two-dimensional with no change to the analysis or results, because all of the averages taken are ensemble averages over a number of statistically equal arrays, not spatial averages. The element pattern is removed from the above, since it is assumed to be the same for all elements and plays no part in the statistical process. It can be included to modify the results at a later stage if desired.

7.2.1 Average Pattern Characteristics

The following development is an outline only because the details are included in the chapter by Skolnik of the text edited by Collin and Zucker [5] and elsewhere [6–8]. In all cases, it is assumed that all correlated errors have been removed and only random errors remain. The average pattern characteristics that are the results of the analysis do not pertain to any one antenna, but describe the observed results averaged over a large number of arrays that have the same statistical phase and amplitude errors.

In the treatment by Skolnik, it is assumed that the phase error Φ_n is described by a Gaussian probability density function with zero mean and variance $\overline{\Phi^2}$. The amplitude errors δ_n have variance $\overline{\delta^2}$ and zero mean, and the failed elements are randomly distributed as noted earlier. Under these conditions, Skolnik [5] shows the average power pattern to be

$$\overline{|F(\theta, \phi)|^2} = P^2 \exp(-\overline{\Phi^2})|F_0(\theta, \phi)|^2 + [(1 + \overline{\delta^2})P - P^2 \exp(-\overline{\Phi^2})] \sum A_n^2 \quad (7.2)$$

This expression shows that the effect of random errors produces a radiation pattern consisting of the ideal pattern $|F_0(\theta, \phi)|^2$ reduced by factors that account for failed elements and phase error, plus another term that is a constant with no angular dependence.

It is convenient to normalize the above to the peak of the resulting pattern, which is $P^2 \exp(-\overline{\Phi^2})|F_0(\theta, \phi)|_{max}$. The result is

$$\overline{|F_N(\theta, \phi)|^2} = \overline{|F_{on}(\theta, \phi)|^2} + \{(1 - P) + \overline{\Phi^2} + \overline{\delta^2}\} \frac{1}{Pg_A}$$

where

$$g_A = \frac{(\sum A_n)^2}{\sum A_n^2} = N\epsilon_T \quad (7.3)$$

is the directivity of the ideal pattern with isotropic element patterns, or N times the array taper efficiency ϵ_T (defined in Chapter 2), and F_{on} is the F_0 normalized as noted above.

In this form, the normalized sidelobe level $\overline{\sigma^2}$ is given by

$$\overline{\sigma^2} = \frac{\overline{\epsilon^2}}{Pg_A} \tag{7.4}$$

with $\overline{\epsilon^2}$ the error variance given by

$$\overline{\epsilon^2} = \{(1 - P) + \overline{\Phi^2} + \overline{\delta^2}\} \tag{7.5}$$

The average sidelobe level $\overline{\sigma^2}$ is sometimes called the *residual sidelobe level*. It is here normalized to the beam peak, but can be normalized to the isotropic level by recognizing that the ideal or design array directivity is g_A multiplied by the element pattern directivity (see Chapter 2).

$$D_A = g_e g_A \tag{7.6}$$

Recall that for a two-dimensional array of $\lambda/2$-spaced elements, the element gain (directivity) is

$$g_e = \pi \tag{7.7}$$

Multiplying the residual sidelobe level by D_A normalizes that level to the isotropic radiation level. Figure 7.1 [11] shows this residual sidelobe level for an array with no failed elements ($P = 1$). In that case, the sidelobe level relative to the isotropic level is

$$\overline{\sigma_I^2} = \overline{\sigma^2} D_A = g_e \overline{\epsilon^2} = g_e(\overline{\Phi^2} + \overline{\delta^2}) \tag{7.8}$$

which is a circle. This result pertains to a one-dimensional array, where it describes average sidelobes in the plane including the array axis, and to a two-dimensional array, where it describes radiation in all space (and here $g_e = \pi$ for $\lambda/2$ spacing). It is often convenient to evaluate the total error variance and then apportion the relative phase and amplitude error variances according to which is easier to control in the design.

The symmetrical form in which errors enter the above equation suggests the convenience of converting the amplitude error (expressed as a ratio) to an equivalent phase error (in radians or degrees). This aid to perspective helps in the tradeoff between amplitude and phase error to determine how much of the error variance

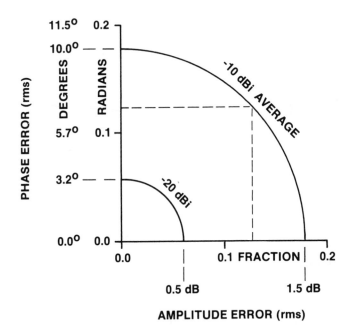

Figure 7.1 Array average (residual) sidelobes (relative to isotropic radiation) due to phase and amplitude error. Element gain π assumed (after [11]).

to allot between the two. Using an expansion of the logarithm valid for small δ, the amplitude error (in decibels) is

$$\delta_{dB} = 20 \log_{10}(1 + \delta) \sim 8.68\delta \qquad (7.9)$$

and the equivalent phase error (degrees) is

$$\Phi_\delta(\deg) = 6.6\delta_{dB} \qquad (7.10)$$

As an example, a 0.5-dB average (or rms) error is thus roughly equivalent to a 3.3-deg rms phase error.

Since the residual sidelobe level, when normalized to the array factor gain, is independent of array size, the foregoing equations point out that for any given array variance, increasing the array size lowers the actual value $\overline{\sigma^2}$ of the residual sidelobes.

Figure 7.2 shows the residual sidelobe level as a function of array directivity normalized to the beam peak for arrays with phase error only. The solid lines

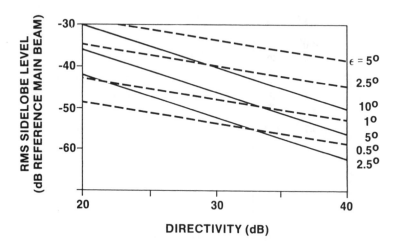

Figure 7.2 Array average sidelobes due to phase error (relative to beam peak). Element gain π assumed. Solid line for two-dimensional scan, dashed line for array of columns (after [12]).

pertain to the residual sidelobes of a two-dimensional array of $\lambda/2$-spaced elements. For comparison, the dashed lines of Figure 7.2 are the residual sidelobes in the principal plane of a linear array of columns, having the same square aperture as the array represented by the solid lines. In the column array case, it is assumed that each column subarray has no phase error, but the lines and phase shifters behind each column have randomly distributed errors. This arrangement reduces the number of degrees of freedom and makes the phase tolerance far more critical. For the array of columns, the sidelobe level of the arrayed phase errors is that of a one-dimensional array and given by (7.6) with g_e the column directivity g_c. If the array is square and the spacing $\lambda/2$ in each plane, the column array factor directivity is given approximately below.

$$g_c = (D_A/\pi)^{1/2} \tag{7.11}$$

Relative to the beam peak, the average sidelobe level is

$$\overline{\sigma^2} = \frac{(\pi)^{1/2}\overline{\epsilon^2}}{D_A^{1/2}P} \tag{7.12}$$

Figure 7.2 demonstrates the requirement for increased precision in the column array case because of the loss of degrees of freedom. The figure also emphasizes the tradeoff between array size and required tolerance in order to meet the desired residual error. For example, residual errors of -50 dB are achievable with an array of 30-dB directivity with about 0.7-deg rms phase errors for a column array, but for a planar array the errors can be 3.5 deg. If the gain were 40 dB, these conditions can be met with about 1.5 and 10 deg for the column and planar arrays.

7.2.2 Directivity

The reduction in directivity due to these residual errors is given by Skolnik [5] as approximately

$$\frac{D}{D_0} = \frac{P}{(1 + \overline{\delta^2}) \exp(-\overline{\Phi^2})} \doteq \frac{P}{1 + \overline{\delta^2} + \overline{\Phi^2}} \tag{7.13}$$

where D is the directivity of the array with errors and D_0 is the directivity of the error-free array. The reduction in directivity is not a function of array size, only of error variance.

Reduced directivity due to excitation errors is not often the driving concern for most array systems because unless the array is quite large, sidelobe distortion becomes severe before there is any major directivity reduction. For example, an rms phase error of 15 deg leads to only a 0.3-dB loss in directivity, while (7.4) shows that this results in an rms sidelobe level of only 11.6 dB below the array factor isotropic level; that is, -31.6 dB below the beam peak for an array factor directivity g_A of 100 (20 dB), or -41.6 dB below beam peak for g_A of 1,000 (30 dB).

7.2.3 Beam Pointing Error

Several authors [13,14] have looked at the issue of beam pointing error due to array phase and amplitude error. Steinberg [14] shows that for a symmetrical array excitation, the variance of beam pointing deviation is given by

$$\overline{\Delta^2} = \overline{\Phi^2} \frac{\Sigma \, I_i^2 x_i^2}{(\Sigma \, I_i x_i^2)^2} \tag{7.14}$$

where I_i is the amplitude of ith element excitation; x_i is the element position divided by interelement spacing d; and $\overline{\Phi^2}$ is the phase error variance. For an array of N elements with uniform amplitude ($I_i = 1$),

$$\overline{\Delta^2} = \frac{12}{N^3} \overline{\Phi^2} \tag{7.15}$$

7.2.4 Peak Sidelobes

It is often important to know the peak sidelobes associated with errored phase and amplitude. Detailed considerations of the peak sidelobe behavior are based on the statistics of the error sidelobes. It can be shown [1,2,4,7,8] that at any angle the amplitude $F(\theta, \phi)$ of the far-field pattern of an ensemble of arrays with the same statistics is given according to:

$$p(F) = (2F/\overline{\sigma^2})I_0[2FF_0/\overline{\sigma^2}] \exp[-(F^2 + F_0^2)/\overline{\sigma^2}] \tag{7.16}$$

where

$\overline{\sigma^2}$ = the variance of an ensemble of array sidelobes (sometimes called the *residual* or *average sidelobe level*);

F = the value of the ensemble pattern, including the design (ideal) pattern and the average or residual pattern;

F_0 = the design (ideal) pattern level at some given angle; and

I_0 = the modified Bessel function.

In this expression, $p(F)$ is the probability that at any angle the field intensity will be between F and $F + dF$. The pattern value F is composed of an ideal pattern (design pattern) with value F_0 (at that point in space) and an average or residual pattern of rms value σ. This type of distribution is often called a *Ricean distribution* [15].

For small errors, or where the design pattern level is relatively large compared to the statistical error pattern, as in the main beam or first sidelobe region of some patterns, $F_0^2 \gg \overline{\sigma^2}$ and the distribution becomes the Gaussian probability function. However, when the errors are large compared to the errorless pattern, as in a low-sidelobe or nulled region, the design pattern contribution is neglected, and the above becomes the Rayleigh density function.

$$p(F) = \frac{2F}{\overline{\sigma^2}} \exp(-F^2/\overline{\sigma^2}) \tag{7.17}$$

An important statistical parameter relating to peak sidelobes is the cumulative probability, which expresses the likelihood that the field intensity F at any point will be less than any given value S, or that the field intensity F will exceed the value S. These parameters are

$$\text{prob}(F \le S) = \int_{F=0}^{S} p(F/\sigma)dF \qquad \text{prob}(F \ge S) = \int_{S}^{\infty} p(F/\sigma)dF \tag{7.18}$$

In the region of the pattern where the statistical contribution dominates, as in the nulled areas described above, (7.17) can be used to readily compute the cumulative probability that the field intensity exceeds the value S. This gives

$$p = \text{prob}(F \geq S) \int_S^\infty p(F)dF = \exp(-S^2/\overline{\sigma^2}) \qquad (7.19)$$

which says, for example, that there is a 1% probability that the residual sidelobe level will be exceeded by more than a factor of 4.6, or 0.0001 that it will be exceeded by more than the factor 9.2.

Rewriting (7.19) to solve for the error yields

$$\overline{\epsilon^2} = -S^2 g_A/\ln(p) \qquad (7.20)$$

which is a convenient form because it again emphasizes the relationship between the residual sidelobe level and the array factor isotropic directivity level $1/g_A$ below beam peak. If the array has phase error only, then to hold all sidelobes a factor of 100 (20 dB) below the array factor isotropic level with probability 0.01 requires an rms phase error of about 2.6 deg. Holding that sidelobe level to a probability of 0.0001 requires phase error of only 1.9 deg.

These numbers are optimistic. They give the probability of exceeding (or not exceeding) a given sidelobe level at a particular point where the deterministic part of the pattern (the design or ideal pattern) is very small or null. A more realistic assessment of the likelihood of having a large sidelobe is obtained from the cumulative probability of (7.16), which accounts for higher sidelobe areas of the deterministic pattern.

Several recent papers [7–9] give peak sidelobe probability curves. Since we will use Hsiao's results, we introduce his terminology (although with a change of notation). The *designed sidelobe* level is the term used previously, and is written F_0. The *desired sidelobe* level S_d is that value not to be exceeded within a certain probability (the S of the previous section). Hsiao's results, shown in Figure 7.3, relate the value of a parameter

$$(\overline{\delta^2} + \overline{\Phi^2})/E \qquad (7.21)$$

where

$$E = \frac{2S_d(\Sigma A_{mn})^2}{\Sigma A_{mn}^2}$$

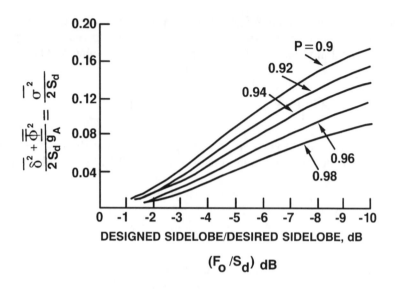

Figure 7.3 Normalized array error, or normalized residual sidelobe level, versus designed to desired sidelobe level ratio (after [7]).

Hsiao shows that the parameter E can be written in terms of the array factor gain as

$$E = 2S_d g_A \tag{7.22}$$

Using (7.4), the ordinate of Figure 7.3 can therefore be written in a form that explicitly includes the residual sidelobe level as (for $p = 1$)

$$(\overline{\delta^2} + \overline{\Phi^2})/E = \frac{1}{2}\frac{\overline{\sigma^2}}{S_d} \tag{7.23}$$

Figure 7.3 [8] gives the required average (residual) sidelobe level as a function of the designed sidelobe level, with both values normalized to the desired (peak) sidelobe level. The figure shows that at any point in space, by increasing the ratio of desired sidelobe to designed sidelobe (moving to the right in the figure), one can increase the ratio of allowed average sidelobes to desired sidelobe level. Thus, for a given desired sidelobe level at one point in space and for some given probability line, one can relax tolerance (and thus allow residual sidelobes to rise) by requiring a larger ratio of desired-to-design sidelobes. If an array were designed with sidelobes within a decibel or so of the required level, the necessary residual level and mean square error would need to be extremely small. However, by designing the array for sidelobe levels 5 to 10 dB below the required levels, one can significantly relax the required tolerance, while allowing the average (residual) sidelobes to rise. This process of over-designing the array to relax tolerance is a well-known procedure, and the figure gives the required data to facilitate this tradeoff.

The results cited above pertain to the probable peak sidelobe level at any point in pattern space. Several authors have gone beyond this to estimate the number of probable times the specification level is exceeded in all of the pattern space. Allen [4] related this likelihood to the number N of pattern beamwidths within the region. He argued that if the pattern had N sidelobes, then the likelihood of one exceeding the threshold is N times that for a single point, so a 100-element array with 100 sidelobes would need 100 times lower probability p of exceeding the threshold $(1 - p)$ at a particular point. Thus, for a 100-element array to have a 99% probability of not exceeding the given threshold anywhere in the pattern, the probability p of exceeding at a single point needs to be $(1 - 0.99)/100$ or 0.0001. Using this number in (7.20) for the case of holding all pattern sidelobes -20 dB below the isotropic array gain factor with probability 0.99 leads to the required phase error $\epsilon_{deg} = 1.88$ deg.

Kaplan [9] also estimated the likelihood of exceeding the threshold by estimating the number of likely "pop-ups" or points in pattern space where the threshold (in this case, the specified-to-residual ratio) is exceeded. Again using the number N as the number of beamwidths in some particular region of pattern space, and under the assumption of some ideal pattern with nearly equal sidelobe levels within the chosen region, Kaplan uses the following binomial expression for the probability of exceeding the threshold a given number of times (k):

$$\text{prob(lobes exceeding threshold} \leq k) = \sum_{n=0}^{k} C_n^N P_0^{N-n}(1 - P_0)^n \quad (7.24)$$

where $C_n^N = N!/n!(N - n)!$

For example, if the number of sidelobes (N) in some given region is 10, and the probability P_0 of exceeding the threshold at *any point* is 0.9, then the probability of exceeding the threshold k times within the region is given from the above (see Table 7.1).

7.3 GRATING LOBES AND AVERAGE SIDELOBE LEVELS DUE TO PERIODIC PHASE, AMPLITUDE, AND TIME-DELAY QUANTIZATION

The practical issues of cost, volume, and manufacturability of array antennas lead to choices that directly influence the array characteristics. These considerations

Table 7.1
Number and Probability of Peak Sidelobes
Exceeding Threshold K

Number of Lobes Exceeding Threshold	Probability of Having $\leq k$ Lobes Exceeding Threshold
k	
0	0.349
1	0.736
2	0.93
3	0.987

result in the production of phase shifters with three or four (or sometimes more) discrete bits, instead of a continuum of available phase, in the construction of power distribution networks that have fixed, quantized levels, and in the use of time-delay units to feed wide-band phase steered subarrays instead of using one time-delay unit per element. Each of these choices results in periodic phase or amplitude errors across the array as if the array were constructed of subarrays with the quantized state defined for each subarray. Since the errors are highly correlated, they result in large, well-defined sidelobe or grating-lobe-type pattern errors.

Figure 7.4(a–c) shows three types of subarrays representing the contiguous levels. Figure 7.4(a) shows several patterns and the phase of an array with phase shifters having discretely quantized phase shifter states. The array taper amplitude is constant. The figure shows the array pattern as constrained by the (dashed) subarray pattern, which is fixed in space, and is the pattern of a uniformly illuminated, constant phase aperture of length equal to the distance between adjacent phase states. Figure 7.4(b) shows the very different characteristics of the pattern due to an array that has a quantized amplitude taper. Figure 7.4(c) shows the characteristic grating lobe structure of an array with time delay at the subarray level, when the array is operated at a frequency away from the design center frequency. The common features of each of these quantized illuminations allow them to be analyzed by the same method, and this procedure [16] for obtaining estimates of all resulting grating lobe peaks will be outlined in the following sections.

The peak sidelobe or grating lobe characteristics can often be reduced by disrupting the total periodicity that leads to the large grating lobes. Thus, it is common engineering practice to randomize the phase taper error in an array steered by discrete bit phase shifters. This practice does not reduce the average characteristics of the errored distribution, however, so the average sidelobe level becomes the ultimate pattern limitation. Approximations of the average sidelobe levels of arrays with discrete phase shifter states are also given in the following section.

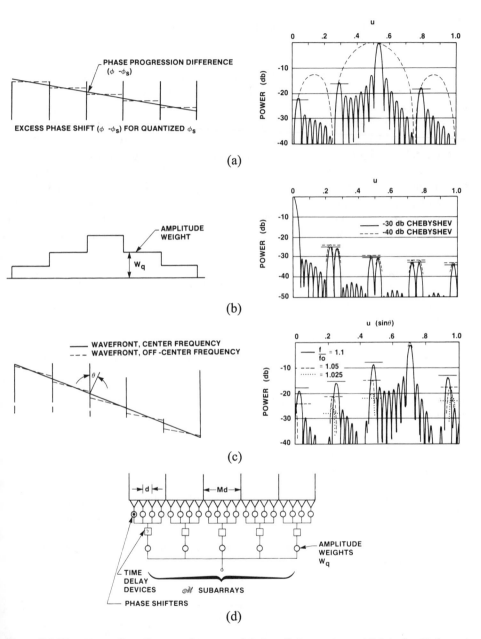

Figure 7.4 Three types of contiguous subarrays and their radiation patterns. All data for 64-element arrays with λ/2 spacing between elements and 8 elements per subarray (after [16]): (a) case 1: discrete phase shifter states (equal amplitude weights); (b) case 2: taper at subarray input ports (ideal phase progression); (c) case 3: time delay at subarray ports (equal amplitude weights); (d) array of contiguous subarrays.

7.3.1 Characteristics of an Array of Uniformly Illuminated Contiguous Subarrays

The common feature of each of the periodic quantization errors discussed above is that, because of the quantization, the array can be considered as divided into subarrays with one quantized state throughout each subarray. Figure 7.4(d) shows this configuration and indicates that each of the quantized illuminations shown at left of Figure 7.4(a–c) can be modeled by appropriately quantized phase, amplitude, or time-delay weights applied to this array of contiguous subarrays.

For each of the configurations shown, the array pattern is written as the product of an array factor and a subarray pattern. In the most general case treated here, the pattern of an array of m subarrays of M elements, each is given below for an array with subarrays steered to u_s by phase shifters, and with time delay between the subarray ports to steer the beam to the desired scan angle u_0.

$$F(u) = A(Z)f(z) \tag{7.25}$$

where

$$A(Z) = \frac{1}{m} \sum_{q=-(m-1)/2}^{(m-1)/2} w_q \exp[jqMZ] \qquad f(z) = \left\{ \frac{1}{M} \sum_{i=-(M-1)/2}^{(M-1)/2} \exp(jiz) \right\}$$

and where $\Sigma |w_q| = m$ (to normalize pattern) and $z = (2\pi ud)/\lambda - \Delta\phi_s = 2\pi d(u/\lambda - u_s/\lambda_0)$ and

$$Z = \frac{2\pi d}{\lambda}(u - u_0)$$

At center frequency and when there is no phase quantization error, $u_s = u_0$. The indicated sums over i and q are over integers for M or m odd, and half integers for M or m even.

Since all subarray ports are the same size, with constant illumination, the subarray pattern $f(z)$ is given as

$$f(z) = \frac{\sin(Mz/2)}{M \sin(z/2)} \tag{7.26}$$

Since these subarrays are generally several wavelengths across, there will occur grating lobes of the array factor $A(Z)$ at the direction cosines

$$u_p = u_0 + \frac{p\lambda}{Md} \qquad p = (\pm 1, \pm 2, \ldots) \qquad (7.27)$$

for all u_p in real space.

The value of the array pattern at or near each of these grating lobe peaks is just the value of the subarray pattern, so it is convenient to define a local coordinate δu centered at the center of any pth grating lobe at wavelength λ.

$$u = u_0 + \frac{p\lambda}{Md} + \delta u \qquad (7.28)$$

The subarray pattern for this generalized case is given in terms of the localized coordinate δu as

$$f(z) = (-1)^p \frac{\sin\left\{\pi Md \dfrac{u_0}{\lambda_0} \dfrac{\Delta f}{f_0} + \dfrac{\pi d}{\lambda} M\delta u + \dfrac{\pi Md}{\lambda_0}(u_0 - u_s)\right\}}{M \sin\left\{\dfrac{\pi u_0 d}{\lambda_0} \dfrac{\Delta f}{f_0} + \dfrac{\pi p}{M} + \dfrac{\pi d}{\lambda}\delta u + \dfrac{\pi d}{\lambda_0}(u_0 - u_s)\right\}} \qquad (7.29)$$

This expression will be used to evaluate the grating lobe power for arrays with quantized distributions. It should be noted, however, that isotropic element patterns have been assumed and that restriction can be removed by reducing all sidelobes by the element power pattern.

7.3.2 Phase Quantization in a Uniformly Illuminated Array

Miller [17] published the first detailed analysis of the adverse effects of using phase shifters with discrete phase states. Although the array is required to produce a smooth phase taper, an N-bit phase shifter has phase states separated by the least significant bit:

$$\phi_0 = \frac{2\pi}{2^N} \qquad (7.30)$$

Figure 7.5 shows that this discretization allows only a staircase approximation of the continuous progressive phase shift required for the array. The staircase phase front results in a periodic triangular phase error that produces the pattern with grating-lobe-like sidelobes, shown in Figure 7.4(a). Miller evaluated the peak first

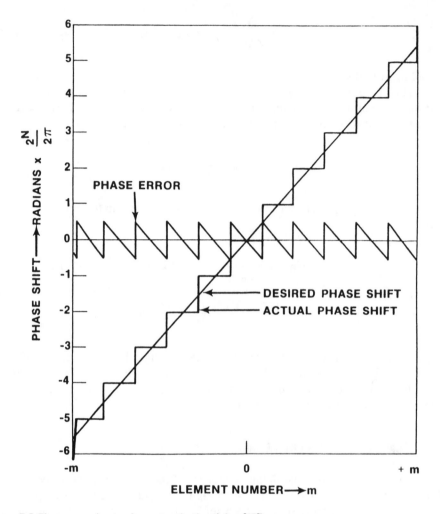

Figure 7.5 Phase error due to phase quantization (after [17]).

grating lobe level for this phase distribution by assuming that the array current distribution was a continuous function (not a discrete set of elements). With this approximation, the first grating lobe level is given as

$$P_{GL} = \frac{1}{2^{2N}}$$ (7.31a)

or

$$P_{GL}(\text{dB}) = -6N \tag{7.31b}$$

This result is shown in Figure 7.6.

Miller also evaluated the average sidelobe level due to this triangular phase error. The mean square error $\overline{\Phi^2}$ is obtained:

$$\overline{\Phi^2} = \frac{1}{3} \frac{\pi^2}{2^{2N}} \tag{7.32}$$

and the average sidelobe level due to quantization error alone is given as

$$\overline{\sigma^2} = \frac{1}{3g_A} \frac{\pi^2}{2^{2N}} \tag{7.33}$$

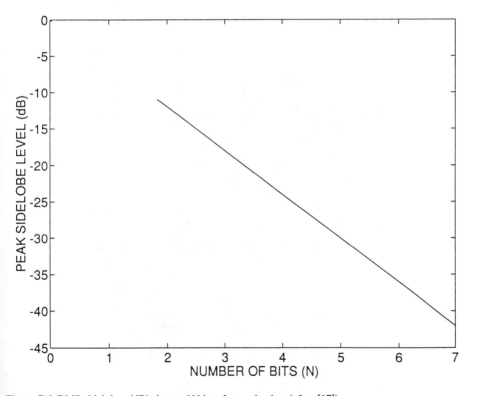

Figure 7.6 RMS sidelobes (dB) due to N bits of quantization (after [17]).

Figure 7.7 shows this average sidelobe level due to phase quantization and its dependence on the number of array elements. In deriving these results, Miller used an expression equivalent to (7.33), with $g_A = N$ for uniform illumination, but reduced the array factor directivity by 2 dB to account for scan and taper losses. Figure 7.7 therefore shows Miller's data with sidelobes approximately 2 dB higher than those given by (7.33).

Miller also gives an expression for the beam deviation (pointing error) due to periodic phase shifter quantization. For a uniformly illuminated array, the pointing error normalized to the array beamwidth is

$$\Delta = \frac{\pi}{4} \frac{1}{2^N} \text{ beamwidths} \tag{7.34}$$

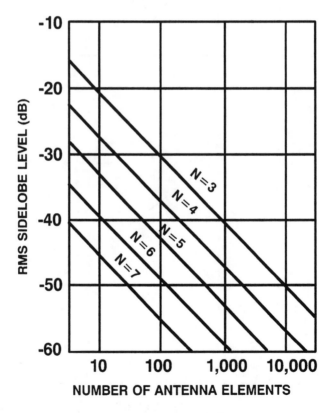

Figure 7.7 Peak sidelobes due to phase quantization (after [17]).

This result does not follow from (7.15) because the phase errors here are periodic, not random, and were obtained by evaluating the pattern slope of a uniformly illuminated continuous aperture with periodic phase steps.

Miller's approximation (7.31) underestimates the actual peak grating lobe level. This was first pointed out by Cheston and Frank [18] and is due to Miller's continuous array approximation. In fact, one can show that the error in Miller's estimate is small when the spatial period of the error is large (large M in our subarray model), and can be quite large when the error period is short. This is as expected because the continuous array approximation is primarily valid for large subarrays.

More accurate descriptions of grating lobe levels due to phase quantization are given by Hansen [19], Mailloux [16], and others. The presentation that follows is from [16]. If the array is to form a beam at u_0, the phase difference between elements should be (at $\lambda = \lambda_0$)

$$\Delta\phi = \frac{2\pi d}{\lambda_0} u_0 \qquad (7.35)$$

Since the least significant phase bit is ϕ_0, the phase across the array is necessarily in error at many points. The resulting pattern error is most serious if the error is entirely periodic. In Figure 7.5, Miller shows the phase as constant across some section of the array between points where least significant phase bits are added. In the more general case, the required interelement phase shift $\Delta\Phi$ usually exceeds the least significant bit, and so some phase progression $\Delta\Phi_S$ (corresponding to direction cosine u_S) is created using the available phase shifter bits. The remaining error incremental phase shift is $(2\pi d/\lambda_0)(u_0 - u_S)$, and again the array is composed of subarrays, each with its maximum pointing at u_S. In this case, the distance between subarray phase centers is such that the total phase error increment between subarrays is equal to the phase of the least significant bit, or

$$\left| M\Delta\phi - M\Delta\phi_s \right| = \frac{2\pi d}{\lambda} M(u_0 - u_s) \qquad (7.36)$$
$$= \frac{2\pi}{2^N}$$

This expression defines the subarray size Md for given scan angles. In practice, since the pattern is repetitive in $(u_0 - u_s)$, one only needs to let $u_0 - u_s$ vary from 0 to $(\lambda/2^N d)$, which corresponds to a phase progression of the minimum bit. Furthermore, since $M = 1$ means one element per subarray and the spacing d is selected to suppress grating lobes, the minimum value of M that need be considered is two. Thus, M can have values from 2 to half the number of array elements.

Using these relationships in (7.29), one can solve for the peak value of the array grating lobes as the subarray pattern amplitude. The normalized power of these grating lobes is written approximately as

$$P_{GL} = |f|^2 = \frac{(\pi/2^N)^2}{[M \sin(p'\pi/M)]^2} \tag{7.37}$$

where $p' = p + (1/2^N)$.

The factor $[M \sin(p'\pi/M)]^{-2}$ is the envelope of the subarray pattern peak power sampled at the pth grating lobe point. This factor also occurs in a later expression, and is therefore plotted in Figure 7.8 for $M > |p'|$ (the near grating lobes). The general expression for power at the peak of the pth grating lobe is written in terms of this envelope function as

$$GL = 10 \log P_{GL} = \text{envelope(dB)} + 9.94 - 6.02N \tag{7.38}$$

where envelope(dB) $= 10 \log\{[M \sin(p'\pi/M)]^{-2}\}$.

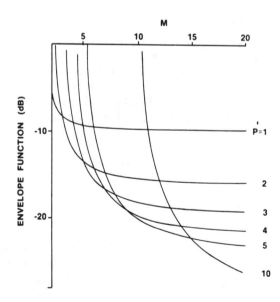

Figure 7.8 Envelope factor (dB) vs. number of elements in subarray with p' as parameter (after [16]).

Only data for integer values of p' are shown in Figure 7.8, but p' is not generally an integer. For a 3-bit phase shifter, the relevant values of p' are all at $p + 0.125$ for $p = \pm 1, \pm 2, \ldots$, and for larger numbers of bits they come closer to the grating lobe index values (p).

Figure 7.4(a) shows a typical case of a uniformly illuminated 64-element array of isotropic elements with $\lambda/2$ spacing. With 3-bit phase shifters (least bit 45 deg), the array forms a perfectly collimated beam at $u_S = u_0 = 0.5$ ($\theta_0 = 30$ deg) using 90-deg phase shift between elements, but at 32.1 deg ($u_0 = 0.53125$), with phase shifters set to the least significant bit phase gradient, there is an excess phase shift of 45 deg across each set of 8 elements. The pattern shows that grating lobes at various levels between -16 and -23 dB result from this periodic phase error. The figure also shows the subarray pattern of the 8-element subarray scanned to $u_S = 0.5$, and indicates how the product of subarray and array factor limits the grating lobe heights. The solid horizontal lines are the estimates of grating lobe height as evaluated from (7.37) or (7.38).

Depending on which scan angles are required, other size subarrays are formed at different scan angles. For example, at $u_0 = 0.5156$, $\theta_0 = 31.04$, and the excess of 45-deg phase shift spans 16-element subarrays. In each case, (7.36) is used to evaluate the subarray size Md.

In the limiting case of M large (e.g., the 16-element subarray noted above), the envelope curves tend to an asymptote and the grating lobe power is

$$P_{GL} \sim [1/(p'2^N)]^2 \tag{7.39}$$

of which a special case for $p = 1$ (the largest lobe). Here the envelope factor is $(1/\pi)^2$ or -9.94 dB, and (7.38) reduces to (7.31), which was obtained by Miller using the continuous triangular error approximation.

It is significant to note, however, that Miller's result underestimates the size of this maximum grating lobe for smaller values of M. In fact, one can obtain a bound on the grating lobe level, since the upper bound of the envelope function is $(1/M)^2$. Thus,

$$GL < -20 \log M + 9.94 - 6.02N \text{ for } M \geq 2 \tag{7.40}$$

For $M = 2$, and using a 3-bit phase shifter, one can show that the level is nearly approached and exceeds Miller's ($-6.02N$) number by about 4 dB.

7.3.3 Reduction of Sidelobes Due to Phase Quantization

As indicated in the previous section, phase quantization leads to unacceptable sidelobe levels because it introduces a large periodic phase error. Although the average error cannot be reduced, it is possible to break up the periodicity of the quantization error and hence reduce the peak sidelobes.

Miller [17] recognized this and suggested that space feeding the array from a common feed horn adds a quadratic phase offset at each element, as indicated in Figure 7.9. When the phase shifters are programmed to correct for the phase offset and scan the array, the resulting error due to quantization no longer possesses the periodic characteristic that resulted in the well-defined grating lobes, but now the sidelobes are distributed so that peaks are reduced to levels approaching that of arrays with random errors.

Smith and Guo [20] presented a detailed comparison of several methods that have been used to reduce the peak level. Using the continuous distribution approximation like that of Miller, Smith and Guo compared a number of different techniques in regard to sidelobe level and beam degradation.

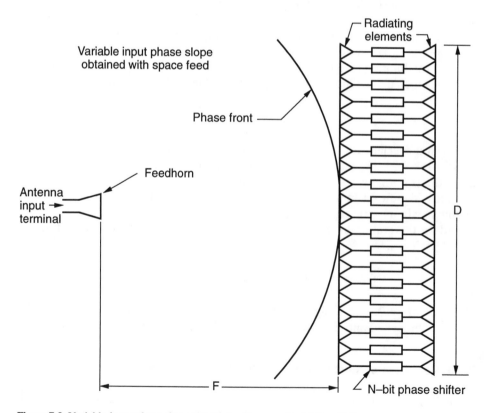

Figure 7.9 Variable input phase slope obtained with space feed (after [17]).

As a baseline, Smith and Guo used the procedure described by Miller, and which they refer to as *rounding off*, because the required phase is rounded to the nearest bit. The peak sidelobes are $-6N$ dB, and the average phase error variance (as given in (7.32)) is

$$\overline{\Phi^2} = \Delta_\Phi^2/3 \tag{7.41}$$

where Δ_Φ is one half of the least bit

$$\Delta_\Phi = \pi/2^N \tag{7.42}$$

The other techniques evaluated by Smith and Guo include a procedure proposed by Aronov [21] and called *mean phase error equal to zero*, another procedure called the *phase added method*, and several procedures called the *two* and *three probable value* methods. These techniques are compared in Table 7.2.

The rounding off technique, with triangular error described earlier, is summarized in the first row of the table. The second row gives particulars for the mean phase error equal to zero method and shows far lower peak sidelobes, at $-12N$, but twice the average error. This procedure is carried out by rounding off the phase or using the next state, depending on the fraction of phase that cannot be set up by the digital phase shifters. This procedure also has a nonzero beam pointing error.

One of the simpler and more successful procedures is the phase added method, which is implemented by adding a random phase offset at each element. These offset phases are included in the calculation of final phase shifter states. This procedure eliminates the parasitic lobes and the beam pointing error without changing the average phase error level.

The two and the three probable value methods trade off increased average sidelobe level for reduced peak sidelobes. Instead of rounding off the phase, the two probable value method uses a statistical algorithm to select one or the other of the nearest phase states while maintaining the mean error equal to zero. These procedures eliminate beam pointing error and reduce the value of the peak sidelobe to $-12N$ dB. The three probable value method uses the three nearest phase states, with probabilities judiciously chosen. Smith and Guo show that this procedure eliminates the pointing error and the peak sidelobes, but has average sidelobes $(\overline{\sigma^2})$ about 4.8 dB higher than the phase added method.

Another practical method of randomizing the periodic errors is called *phase dithering* [22]. This technique is a radar system solution rather than an antenna solution in that it requires averaging over a series of radar pulses. Before each pulse is received, a phase offset is added to each phase shifter command. The phase shifter settings are then determined according to the roundoff method, and

Table 7.2
Comparison of Five Methods Used to Reduce Parasitic Sidelobes

	Mean Pointing Deviation	Mean Maximum Parasitic Sidelobe Level (dB) (large N)	Variance of Phase Error σ^2	Additional Array Hardware	Beam Steering Unit Functions
Rounding off	Not zero	$-6N$	$\Delta_\phi^2/3$		Rounding off
Mean phase error equal to zero	Not zero	$-12N$	$2\Delta_\phi^2/3$		Random number generation, test for rounding up or down
Phase added	0	Not present	$\Delta_\phi^2/3$	Random (known) start phases at each element	Memory of start phases, rounding off
Two probable value	0	$-12N$	$2\Delta_\phi^2/3$		Random number generation, test for rounding up or down
Three probable value	0	Not present	Δ_ϕ^2		Random number generation, test to choose one of three values

so the array has a different triangular error distribution for each pulse and the average suppresses the peak sidelobes.

7.3.4 Subarrays With Quantized Amplitude Taper

A phase-steered array, organized into equally spaced, uniformly illuminated subarrays with different amplitude weights at each subarray, has its grating lobes located at the null points of the subarray pattern. If the whole array is uniformly excited, its beamwidth is narrow and the subarray nulls completely remove the grating lobes. When the excitation amplitude at the subarray input ports is weighted for array factor sidelobe reduction, the beamwidth broadens, and at the grating lobe angles, there occur split (monopulse-like) beams as shown in Figure 7.4(b). The beams are split because of the subarray pattern null.

The height of these split beams is clearly only related to the width (and local shape) of the array factor pattern. As the array factor sidelobes are lowered, the beamwidth broadens and the subarray pattern nulls do not completely remove the unwanted lobes.

To evaluate the power level of these split grating lobes, it is convenient to use a general expression for the array factor $A(Z)$ in (7.25) in the vicinity of each pth grating lobe. At center frequency, and with each subarray scanned to θ_0 (so that $\theta_s = \theta_0$), the array factor grating lobe is centered on the subarray pattern null. In the localized region from the beam peak to somewhat beyond the -3-dB point, the shape of each pth grating lobe of the array factor is approximated by

$$A(z) = \frac{B_b \sin\{[Mm\pi d/(B_b\lambda_0)]\delta u\}}{Mm(\pi d/\lambda_0)\delta u} \tag{7.43}$$

which represents a broadened beam with beam broadening factor B_b, defined such that the beamwidth is given as $0.886\lambda_0 B_b/Mmd$, with B_b the ratio of the beamwidth of the tapered array to the uniform array, M the number of elements in a subarray, and m the number of subarrays. By means of this approximation, it is possible to obtain very general and almost universal applicability without having to specify the taper and general pattern shape.

In the vicinity of the pth grating lobe, the product of the subarray pattern and the array factor is given by

$$A(z)f(z) = \frac{(-1)^p B_b \sin[(Mm\pi/B_b)(d/\lambda_0)\delta u]}{Mm \sin(p\pi/M)} \tag{7.44}$$

This expression has the proper zero at $\delta u = 0$ to produce the characteristic split lobe centered on the pth grating lobe location.

The normalized power at these grating lobes is evaluated at the peak value of the above expression as

$$P_{GL} = \frac{B_b^2}{M^2 m^2 \sin^2(\pi p/M)} \tag{7.45}$$

The grating lobe level can be computed directly from the above, or by using the envelope factor introduced earlier as

$$GL = 10 \log P_{GL} = \text{envelope(dB)} + 20 \log B_b - 20 \log m \tag{7.46}$$

or bounded as before in the limit of m as:

$$GL < -20 \log M + 20 \log B_b - 20 \log m \tag{7.47}$$

Figure 7.4(b) shows an example of a 64-element array with $\lambda/2$ spacing grouped into 8-element subarrays and illuminated at the subarray input ports with -30- and -40-dB Chebyshev tapers. Based on beam broadening factors of 1.29 and 1.43 for the Chebyshev patterns, evaluation of the above expression shows that the -40-dB pattern should have about 0.9 dB higher grating lobes than the -30-dB pattern. The horizontal lines computed from the above are again an excellent approximation of the grating lobe, as seen by comparison with the actual pattern in the figure.

7.3.5 Time Delay at the Subarray Ports

In the limit of a very small frequency excursion for a large array, it may be advantageous to use time delay at the subarray ports. This economy is not achieved without some penalty, however, since the periodic phase error introduced can cause significant sidelobes at frequencies away from the center frequency. In this case, the grating lobe peak is not split, and the peak grating lobe values are given directly by the subarray pattern envelope, as in the discrete phase shifter case.

Using a small angle expansion for the numerator of (7.29), the normalized power in the pth lobe is

$$P_{GL} = \frac{\pi^2 X^2}{\sin^2 \pi[X + p/M]} \tag{7.48}$$

where

$$X = \frac{u_0 d}{\lambda_0} \frac{\Delta f}{f_0}$$

Note that $|X| < 1/M$ so that the main beam does not "squint" out to a grating lobe location. This ensures that P_{GL} never becomes singular. Note also that in this case M is the actual number of elements in the subarray and is directly dictated by the geometry. A plot of grating lobe level versus the variable X is given in Figure 7.10 for various P/M ratios. Figure 7.4(c) shows an example of a uniformly illuminated array with time-delay steering at the subarray level. The results of (7.48) are plotted as horizontal lines and are clearly quite accurate representations of the computed grating lobe levels for various f/f_0 levels.

7.3.6 Discrete Phase or Time-Delayed Subarrays With Quantized Subarray Amplitudes

Figure 7.11 shows the grating lobe structure of a 64-element array with -40-dB Chebyshev illumination at the input ports of 8-element subarrays. The array is scanned using time delay at the subarray input ports, and phase shifters within the subarrays. The solid horizontal lines show the grating lobe levels computed using (7.48) (or Figure 7.10), with $f/f_0 = 1.05$. The figure clearly indicates that the results for the time-delayed subarrays can be extended to include a situation in which

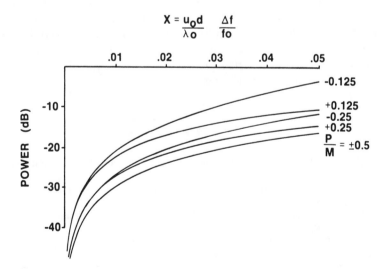

Figure 7.10 Grating lobe power for array with time delay at subarray ports (after [16]).

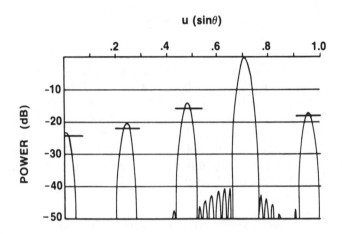

Figure 7.11 Power pattern for array with time delay at subarray ports and a 40-dB Chebyshev taper $f/f_0 = 1.05$ (after [16]).

there is pattern distortion due to quantized amplitude taper in addition to time delay. The reason for this more general result is that (7.47) was derived on the basis of the subarray pattern envelope and since the subarray pattern null does not fall at the grating lobe angle, the lobes are not split and the beam broadening factor argument used in the quantized amplitude case does not apply. So, if the grating lobes that result from phase shifter quantization or time-delay quantization are large, then the grating lobes are sampling subarray patterns far from the nulls, and the quantized amplitude taper has little effect on the validity of the approximations. The analysis of phase and time-delay quantization can be applied in many situations, even when the amplitude taper is quantized.

REFERENCES

[1] Ruze, J., "Physical Limitations on Antennas," Research Laboratory for Electronics, MIT, 30 Oct. 1952.

[2] Ruze, J., "The Effect of Aperture Errors on the Antenna Radiation Pattern," *Nuovo Cimento* (Suppl.), Vol. 9, No. 3, 1992, pp. 364–380.

[3] Elliott, R. E., "Mechanical and Electrical Tolerances for Two-Dimensional Scanning Antenna Arrays," *Trans. IRE*, PGAP, Vol. AP-6, 1958, pp. 114–120.

[4] Allen, J. L., "The Theory of Array Antennas," MIT Lincoln Laboratory Technical Report, No. 323, 1963.

[5] Skolnik, M. I., "Nonuniform Arrays," Ch. 6 in *Antenna Theory*, Collin and Zucker, eds., New York: McGraw-Hill, 1969, pp. 227–234.

[6] Moody, H. J., "A Survey of Array Theory and Techniques," RCA Victor Report No. 6501.3, Nov. 1963, RCA Victor Co., Research Labs., Montreal, Canada.

[7] Hsiao, J. K., "Array Sidelobes, Error Tolerance, Gain and Beamwidth," NRL Rept. 8841, Naval Research Laboratory, Washington, D.C., 28 Sept. 1984.

[8] Hsiao, J. K., "Design of Error Tolerance of a Phased Array," *Elect. Letters*, 12 Sept. 1985, Vol. 21, No. 19, pp. 834–836.

[9] Kaplan, P. D., "Predicting Antenna Sidelobe Performance," *Microwave J.*, Sept. 1986, pp. 201–206.

[10] Allen, J. L., "Phased Array Radar Studies," MIT Lincoln Lab. Tech. Report, No. 236, 1960.

[11] Ruze, J., "Pattern Degradation of Space Fed Phased Arrays," Project Rept. SBR-1, MIT Lincoln Laboratory, 5 Dec. 1979.

[12] Mailloux, R. J., "Periodic Arrays," Ch. 13 in *Antenna Handbook, Theory, Applications and Design*, Y. T. Lo and S. W. Lee, eds., New York: Van Nostrand Reinhold, 1988.

[13] Carver, K. R., W. K. Cooper, and W. L. Stutzman, "Beam Pointing Errors of Planar Phased Arrays," *IEEE Trans.*, Vol. AP-21, March 1973, pp. 199–202.

[14] Steinberg, B. D., "Principles of Aperture and Array Systems Design," New York: John Wiley and Sons, 1976.

[15] Rice, S. O., "Mathematical Analysis of Random Noise," Bell System Tech., Vol. 23, 1944, p. 282; also Vol. 24, 1945, p. 40.

[16] Mailloux, R. J., "Array Grating Lobes Due to Periodic Phase, Amplitude and Time Delay Quantization," *IEEE Trans.*, Vol. AP-32, No. 12, Dec. 1984, pp. 1364–1368.

[17] Miller, C. J., "Minimizing the Effects of Phase Quantization Errors in an Electronically Scanned Array," *Proc. 1964 Symp. Electronically Scanned Phased Arrays and Applications*, RADC-TDR-64-225, RADC Griffiss AFB, Vol. 1, pp. 17–38.

[18] Cheston, T. C., and J. Frank, "Array Antennas," Ch. 11 in *Radar Handbook*, M. E. Skolnik, ed., McGraw Hill, 1990.

[19] Hansen, R. C., "Linear Arrays," Ch. 9 in *The Handbook of Antenna Design*, Vol. 2, Peter Peregrinus, pp. 104–106.

[20] Smith, M. S., and Y. C. Guo, "A Comparison of Methods for Randomizing Phase Quantization Errors in Phased Arrays," *IEEE Trans.*, Vol. AP-31, No. 6, Nov. 1983, pp. 821–827.

[21] Aranov, F. A., "New Method of Phasing for Phased Arrays Using Digital Phase Shifters," *Radio Eng., Electron. Physics*, Vol. 11, 1966, pp. 1035–1040.

[22] Brookner, E., "Antenna Array Fundamentals," Ch. 3 in *Practical Phased-Array Antenna Systems*, E. Brookner, ed., Dedham, MA: Artech House, pp. 3–25.

Chapter 8

Special Array Feeds for Limited Field-of-View and Wide-Band Arrays

Most phased array antennas discussed in this text are designed for wide-angle scanning. This chapter, however, addresses a specialized group of array systems that take advantage of restrictions in the scan coverage in order to produce a very-high-gain scanning system with relatively few phase controls, or that provide wide-band, wide-angle scanning performance for large apertures without an accompanying large number of time-delay controls. Many of these systems are based on the multiple-beam properties of reflectors and lens systems, and so obtain their high gain from the collimation provided by these quasi-optical systems. They achieve some restricted scan coverage by means of a complex feed. Several of the techniques, however, are strictly array systems, where again the scan tradeoffs are used to reduce the number of array controls.

The chapter is introduced by a section on multiple-beam systems because these fundamental beam formers are the basis of many limited scan and wide-band systems.

8.1 MULTIPLE-BEAM SYSTEMS

While phased arrays have a single output port, multiple-beam systems have a multiplicity of output ports, each corresponding to a beam with its peak at a different angle in space. Typical systems needing simultaneous, independent beams include multiple-access satellite systems and a variety of ground-based height-finding radars. Figure 8.1 shows a schematic diagram of a multiple-beam antenna with a number of input ports and a switching network that selects a single beam or a group of beams as required for specific applications. Figure 8.2 illustrates the use of generic lens or reflector apertures in a multiple-beam system.

Many antenna requirements emphasize high gain with low sidelobes. In addition, it is often important that the system have a high *beam crossover level* so that nearly the full system gain is available within any point in the antenna field of

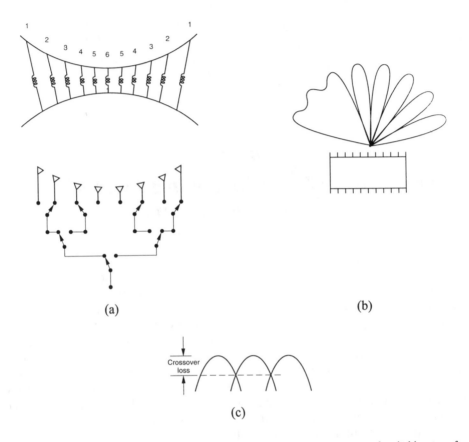

Figure 8.1 Multiple beam antenna systems: (a) basic multiple-beam antenna and switching tree for eight-beam system; (b) combined shaped and pencil beam system; (c) beam crossover.

view. The beam crossover level is shown in Figure 8.1 as the relative gain of either of two adjacent beams at the point of their intersection. Typical crossover levels can range from about 4 dB (actually 3.9) below the beam peak for the beams used in the Woodward-Lawson [1] synthesis procedure to much higher or lower levels, depending on the desired sidelobes and system loss. Another critically important feature of multiple-beam forming networks is that they should be lossless, or have minimal loss, in order that the reduced gain not render the system impractical.

Other applications for multiple-beam arrays include their use in the synthesis of shaped patterns, where the beams are the constituent beams that combine to make up the shaped pattern, as in the Woodward-Lawson procedure. In this case, the sidelobes are often not so important, but it is necessary that the crossover levels be relatively high in order to have a smooth approximation of the desired pattern,

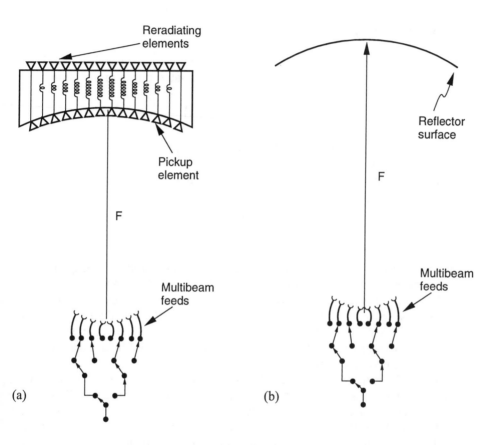

Figure 8.2 Generic lens (a) and reflector (b) multiple-beam systems.

and it is also necessary that the loss is minimized. A procedure for shaped pattern synthesis with multiple beams is given by Ricardi [2]. A recent paper [3] discusses the design of shaped beam patterns using minimax gain optimization, and Galindo-Israel et al. [4] and Thomas [5] demonstrate the synthesis of contoured patterns and low-sidelobe patterns using orthogonal constituent beams.

In still other cases, multiple-beam arrays are used as one component of scanning systems. An example is in the use of a multiple-beam array feed for a reflector or lens system. Such systems are a special case, and their characteristics are discussed in Section 8.2 on limited scan (or limited field-of-view) systems. The following sections describe some of the principal characteristics of multiple-beam systems.

8.1.1 Beam Crossover Loss

The Woodward-Lawson synthesis technique makes use of beams radiated by a uniformly illuminated linear array with uniquely related phase progressions:

$$a_m = \exp[-jkd_x u_i m] \tag{8.1}$$

for $u_i = (\lambda/Nd_x)i$ and $i = \pm(1/2, 3/2, 5/2, \ldots)$ for N-even and $i = \pm(0, 1, 2, 3, 4, \ldots)$ for N-odd.

The set of beams formed by this excitation has the familiar

$$f_i(u) = \frac{\sin[N\pi(ud_x/\lambda - i/N)]}{N \sin[\pi(ud_x/\lambda - i/N)]} \tag{8.2}$$

angular dependence, with a broadside beam for N odd (Figure 8.3(a)) and with symmetrically spaced beams displaced one half of the null beamwidth from broadside (Figure 8.3(b)) for N even. Note that these beam peaks move with frequency to form a contiguous set of beams that cross at the 4-dB (actually 3.92-dB) point. With increased frequency, the beams narrow and thus each moves toward broadside. The Woodward-Lawson beams are thus ideally excited by a phase-shift network, not a time-delay network.

Throughout the chapter, these beams will often be referred to as the sin x/x beams, a liberty that alludes to the form of the pattern for a continuously illuminated uniform aperture. This near equivalence is discussed in Chapter 2. These beams are known to have the narrowest beamwidths and highest directivity of any but superdirective illuminations, and, furthermore, they are orthogonal in space over the region $-\lambda/2d_x \leq u \leq \lambda/2d_x$. The adjacent beams have relatively high crossover levels and so provide good pattern coverage for all angles. Since these beams are orthogonal, they can be excited by lossless networks [6,7], as will be described in the next section.

For some applications, the 4-dB crossover points of these adjacent beams might be considered too much loss, but it is not possible to simply crowd the patterns close together with any passive feed network without suffering excessive *orthogonality loss*. This characteristic is described in the next section.

It should be noted that, in principle, one can regain 3 dB of the crossover loss by summing two adjacent beams at the crossover point, or use a variable power divider network to properly weight contributions from two adjacent beams at arbitrary points between the beam centers, should this increase in complexity be warranted [2].

For a two-dimensional grid of beams, the problem is yet more severe. Arrays with rectangular grids, and typically formed by orthogonal beam matrices as described later in the chapter, have sin x/x beams arrayed in two dimensions. If

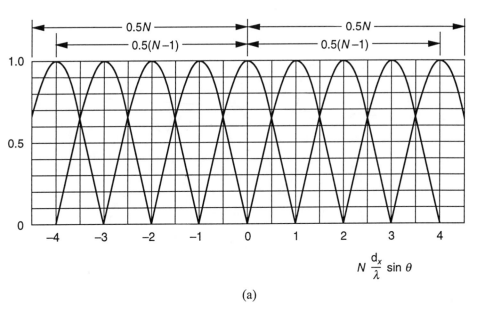

(a)

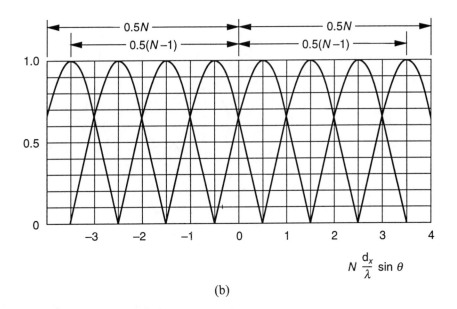

(b)

Figure 8.3 Woodward-Lawson beams: (a) odd numbers of elements (shown for $N = 9$); (b) even numbers of elements (shown for $N = 8$).

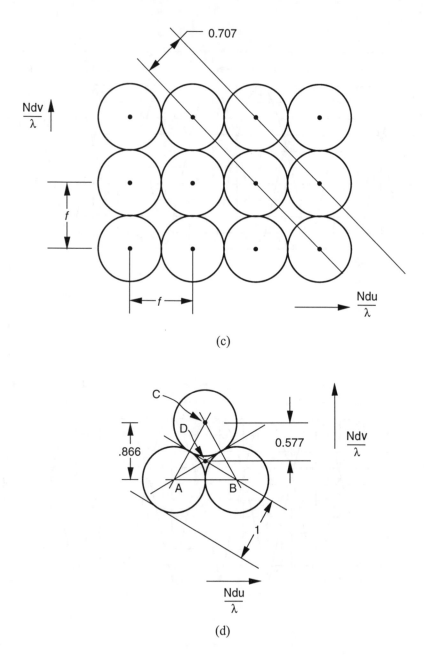

Figure 8.3 (cont.) Woodward-Lawson beams: (c) beam configuration for square grid (circles at -3.92 dB) (aperture length Nd in both planes); (d) beam configuration for isosceles triangular grid (circles at -3.92 dB) (aperture length Nd in both planes).

the beams are located in a square grid (Figure 8.3(c)) and have equal beamwidths in both planes, adjacent beams in each principal plane can have the 4-dB crossover points (for the orthogonal sin x/x beams), but the locations on the diagonal between beams have quite low crossovers. The circles in Figure 8.3(c) are plotted on the coordinates $u(d/\lambda)$ and $v(d/\lambda)$, where d is the interelement spacing, so the peak-to-null distance is unity.

In this case, the beams at locations (u_i, v_j) have the form shown below:

$$\frac{\sin[N\pi(ud/\lambda - i/N)]}{[N\pi(ud/\lambda - i/N)]} \frac{\sin[M\pi(vd/\lambda - j/M)]}{[M\pi(vd/\lambda - j/M)]} \tag{8.3}$$

This arrangement of beams, shown in Figure 8.3(c) for $M = N$, has very low crossover levels $(-8.8$ dB$)$ along the diagonal plane at $u = v$, and the crossover can be improved by selecting a triangular grid of beams. When this is done, the beams are no longer orthogonal and it becomes necessary to trade crossover level for orthogonality loss.

Triangular grids of beams offer advantages. An isosceles triangular lattice (Figure 8.3(d)) with adjacent beams spaced one unit apart in azimuth and 0.866 in elevation has its lowest crossover, not along the diagonal lines AC or CB where the center is at $\Delta/2$ from point A, but instead at point D, which is a distance (line CD) 0.577 from the nearest beam center. Here the crossover level is at -5.4 dB below the beam peak.

Beam crossover loss is thus seen to be an important factor in the design of multiple-beam systems. Ideally, one would like to produce low-sidelobe beams and stack them close together so that the crossover levels are only a decibel or two below the beam peaks. When implemented with a passive, lossless beam former, this condition leads to excessive network loss because of the nonorthogonality of the closely packed beams. This subject is addressed in the next section.

8.1.2 Orthogonality Loss and the Stein Limit

The sin x/x type beams unfortunately have high sidelobes $(-13$ dB$)$, and so there has been considerable interest in the synthesis of multiple beams with lower sidelobe levels. Allen [8] showed that requiring a network to excite two or more independent radiating beams without loss requires that the radiated beams in space be mutually orthogonal over one period of the pattern from $u = -\lambda/(2d_x)$ to $\lambda/(2d_x)$. An integral over any number of pattern periods would also exhibit orthogonality. This condition was a direct result of two well-known properties of the scattering matrix of a lossless reciprocal network. Reciprocity dictates that the scattering matrix be symmetrical, and the lossless character of the matrix dictates that the matrix be unitary. Allen's proof began with the assumption that a lossless reciprocal network could exist that would form the required beams, and then the symmetric and unitary conditions dictated the resulting aperture fields, which Allen showed were orthogonal over each period. Kahn and Kurss [9] extended these conclusions and showed that if the array is required to form N similar uncoupled beams with a lossless network, then the angular spacing between the beams is fixed (and equal to $\lambda/(Nd_x)$ in sine space, but that if the requirement for forming N beams were removed, then one could combine beam input ports to obtain lower sidelobes.

White [10] derived extremely general relationships that extended Allen's results to arbitrary multiple-beam antenna systems, whether arrays or quasi-optical beam formers. White showed that for reciprocal or nonreciprocal lossless networks radiating multiple beams from a common aperture, the beams must be orthogonal in space, and so *the radiation pattern and crossover levels cannot be specified independently.* White showed that by combining adjacent beam ports in phase, one can obtain beams with a cosine amplitude distribution across the array, and hence -23-dB sidelobes; but then the interbeam spacing for orthogonality is $2\lambda/d_x$, and this corresponds to crossover levels of -9.5 dB. Similarly, still lower sidelobes can be formed by suitably combining the $\sin[\ ]/N \sin[\ ]$ beams to obtain a $\cos^2$ amplitude illumination or a $\cos^2$ over a pedestal illumination across the array, but this illumination is only orthogonal if the interbeam separation is $3\lambda/d_x$. Moreover, the crossover levels are still lower. Alternatively, White showed that if one forced the beam spacings to be less than the orthogonal spacing, then the beams would be necessarily coupled and the feed network lossy.

Formation of the several beams can be shown simply. Consider input ports i and j that form array excitations $\exp[-jknd_xu_i]$ and $\exp[-jknd_xu_j]$ for

$$u_p = p\lambda/(Nd_x) \tag{8.4}$$

for integer p.

Superimposing the excitations of the two adjacent beams produces the excitation below at the nth element (assume $j = i + 1$), and that $u_j = u_i + \delta$:

$$\exp[-jnkd_xu_i] + \exp[-jknd_xu_j] = 2\exp[-jknd_x(u_i + \delta/2)]\cos[knd_x\delta/2] \tag{8.5}$$

Note that the beam angle is at the point between the two constituent beams and that the element amplitude has the cosine dependence.

Similarly, a judicious superposition of three adjacent beams with amplitudes $1/4$, $c + 1/2$, and $1/4$ and located at $u_i - \delta$, u_i, and $u_i + \delta$ gives

$$\exp[-jnkd_xu_i]\{+1/4\exp[+jnkd_x\delta] + 1/4\exp[-jnkd_x\delta] + (c + 1/2)\}$$
$$= \exp[-jnkd_xu_i][\cos^2(nkd_x\delta/2) + c] \tag{8.6}$$

This illumination is known as a $\cos^2$ on a pedestal function, and the pedestal height can be varied to produce a low-sidelobe pattern. In the limit, the pattern sidelobes can be -43 dB for a pedestal height of 0.08.

White presented these examples and illustrated several others showing how low-sidelobe beams with high crossover levels can be decoupled using additional apertures by resistive (lossy) decoupling networks or by introducing active amplifiers to recover the signal-to-noise ratio on receive.

Other low-sidelobe patterns can also be synthesized using the orthogonal $\sin x/x$ patterns. Thomas [5] illustrated the synthesis of Taylor patterns by proper superposition of beams. However, the Thomas procedure was one of synthesis of a single low-sidelobe beam with orthogonal beams, not the formation of multiple low-sidelobe beams. Consequently, the orthogonality condition has no meaning in this case.

Stein [11] derived the conditions for maximum efficiency from multiple-beam networks and obtained relations for evaluating this maximum efficiency in terms of *beam coupling factors*. This maximum efficiency is often termed the *Stein limit*. Consider the linear multiple-beam network of Figure 8.1(a) radiating M beams. Using Stein's notation, the signals y_k reflected in each of the ports are related by the linear relation

$$ y_k = \sum_{m=1}^{M} S_{km} x_m \qquad \text{or} \qquad \mathbf{y} = \mathbf{S}\mathbf{x} \tag{8.7} $$

where S_{K_m} is the unspecified scattering matrix, and $\mathbf{x}$ and $\mathbf{y}$ are column matrices.

If the kth input port is excited with a signal of unity power, the antenna system vector far field is given by

$$ \mathbf{E_k}(\theta, \phi) = q_k \mathbf{R_k}(\theta, \phi) \frac{e^{j2\pi r/\lambda}}{r} \tag{8.8} $$

Here the $\mathbf{R_k}$ is called the *beam pattern* and normalized so that the integral of the following dot product is unity:

$$ \frac{1}{2Z_0} \int_\Omega \mathbf{R_k^*}(\theta, \phi) \cdot \mathbf{R_k}(\theta, \phi) d\Omega = 1 \tag{8.9} $$

where $d\Omega = \sin \theta \, d\theta \, d\phi$ and Z_0 is the free-space impedance. With this normalization, the total radiated power for the kth beam in the far zone is

$$ P_k = \int_\Omega \frac{r^2}{2Z_0} \mathbf{E_k^*}(\theta, \phi) \cdot \mathbf{E_k}(\theta, \phi) d\Omega = q_k^* q_k = |q_k|^2 \tag{8.10} $$

Since unit power is incident upon the junction, $|q_k|^2$ is the radiation efficiency for this beam, and $1 - |q_k|^2$ represents losses in the network and the waves reflected back into all the feed ports.

For a lossless system, one can simply measure these efficiencies using the power reflected into the entire set of feed lines:

$$|q_k|^2 = 1 - \sum_{i=1}^{M} |S_{ik}|^2 \tag{8.11}$$

Stein defines a parameter related to beam overlap as

$$\beta_{kj} = \frac{1}{2Z_0} \int \mathbf{R}_k^*(\theta, \phi) \cdot \mathbf{R}_j(\theta, \phi) d\Omega \tag{8.12}$$

where, from the previous normalization,

$$\beta_{kk} = 1$$

Note also that $\beta_{kj} = \beta_{jk}^*$ and that $|\beta_{kj}| \leq 1$. The term β_{kj} defined above is called the *beam coupling factor*, and the square matrix β is the *beam coupling matrix*. The off-diagonal terms of this matrix imply coupling between the various beams and, if zero, define an orthogonality relationship between the beams.

If all the input ports are excited, the total radiated power is given by

$$\begin{aligned} P_{RAD} &= \int \frac{r^2}{2Z_0} \mathbf{E}^* \cdot \mathbf{E} \, d\Omega \\ &= \sum x_k^* q_k^* \beta_{kj} x_j q_j \end{aligned} \tag{8.13}$$

which can be written in terms of a new matrix Γ as

$$P_{RAD} = \sum_{k,j=1}^{M} x_k^* \Gamma_{kj} x_j = \mathbf{x}^\dagger \Gamma \mathbf{x} \tag{8.14}$$

where

$$\Gamma_{kj} = q_k^* \beta_{kj} q_j \tag{8.15}$$

This new matrix has eigenvalues α_k as given by the equation

$$\boldsymbol{\Gamma}\mathbf{x} = \alpha\mathbf{x} \tag{8.16}$$

or the characteristic equation $\det\{\boldsymbol{\Gamma} - \alpha\mathbf{I}\} = 0$ for $\mathbf{I}$, the identity matrix.

Stein's limit, based on the Hermitian and positive semidefinite properties of this matrix, states that the largest of the eigenvalues of the $\boldsymbol{\Gamma}$ matrix cannot exceed unity:

$$(\alpha_k)_{\max} \leq 1 \tag{8.17}$$

A simple, but perhaps the most intuitively meaningful, example of the utility of Stein's limit occurs when all the beams have equal radiation efficiencies, $q_k = q$ for all K.

$$\Gamma_{kj} = |q|^2\beta_{kj} \tag{8.18}$$

It follows that the eigenvalue equation takes on the simplified form

$$|q|^2\boldsymbol{\beta}\mathbf{x} = \alpha\mathbf{x} \tag{8.19}$$

and the eigenvalues α_k of $\boldsymbol{\Gamma}$ are clearly related to the set of eigenvalues β_k of the matrix $\boldsymbol{\beta}$ by the linear relation

$$\alpha_k = |q|^2\beta_k \tag{8.20}$$

In this case, this Stein limit becomes

$$|q|^2 \leq 1/(\beta_k)_{\max} \tag{8.21}$$

This far-reaching conclusion states that the efficiency $|q|^2$ is less than the inverse of the maximum eigenvalue of the beam coupling matrix, and the limitation pertains because of the overlap of the beams in space, without explicit reference to the network that forms the beams. This form is particularly simple for computation, because the coupling matrix $\boldsymbol{\beta}$ is readily obtained from (8.12) (most often by transforming the pattern expressions into aperture fields and making use of convolution-type integrals) and the eigenvalues β_k found by traditional methods.

A further result is that since all the diagonal elements β_{kk} are unity, the sum of the diagonal elements (the trace of $\boldsymbol{\beta}$) is M. However, for any Hermitian matrix, the sum of all the eigenvalues equals the trace of the matrix, so the eigenvalue sum

$$\sum_{k=1}^{M} \beta_k = M \tag{8.22}$$

and the largest eigenvalue β_k must be less than or equal to unity. Thus, $|q|^2 \leq 1$ with $|q|^2 = 1$ possible only if all the eigenvalues β_k are equal, which requires that all off-diagonal elements of $\boldsymbol{\beta}$ vanish, and if the beams are all mutually orthogonal, as had been previously pointed out by Allen and White.

Stein gives examples showing this coupling factor for several types of overlapped beams. An important simple case is that of two identical beams. Here (see (8.19)), the eigenvalues of concern are simply those of the β matrix

$$\boldsymbol{\beta x} = \beta \mathbf{x} \tag{8.23}$$

where the matrix is in bold type and the eigenvalue in normal type. The eigenvalues β are obtained as

$$\begin{vmatrix} 1 - \beta & \beta_{12} \\ \beta_{12} & 1 - \beta \end{vmatrix} = 0 \tag{8.24}$$

and are given by

$$\begin{aligned} \beta_1 &= 1 - \beta_{12} \\ \beta_2 &= 1 + \beta_{12} \end{aligned} \tag{8.25}$$

From (8.21), the upper bound of the radiation efficiency is

$$|q_{max}|^2 \leq \frac{1}{\beta_{k\ max}} = \frac{1}{1 + |\beta_{12}|} \tag{8.26}$$

Figure 8.4 shows the efficiency $|q^2|$ and beam coupling factor for two beams of a uniformly illuminated aperture as the interbeam spacing is increased. The beams chosen in this figure follow the example of White [10], who assumed for simplicity a set of $\sin x/x$-type patterns and a very narrow beam array so that the integral β_{kj} has the approximate form of the infinite integral below:

$$\int_{-\infty}^{\infty} \frac{\sin x}{x} \frac{\sin(x + t)}{x + t} \, dx = \pi \frac{\sin t}{t} \tag{8.27}$$

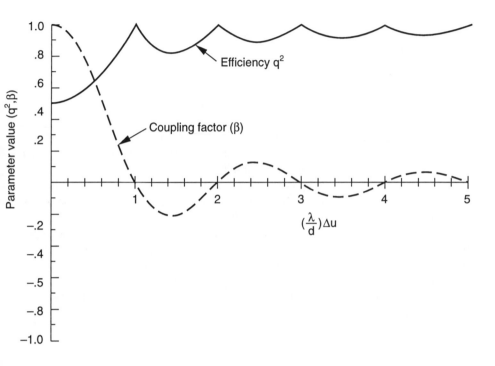

Figure 8.4 Orthogonality loss in two-beam system.

Choosing the normalized form below, and assuming a one-dimensional aperture distribution, the (scalar) form for the R_j is given as

$$R_k = (2Z_0 d/\lambda)^{1/2} \frac{\sin[(\pi d/\lambda)u_k]}{(\pi d/\lambda)u_k} \tag{8.28}$$

Choosing u_j displaced from the u_k by some increment Δ in sine space so that

$$u_k = u_j + \Delta u \tag{8.29}$$

one obtains

$$\beta_{kj} = \frac{\sin[(\pi d/\lambda)\Delta u]}{(\pi d/\lambda)\Delta u} \tag{8.30}$$

as the beam coupling factor. Equation (8.26) gives the upper bound of the radiation efficiency for this case as

$$|q|^2_{max} = \frac{1}{1 + |\beta_{jk}|} \tag{8.31}$$

The dashed curve of Figure 8.4 shows the coupling factor to be unity when the beams are coincident ($\Delta u = 0$) and decreasing as the spacing between beams increases. The curve shows that the coupling factor is zero for spacings of any nonzero multiple of λ/d, which corresponds to the orthogonal spacings. The envelope of the curve decreases with increasing spacing. The solid curve shows the maximum efficiency (often called the *Stein limit*) as starting at 0.5 for coincident beams. As the interbeam spacing is increased, the unity efficiency case repeats periodically (at the orthogonal spacings), and the envelope of the efficiency curve increases monotonically as the beam coupling decreases with spacing.

A second case of significant importance is that of a linear array of omnidirectional elements with half-wave spacing. Here the array pattern is given in the usual form (assumed scalar)

$$R_k(u) = \sum_{-(N-1)/2}^{(N-1)/2} a_n e^{jknd_x(u-u_k)} \tag{8.32}$$

and the coupling factor β_{kj} by

$$\beta_{kj} = \frac{1}{2Z_0} \int_{-1}^{1} du \left\{ \sum_{-(N-1)/2}^{(N-1)/2} a_n e^{jknd_x(u-u_k)} \right\} \left\{ \sum_{-(N-1)/2}^{(N-1)/2} a_m^* e^{-jkmd_x(u-u_j)} \right\} \tag{8.33}$$

After changing the order of integration and summation, this can be written in terms of the sinc function sin t/t as

$$\beta_{kj} = \frac{1}{Z_0} \sum_n \sum_m a_n a_m^* e^{-jk(nu_k - mu_j)d_x} \, \text{sinc}[k(n - m)d_x] \tag{8.34}$$

where the limits have been left off the summations for convenience.

For half-wave spacing, the sinc expression is zero unless $k = j$, and so the summation reduces to the form

$$\beta_{kj} = \frac{1}{Z_0} \sum_{-(N-1)/2}^{(N-1)/2} a_n a_n^* e^{-jkn(u_k - u_j)d_x} \tag{8.35}$$

This fairly general expression can be used to evaluate the coupling for arbitrarily tapered arrays and so is very convenient for evaluating the coupling of low-sidelobe arrays.

In the limit of a uniformly illuminated array ($a_n = 1$), this summation is readily accomplished and leads to the form

$$\beta_{kj} = \frac{\sin(N\pi(u_k - u_j)d_x/\lambda)}{\sin(\pi(u_k - u_j)d_x/\lambda)} \tag{8.36}$$

Comparing this result with (8.30) shows that the continuous and discrete apertures have the expected similarity. The array patterns demonstrate orthogonality for beam spacings

$$u_k - u_j = \frac{Q}{Nd_x} \tag{8.37}$$

for any integer Q, and so Figure 8.4 is also a good qualitative description of coupling and orthogonality loss for linear array antennas.

Stein gives curves of efficiency for several beams and clusters of beams, choosing the circularly symmetric forms of the uniform, Gaussian, and several tapered illuminations. Johansson [12] presents a detailed catalog of efficiencies for multibeam circular arrays with beams arranged in square or hexagonal (triangular) grids. In all of these cases, the requirement for low-sidelobe beams leads to either low radiation efficiency if high crossover levels are required or increased spacing and low crossover levels with improved efficiency. The tradeoff of crossover level and efficiency between these two extremes is a primary consideration in multiple-beam system design.

In 1985 Dufort [13] considered the case of equal multiple-beam patterns from a large array with beam separation in the characteristic Butler matrix (Hansen-Woodward) directions, but with a tapered aperture illumination. Dufort obtained the following reduced form for Stein's limit in this case: "The maximum efficiency possible is the ratio of the average to the peak value of the aperture power distribution."

This powerful and useful result allows the immediate conclusion that the maximum efficiency is unity for a uniform illumination, and is 1/2 for a cosine taper. It also explains how, for most low-sidelobe distributions, a loss of 3 dB or more must be accepted. Dufort showed that for passive networks the use of attenuators to control aperture taper can produce optimum results. Dufort also showed that for lens-type low-sidelobe multiple-beam antennas, where the loss may be shared between aperture and spillover, the Stein limit is achieved by a combination of attenuation and feed distribution without overlapping the feed networks.

At this point it should also be remarked that the use of *digital beam forming* on receive completely avoids the orthogonal spacing problem. Since the adjacent beams are formed completely by digital processing, one can form arbitrarily low sidelobe beams with any selected beam separation. Several references on digital beam forming are given in Chapter 3.

8.1.3 Multiple-Beam Matrices and Optical Beam Formers

Figure 8.5(a) shows a beam forming circuit due to Butler [6] that forms eight beams using a combination of microwave power dividers and phase shifts. Other networks have been devised by Butler and by Shelton and Kelleher [7]. Other variations and circuits are shown in Volume 3 of *Microwave Scanning Antennas* [14]. The *Butler matrix* (as the network is popularly termed) is the analog implementation of the fast Fourier transform, and as such requires $N \log N$ signal combinations (sums, differences) to excite N beams of an N-element array from N input ports. Butler matrices have been built with excellent phase tolerance for up to 64 beams. A recent study of high-power waveguide Butler matrices and fabrication of an eight-element matrix [15] achieved maximum phase error of 6 deg with rms error less than 3 deg and power dissipation of 0.4 dB. Computer studies of a 32-element matrix indicated that very similar performance can be obtained.

Section 8.2.3 gives a more detailed analysis of relationships between the input and output signals of a Butler matrix. There are, however, several important features of the beam forming network that contribute to the discussion at hand. The basic Butler matrix produces ideal (symmetrical) orthogonal beams of the type used in Woodward-Lawson synthesis (Chapter 3). The beam maxima u_i are at

$$u_i = (\lambda/L)i = i\lambda/(Nd_x) \tag{8.38}$$

for $i = \pm 1/2, \pm 3/2, \pm 5/2, \ldots, (N-1)/2$; and the phase progression between elements is

$$\delta_i = (2\pi d_x/\lambda)u_i = 2\pi i/N \tag{8.39}$$

Figure 8.3(a,b) shows the location of this set of beams plotted against the normalized coordinate $Nd_x u/\lambda$. In Figure 8.3 the beams are shown only to their first zeros, but Figure 8.5(b) shows the complete pattern for two beams of the set of eight beams.

For an aperture of N elements and "length" $L = Nd_x$, the N beams will fill a sector of width $N(\lambda/Nd_x) = \lambda/d_x$ in u-space to the -4-dB point. The Butler matrix is thus ideal for synthesizing a shaped pattern over such an extended region, since N switches can clearly be used to create any realizable pattern over the given region

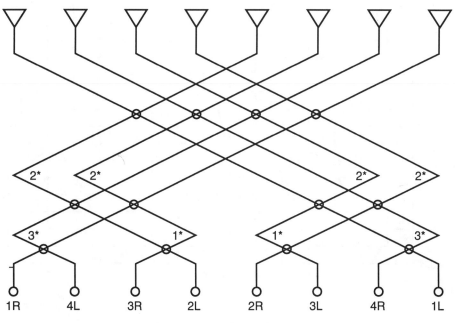

8–beam 8–element Butler matrix

* Units of phase shift are π/8 radians

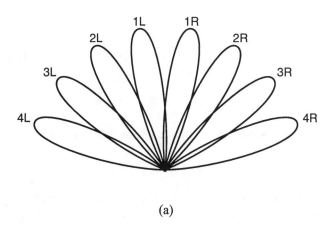

(a)

Figure 8.5 Constrained circuits for forming multiple beams: (a) eight-beam, eight-element Butler matrix.

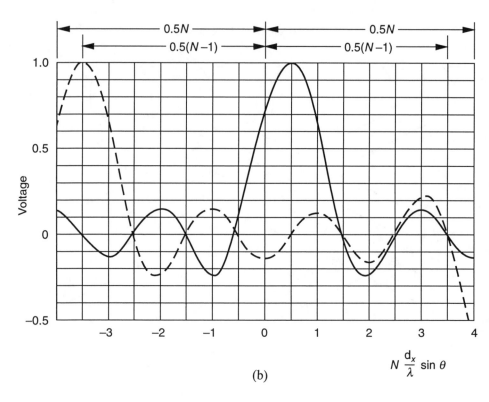

(b)

Figure 8.5 (cont.) Constrained circuits for forming multiple beams: (b) two orthogonal beams (plotted for $N = 8$) with $i = 1/2$ and $i = -7/2$.

by combining N pencil beams. The outermost beam of the set has its peak value at

$$u_{max} = (\lambda/2d_x)[N - 1]/N \tag{8.40}$$

and the phase progression between elements for this beam is

$$\delta_{max} = \pi[1 - 1/N] \tag{8.41}$$

There can be no beam with u larger than this, because the outermost beam is one-half beamwidth from $u = 0.5\lambda/d_x$. If there were another beam at $u = (\lambda/2d_x)(N + 1)/N$, its phase progression would be

$$\delta = \frac{2\pi d_x}{\lambda}(\lambda/(2d_x))\frac{(N + 1)}{N} = \pi(1 + 1/N) \tag{8.42}$$

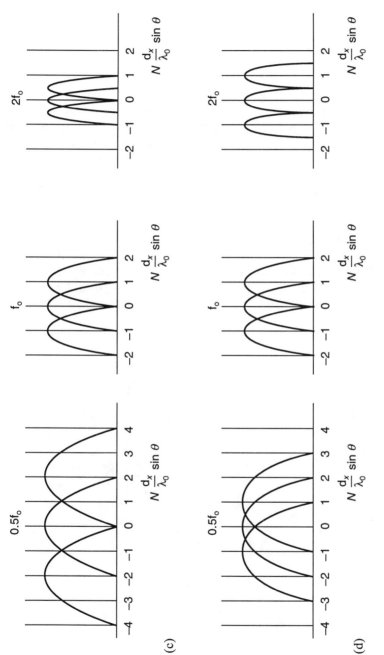

Figure 8.5 (cont.) Constrained circuits for forming multiple beams: (c) beam cluster motion as function of frequency; (d) beam cluster geometry for time-delayed beams.

which is the same as

$$\delta = \pi(1 + 1/N) - 2\pi = -\pi(1 - 1/N) \qquad (8.43)$$

and so is the phasing for the beam at the left of the set.

In this case, the frequency dependence has been retained and signifies that the beam angles u_i vary with frequency, because the δ_i are independent of frequency. The Butler matrix thus forms phase-steered beams which *squint* with frequency. The fan of beams is narrowed at the highest frequency and broadened at the lowest frequency, but the beams remain orthogonal (Figure 8.5(c)). If the beams were time-delay steered instead of phase steered, Figure 8.5(d) indicates that the beams would overlap at low frequencies and have low crossover points at the higher frequencies.

Figure 8.6 shows the Blass matrix [14,16], an alternative constrained network for multiple-beam forming. The Blass matrix uses relative line lengths to provide steering phases and power dividers (directional couplers) to excite the multiplicity of beams. The circuit suffers loss from the coupling network, even for orthogonal beams. The circuit [14] produces true time-delayed beams that do not squint with frequency.

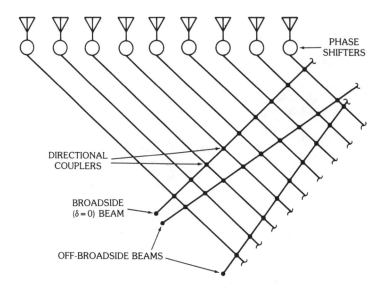

Figure 8.6 Constrained Blass true time-delayed multiple-beam forming circuit.

Multiple-beam lenses and reflectors are true time-delay devices, designed to scan on the basis of optical path lengths, and their radiated beams are essentially fixed in space. The individual beams broaden at the low frequencies and narrow at the high frequencies while remaining fixed in position, as indicated in Figure 8.5(d), so the interbeam spacing changes with frequency and the beams are not orthogonal except at a single design frequency. Mathematically, one can write the beams in the same format as (8.1), but with beam maximum locations fixed at values u_i that are fixed in location, independent of frequency. For an array or lens or reflector with true time-delayed beams at some center frequency designated by the wavelength λ_0, at which the beams are chosen to be orthogonal, one selects

$$u_i = (\lambda_0/L)i \tag{8.44}$$

Since the peaks of time-delayed beams are fixed in space at all frequencies, and the width of the beams as measured to the four-dB point is $\lambda/2Nd_x$, the beams narrow with increasing frequency, and the extent of their overlap changes. At the high frequencies, the crossover levels are very low (the beams overlap very little), while at the low frequencies the beams cross at higher levels, and so suffer orthogonality loss. The variation in crossover level is depicted in Figure 8.5(d).

Figure 8.7 shows several lens and reflector geometries used in wide-band multiple-beam systems. Lens and reflector multiple-beam antennas are true time-delay devices and so have good wide-band properties. Lenses offer more flexible design conditions than reflectors because the specular reflection from the reflector surface determines the angle of local radiation (ray path), while with a lens this is a degree of freedom that can be used in the design of the lens scanning characteristics. On the other hand, lenses are bulkier and heavier than reflector multiple-beam systems. The text by Sletten [17] lists a number of reflector and lens scanning systems. Other recent book chapters by McGrath [18] and Lee [19] discuss constrained and optically designed lenses.

The term *constrained lens* refers to the way the electromagnetic energy passes through the lens face. Unlike dielectric lens action, a constrained lens includes a number of radiators to collect energy at the lens back face and to reradiate energy from the front or radiating face. Within the lens, the energy is constrained by transmission lines, and this allows design freedom to tailor the lens scanning characteristics. The constrained lens of Figure 8.7(a) is called *Bootlace* and allows the front- and back-face elements to be displaced to optimize performance. The so-called *Rotman lens* geometry [20] shown in Figure 8.7(b) is a variety of the more general *Gent* bootlace lens. The Rotman lens is a two-dimensional lens with a flat front face. Signals received from a radiating feed are picked up by radiators at the back face of the lens and distributed by transmission lines to radiate at the lens front face. The lens front face radiator locations (y_n) are not the same as those on the back face, and this adds an extra degree of freedom to the design. Rotman

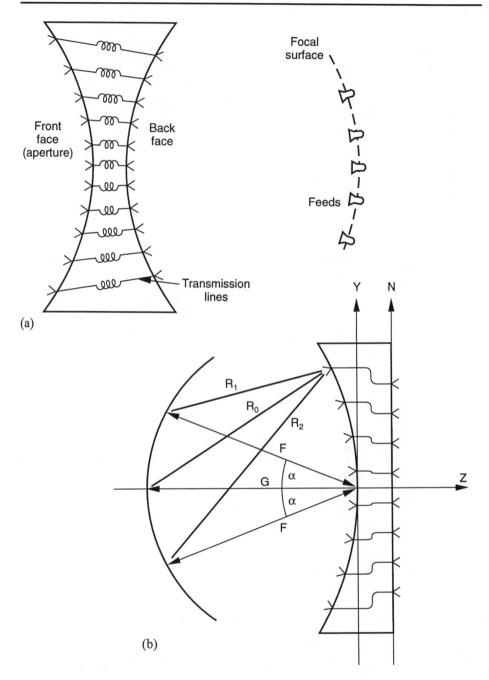

Figure 8.7 Several multiple-beam lens and reflector systems: (a) generic bootlace lens (after [18]); (b) Rotman lens.

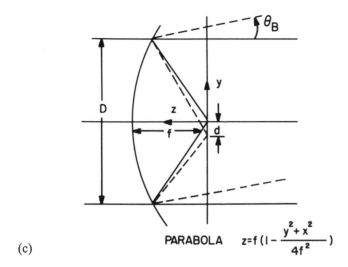

$$\text{PARABOLA} \quad z = f\left(1 - \frac{y^2 + x^2}{4f^2}\right)$$

(c)

Figure 8.7 Several multiple-beam lens and reflector systems: (c) reflector with displaced feed for multiple-beam radiation.

showed that the lens has three points of perfect focus, one on axis and two symmetrically displaced from the axis.

The Rotman lens is an excellent scanning system and has found use in a number of military and civilian systems as a fundamental multiple-beam antenna and as a feed for reflector and lens systems. Here, however, it is but one example of a multiple-beam system. The text by Ajioka [21] gives a description of a number of such multiple-beam systems.

Reflectors Scanned by Off-Axis Feeds

Reflector systems can be scanned by lateral displacement of the feed array from the true focus. Unless special shaping techniques are used, only the on-axis focus is true, and the off-axis beams have a number of aberrations, including defocusing, coma, astigmatism, and higher order aberrations. However, additional feed points can provide adequate scanned beams if some sidelobe deterioration can be tolerated. Reflector systems can therefore be used as multiple-beam systems and as shaped-beam systems by combining clusters of these constituent beams [22].

The beams of a parabola with feed displaced as shown in Figure 8.7(c), though imperfectly formed, are scanned to an angle θ_B related to the feed offset angle $(\tan^{-1} d/f)$ by a factor called the *beam deviation factor* (BDF).

$$\text{BDF} = \frac{\theta_B}{\tan^{-1}(d/f)} \tag{8.45}$$

The beam deviation factor is generally between 0.7 and 0.9 and increases with f/d. Lo [23] gives typical values of the beam deviation factor as a function of f/d.

There have been many studies of the best surface on which to locate the off-axis feed. Ruze [24], using geometrical optics, showed that when astigmatism is neglected, the feed locus for shaped nulls is given by

$$z = -\frac{y^2}{2f} \tag{8.46}$$

a relationship that defines a parabola called the *Petzval* surface. For a feed on this surface, the beam can be scanned a number of beamwidths θ_3 (with a -10.5-dB coma lobe and 1-dB reduced gain) as given by

$$\frac{\theta_B(\text{max})}{\theta_3} = 0.44 + 22(f/d)^2 \tag{8.47}$$

For example, for $f/d = 0.4$, $\theta_B(\text{max})$ is ± 3.96 beamwidths off axis according to this criterion. Other research studies have investigated large lateral feed displacement [25] and the off-axis scanning of feeds on optimized surfaces derived using physical optics, which are close to, but not identical to, the Petzval surface [26].

If a parabola is to be used to form a shaped beam, or multiple beams in one plane, while maintaining an on-axis beam in the other, then other loci define the best positions for a multiple-element feed. The equations for these lines are given in [17], as well as in previous references by Sletten [27] and others. Since it is beyond the scope of this text to detail these elements of off-axis reflector feed design, the reader is referred to the previous references.

In summary, reflector feed displacement produces scanning, accompanied by high sidelobes, and it is not possible to produce low-sidelobe scanned beams by feed displacement alone. These higher sidelobe beams may be perfectly adequate for shaped-beam synthesis, however, and more sophisticated techniques to be described later do provide for high-quality scanning of reflectors.

8.2 ANTENNA TECHNIQUES FOR LIMITED FIELD-OF-VIEW SYSTEMS

A variety of techniques have been developed for special systems that need to scan high-gain antenna patterns over a limited sector of space. These techniques are often discussed in general as beam forming feeds, and as such have application not only for limited field of view systems, but also as wide-band scanning systems. The early survey by Tang [28] addressed the broadband aspects of many of the approaches cited here and reviewed some of the early historical developments of this technology. Later references include specific limited field-of-view applications to precision approach radars [29] and surveys by Ajioka and McFarland [21] and Rusch et al. [22].

These limited field-of-view systems range from arrays of horns or subarrays, to a variety of single and dual reflectors, to single or multiple lens systems, and to systems that combine lenses, reflectors, and arrays. Although this variety admits to comparison on a number of different levels (sidelobes, efficiency, pattern control, etc.), the most basic comparison that relates to system cost is the number of required control elements.

8.2.1 Minimum Number of Controls

Several authors [30–32] have investigated the theoretical minimum number of controls necessary to scan a given antenna pattern over a prescribed sector of space. Perhaps the simplest way [33] to understand the reason for the minimum is to realize that the Woodward-Lawson beams form a complete orthogonal set, and that one can synthesize the scanned beam if the entire set of beams is used. However, since a scanned pattern can be approximated using only those beams that span the entire scan sector, the minimum number of controls is that necessary to access that number of beams. Consider a multiple-beam system with N beams and N input ports. That system spans N beamwidths in space (with -4-dB coverage at the outer scan angle), and the set of beams can be accessed by $N - 1$ switches, as shown in Figure 8.1 for the case $N = 8$ (or it covers $N - 1$ beamwidths to the peak of the outermost beams).

Clearly, the rule is that one needs approximately as many controls as the number of multiple beams required to fill the scan sector. Based on this argument, Patton [30] introduced the term *element use factor*, which is the ratio of the actual number of phase shifters in the control array to the minimum number based on this criterion of beam filling. Patton's expression for the minimum number of controls for a one-dimensional array scanning to $\pm \theta_{max}$ and with beamwidth θ_3 is

$$N_{min} = \frac{\sin \theta_{max}}{\sin(\theta_3/2)} \tag{8.48}$$

and N/N_{min} is the element use factor. For a rectangular array with a rectangular scan sector, the minimum number is

$$N_{min} = \left[\frac{\sin \theta^1_{max}}{\sin(\theta^1_3/2)}\right]\left[\frac{\sin \theta^2_{max}}{\sin(\theta^2_3/2)}\right] \tag{8.49}$$

where $\theta^1_{\max}$ and $\theta^2_{\max}$ are the maximum scan angles in the two planes measured to the peak of each beam, and θ^1_3 and θ^2_3 are the half-power beamwidths in these planes.

Several authors have given derivations of similar minimum criteria, notably Stangel [33] and Borgiotti [34]. Stangel's theorem states that the minimum number of antenna elements to scan a solid angle Ω with $G_0(\theta, \phi)$ the maximum achievable gain in the θ, ϕ direction over the sector is

$$N_{\min} = \frac{1}{4\pi} \int_\Omega G_0(\theta, \phi)d\Omega \qquad (8.50)$$

where the integral is taken over the solid angular surface $d\Omega = \sin \theta d\theta d\phi$.

Figure 8.8 shows the relative number of controls N/G_0 for an array to scan over a conical volume of space. The curve is based on a gain envelope $G_0(\theta, \phi) = G_0 \cos \theta$, where $G_0 = 4\pi A/\lambda^2$ is the gain for a uniform aperture and so includes

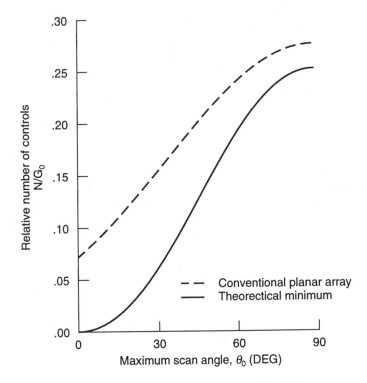

Figure 8.8 Number of controls needed to scan a conical volume (after [31]).

a first-order beam broadening as a function of scan. The comparison is made between the theoretical minimum number and a conventional planar array with half-wave spacing. The figure shows that a significantly reduced number of elements is required if the maximum scan angle is small.

Stangel's theorem has more general application than just the restricted scan case and can be applied to wide-angle scan with tailored gain-scan contours. In the limited scan case, however, and for uniform illumination with $\cos \theta$ scan loss and a rectangular scan sector, one can show that this theorem reduces to the condition given by Patton: for N equal to N_{min}, the element use factor is unity.

One can also relate the element use factor to the array size amd maximum scan. Equation (8.44) can be used for the rectangular array, assuming that the approximate 3-dB points in the planes θ^1 and θ^2 are given by $\theta_3^1 \sim \lambda/N^1 d_1$ and $\theta_3^2 \sim \lambda/N^2 d_2$ for element spacings d_1 and d_2 in the orthogonal planes, and element numbers N^1 and N^2 in those planes. Since the total number of elements is $N = N^1 N^2$ and assuming that each element requires Q controls, one can write the element use factor as

$$N/N_{min} = \frac{0.25Q}{[(d_x/\lambda) \sin \theta_{max}^1][(d_y/\lambda) \sin \theta_{max}^2]} \tag{8.51}$$

Evaluating the array element use factor is thus simply related to how far the array scans in $(d/\lambda) \sin \theta$ space. In this form, however, it is clear that if the array could be made to scan to

$$(d/\lambda) \sin \theta_{max} = 0.5 \tag{8.52}$$

in both planes with only one control per element, it would have an element use factor of unity. This fact is explored further in the next section.

The next sections briefly discuss a number of array techniques for limited sector coverage. In addition to element use factors, these techniques differ widely in their peak and average sidelobe levels and relative complexity of implementation.

8.2.2 Periodic and Aperiodic Arrays for Limited Field of View

Since the coverage sector is limited, it seems reasonable that one could develop a high-gain scanning array by using widely spaced, high-gain elements. The elements would have narrowed patterns, as appropriate to the scan sector, and gain commensurate with their interelement spacing.

A periodic linear array, with element spacing d_x more than one wavelength, has grating lobes in real space, with locations given in Chapter 1. The linear array is to scan over some sector $\theta_{max} \leq \theta \leq \theta_{max}$. For $\theta = \theta_{max}$, the nearest grating lobe is in real space at $\sin \theta_p$ for $p = -1$.

$$\sin \theta_{-1} = \sin \theta_{max} - \lambda/d_x \tag{8.53}$$

With still larger spacing, it may be that many such lobes radiate. Since the array factor is multiplied by the element pattern, the grating lobes are suppressed by the element pattern, but the grating lobe nearest broadside is suppressed very little because it is within the element pattern main beam. Figure 8.9 shows three curves and is intended to illustrate the action of the element pattern in altering the radiated pattern. The upper sketch shows the shape of a typical element pattern (an E-plane horn, or uniformly illuminated aperture) that occupies the entire inter-element distance d_x. The element pattern has its peak at $\sin \theta = 0$ and its nulls at $n\lambda/d_x$ for all n. The array pattern, assuming d_x is several wavelengths, has a main beam and a spectrum of equal grating lobes spaced λ/d_x apart in $\sin \theta$ space. For the array at broadside, each of these lobes is suppressed by the element pattern nulls, and only the main beam contributes to the product of element pattern and array factor. The central sketch in the figure shows the grating lobe spectrum for a main beam moved away from broadside. For this case, the lower sketch shows that the product of element pattern (upper figure) and array factor (central figure) produce a radiation pattern that has some grating lobe suppression for the far grating lobes, but the nearest one to the main beam is within the main beam of the element pattern, and so is suppressed very little. If the array were scanned to $d_x/\lambda \sin \theta = 0.5$, the main beam and grating lobe would be equal.

It is possible, however, to specify an ideal element pattern that will suppress the grating lobe and therefore to use larger element spacings and fewer array elements. Such an ideal pattern [34] (shown in Figure 8.10) would have a nearly constant level out to the maximum scan angle θ_{max} and be zero outside to suppress the grating lobe. This pattern, with its steep edges, allows the maximum element spacing and thus minimizes the number of elements and controls. For the ideal pattern, and for a very large array, the grating lobe is suppressed if it is just outside of the element pattern. This implies that the ideal pattern is constant out to

$$(d_x/\lambda) \sin \theta_{max} = 0.5 \tag{8.54}$$

and zero thereafter. This condition gives the largest spacing d_x consistent with grating lobe suppression, and is precisely the criterion that leads to an element use factor of unity in (8.51). The above can thus be seen as an alternate way of understanding the condition for the minimum number of controls.

Unfortunately, it is not possible to synthesize the ideal element pattern with a single element of width d_x. For example, if the aperture illumination is continuous,

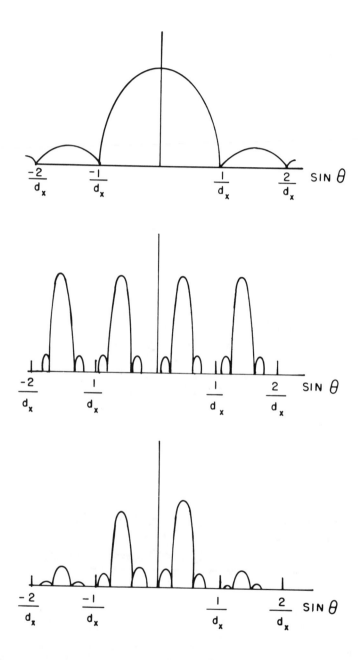

Figure 8.9 Element pattern (top), array factor (middle), and array (bottom) pattern for *E*-plane uniformly illuminated (horn) aperture.

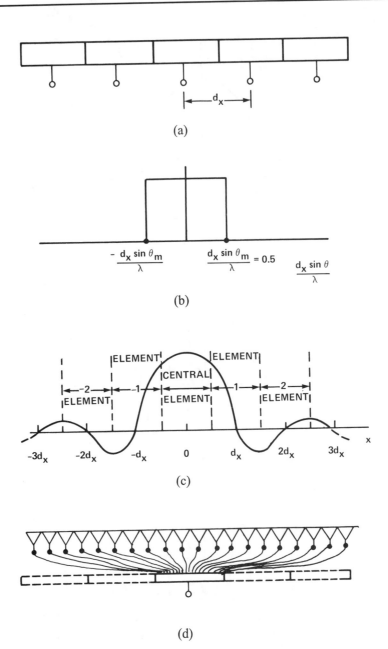

(a)

(b)

(c)

(d)

Figure 8.10 Element or subarray aperture distribution for ideal limited field-of-view scanning system: (a) oversize elements or subarrays for limited scan; (b) ideal element pattern for limited scan system (cos θ suppressed); (c) subarray distribution for scan to $(d_x/\lambda) \sin \theta = 0.5$; (d) overlapped feed distribution network.

one can compute the required illumination as an inverse transform of the ideal pattern. The ideal pattern is

$$f(u) = 1 \qquad -u_{max} \leq u \leq u_{max}$$
$$= 0 \qquad |u| > u_{max} \tag{8.55}$$

for $u_{max} = 0.5/(d_x/\lambda)$. The required illumination is

$$a(x) = \int_{-\infty}^{\infty} f(u)e^{-jux2\pi/\lambda}du$$
$$= u_{max}\frac{\sin[(2\pi/\lambda)xu_{max}]}{(2\pi/\lambda)xu_{max}} \tag{8.56}$$

This distribution is sketched in Figure 8.10(c). Its first zero is at $x = \lambda/2u_{max}$ = d_x and the illumination oscillates with equally spaced zeros. Thus, to synthesize the ideal element pattern requires an amplitude distribution that extends over a large number of elements. Adjacent elements (or subarrays) would have the same aperture illumination as the above, but with peaks at $x = nd_x$ for integer n.

One can synthesize an ideal element pattern only by building a network that connects each port with a subarray of many elements. Since this is so for each input port, the subarrays overlap and can approximate the complex distribution of Figure 8.10(c). The most successful examples of such overlapped subarray synthesis to date have been achieved using space-fed subarrays and will be described in later sections, where they are referred to as *dual transform systems*. Other subarray schemes are also described in later sections. A specific case shows that a conventional filled array with 0.5λ spacing in both planes and scanning over the entire hemisphere ($\theta_{max} = 90$ deg) has an element use factor of unity.

Periodic Horn Apertures

Waveguide-excited horn elements are efficient, high-power radiators and are particularly appropriate for limited field-of-view systems on high-altitude satellites. The array can be scanned within the limits allowed by the grating lobe lattice and the element pattern falloff. Figure 8.11 shows the element patterns for ideal E- and H-plane horn apertures. The antisymmetric patterns shown in the figures are for reference only and are discussed later. For a scanned array, the level of main beam and grating lobes are indicated by the vertical lines on the figures. The first symmetrical E- and H-plane first pattern nulls are at $(d_x/\lambda) \sin \theta = 1.5$ and in the H-plane and $(d_y/\lambda) \sin \theta = 1.0$ in the E-plane. The H-plane grating lobes for broadside scan are at $(d_x/\lambda) \sin \theta = \pm1, \pm2, \ldots$. In the H-plane, these lobes radiate even

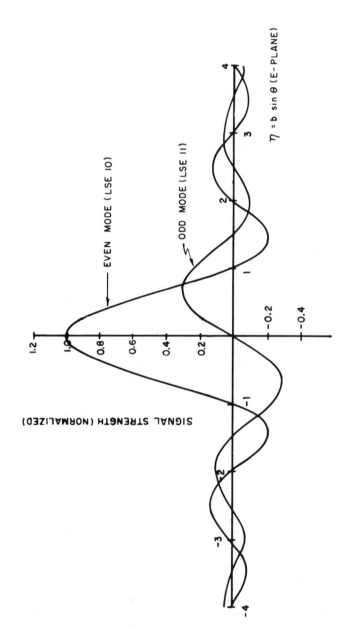

Figure 8.11(a) Waveguide horn element patterns and grating lobe locations: *E*-plane element pattern.

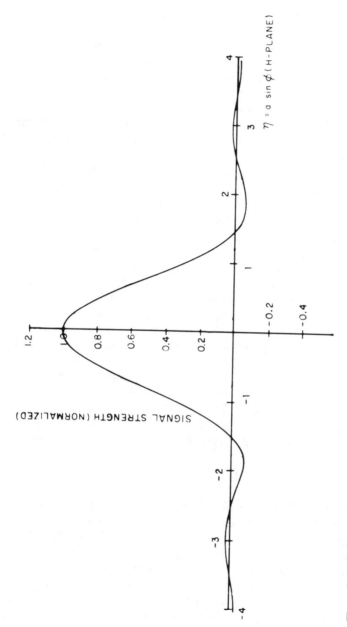

Figure 8.11(b) Waveguide horn element patterns and grating lobe locations: *H*-plane element pattern.

when the array is at broadside and become increasingly severe as the array is scanned. In the E-plane, the broadside grating lobe locations correspond to the element pattern nulls and do not radiate. As the array is scanned, the grating lobes move out of the element pattern nulls and radiate with the levels shown by the dashed vertical lines. The nearest grating lobe increases substantially as it moves up the edge of the main beam of the element pattern. Some improvement of the H-plane pattern can be produced by dielectric loading [35] the H-plane horn edges to increase the illumination at the edges. This narrows the pattern and brings the nulls in closer to broadside to cancel the H-plane grating lobes at broadside scan.

Practical horn elements have also filled nulls in the E-plane (because of phase front curvature), and so it is usually necessary to add a focusing lens at each horn aperture in order to have the E-plane grating lobes nulled for broadside radiation.

From the above, it is clear that single-mode horns always have grating lobes present when scanned, and these lobes get quite large as the scan angle increases. Horn arrays are still useful for satellite-based limited scan systems, however, if the radiating grating lobes are beyond the rim of the earth. In that case, the elements are used with spacings that allow scanning to angles at which the gain becomes appreciable. For 3-dB loss, this is approximately $(d/\lambda) \sin \theta_{max} = 0.44$ in either plane. Amitay and Gans [36] describe the use of an array of oversize horns as feed for a limited field-of-view imaging system. Here again, the array needs to scan over only a limited sector, and so large aperture horns are appropriate.

Multimode horn apertures [37,38] can provide a degree of grating lobe control by producing a null in the horn element pattern to suppress the dominant (first) grating lobe. This is depicted in Figure 8.12(a). The technique is best illustrated by referring to the symmetrical and antisymmetrical patterns of Figure 8.11. The technique combines the symmetrical horn mode with the asymmetrical (odd) mode (in each plane), and, as shown in Figure 8.12(a), results in a shift of the element pattern peak in the direction of scan. This control can produce scan to quite large values of the $(d/\lambda \sin \theta)$ variable in the E-plane. Typically, the horn scans to

$$d_E/\lambda \sin \theta_{max}^E = 0.6 \tag{8.57}$$

for an E-plane horn. This control can be achieved using two phase shifters per horn instead of one and using passive circuits like that shown in Figure 8.12(b).

Although the primary E-plane grating lobe is effectively reduced by this procedure, other grating lobes increase with scan, and the end-of-scan condition is accompanied by other grating lobes at levels from -20 to -13 dB below the beam peak. A collimating lens is also required to suppress unwanted grating lobe radiation for the array at broadside. Figure 8.12(c) shows a sketch of an E-plane horn aperture with the collimating dielectric lens and phase shifter combination to excite even and odd modes with the proper ± 90-deg phase relationship at the horn aperture. Figure 8.12(c) shows a computed pattern for an ideal horn aperture at

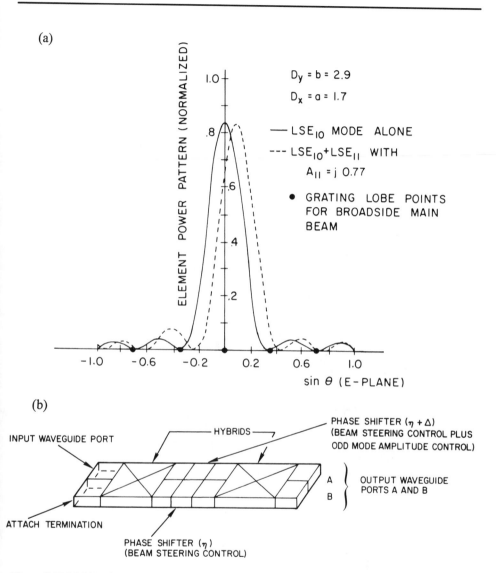

(a)

$D_y = b = 2.9$

$D_x = a = 1.7$

—— LSE$_{10}$ MODE ALONE

--- LSE$_{10}$+LSE$_{11}$ WITH
$A_{11} = j\ 0.77$

● GRATING LOBE POINTS
FOR BROADSIDE MAIN
BEAM

ELEMENT POWER PATTERN (NORMALIZED)

sin θ (E-PLANE)

(b)

INPUT WAVEGUIDE PORT

HYBRIDS

PHASE SHIFTER $(\eta + \Delta)$
(BEAM STEERING CONTROL PLUS
ODD MODE AMPLITUDE CONTROL)

A
B
} OUTPUT WAVEGUIDE
PORTS A AND B

ATTACH TERMINATION

PHASE SHIFTER (η)
(BEAM STEERING CONTROL)

Figure 8.12 Multimode horn apertures for limited field of view: (a) control of element pattern null for grating lobe suppression; (b) circuit for nulling.

broadside (solid) and scanned (dashed) by application of the appropriate level of odd-mode excitation. Motion of the beam peak reduces scan loss, while motion of the first null to the left of broadside ensures good suppression of the first grating lobe. Figure 8.12(d) shows the locus (dashed) of grating lobe levels as the element (and array) are scanned, with the $n = -1$ lobe nulled at all angles.

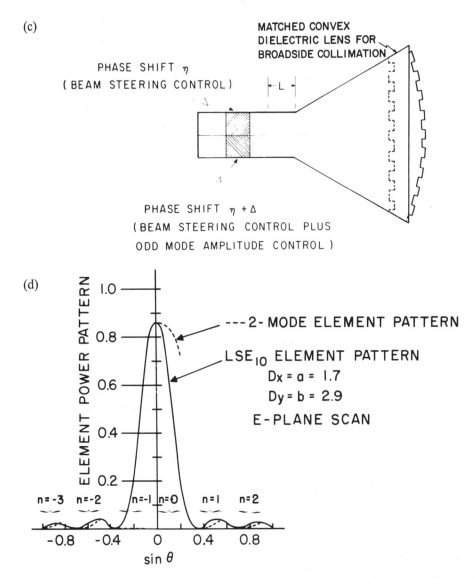

Figure 8.12 (cont.) Multimode horn apertures for limited field of view: (c) *E*-plane horn and lens with odd-mode amplitude control; (d) locus of grating lobe power levels and main beam scan.

In the *H*-plane the element pattern is much broader than in the *E*-plane, and something must be done to suppress the broadside grating lobes, which are at about -9.5 dB below the main beam. The technique of loading the horn edges [35] eliminates this broadside grating lobe, but does not substantially improve the end-of-scan grating lobes, which can be as large as -7 dB relative to the main beam at

$$(d_H/\lambda) \sin \theta_{\max}^H = 0.6 \tag{8.58}$$

Using (8.51) with $Q = 4$ (4 phase controls per element), the element use factor is about 2.8. If a conventional array were to scan over a ± 10-deg sector using conventional 0.5λ spacing, the element use factor would be 8.3, or about 3 times as many controls as for this structure. For ± 5 deg, the element use factor is 33 or nearly 12 times as many as for this array approach. There is a major advantage to these techniques if the scan sector is small and as long as sidelobes are not a consideration.

The problem with all of the horn aperture limited scan techniques is in maintaining sidelobes below a tolerable level. At the scan limit, the primary remaining lobes are at -12 to -14 dB in the E-plane and -7 to -9 dB in the H-plane. The use of random row displacements as described in Chapter 2 is shown to reduce all of the set of lobes (u_p, v_q) for p not equal to zero by the factor given in Chapter 2 (2.54), which can be as large as $1/N_y$ for a uniformly illuminated array, where N_y is the number of rows. For a large array, this can be a 10- to 20-dB reduction in these grating lobes. The set of grating lobes (u_0, v_q) lies along the ridge $u = u_0$ and is unaltered by the row displacement. If the array has a limited field of view in one plane and wide-angle scan in the other, then by using the row displacement and smaller spacings in the wide scan plane, one can suppress all the radiating grating lobes. When limited scan is required in both planes, one must use some other technique to further suppress the (u_0, v_q) grating lobes if they are intolerable.

Angular Filters for Grating Lobe Suppression

One technique that provides sidelobe or grating lobe suppression is the use of angular filters that use dielectric layers [39] or metallic screens [40] in cascade to produce an angular passband-stop band selectivity. Figure 8.13 shows the geometry and characteristics of a dielectric layer angular filter. A full electromagnetic model was used in the design of the filter and in these calculations. In order to achieve a steep angular passband with modest dielectric constants, the grid spacing of one wavelength was used between filter elements. While this did result in good suppression of near grating lobes, it created a second passband beyond about 60 deg in all planes, and this allowed radiation for larger angles to pass unattenuated. Figure 8.14 shows a metal grid angular filter (radome), its approximate transmission line equivalent circuit, where

$$k_x = \frac{2\pi}{\lambda} (1 - v^2 - u^2)^{1/2} \tag{8.59}$$

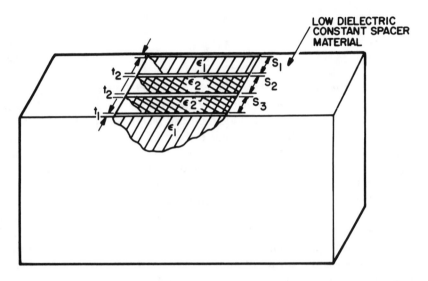

Figure 8.13(a) Dielectric layer angular filter geometry and performance: dielectric layer filter.

and the angular passband characteristics of a particular filter with dimensions given in the reference. The calculation leading to Figure 8.14(c) was a full wave electromagnetic analysis and shows the (UV) filter passband in $u - v$ space (with reflection loss in decibels) to be almost perfectly circular, as would be expected using k_x, a constant in (8.59), for the simplified equivalent circuit. Figure 8.15 shows the scan performance of an eight-element array with and without the use of an early dielectric angular filter [38]. As mentioned earlier, this filter provides extremely good suppression of the near grating lobes, but has a second passband that allows significant radiation at larger angles.

Constrained Overlapped Networks for Limited Field-of-View Arrays

The "ideal" element pattern of Figure 8.10 would enable an array to scan to $(d/\lambda) \sin \theta_{max} = 0.5$, and so achieve an element use factor of unity. The pattern is clearly very narrow compared to a uniformly illuminated aperture of dimension d, which would have nulls at $(d/\lambda) \sin \theta = \pm 1, \pm 2, \pm 3, \ldots$. Thus, it is apparent that no single element can produce the ideal pattern; the required coverage needs an element that is actually larger than the available space and must overlap with adjacent elements.

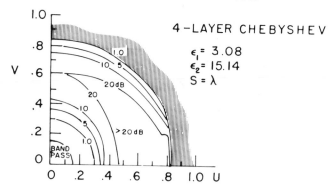

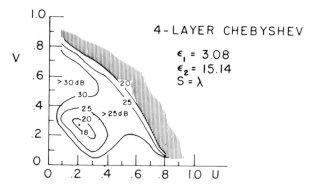

Figure 8.13(b) Dielectric layer angular filter geometry and performance: filter angular transmission characteristics.

Perhaps the simplest network to achieve a degree of overlap consists of a power divider combination shown in Figure 8.16(a). Originally called a phase interpolation network, this circuit has the advantage of being lossless at broadside and very simple to build because it requires conventional sum and difference power dividers, which are inexpensive to produce. For a uniformly illuminated array, the sum hybrids "interpolate" the phase between adjacent signals to produce a signal with phase angle half way between those of the adjacent signals. Since this is done

(a)

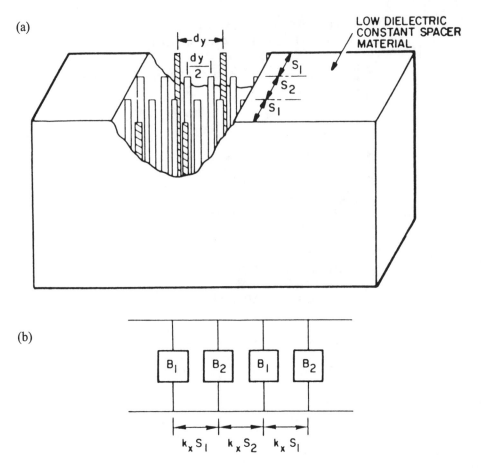

Figure 8.14 Metallic grid angular filter geometry: (a) metal grid filter; (b) equivalent circuit of metal grid filter.

by direct addition when the phase difference is less than 180 deg, there is an amplitude modulation imposed on the array that constitutes an error signal. There is no phase error if the array is uniformly illuminated, but for a tapered array there is a symmetrical phase error that leads to increased sidelobe levels. The signal amplitude at the interpolated ports is shown in Figure 8.16(a) and given by

$$S = \frac{\sin \delta}{2 \sin(\delta/2)} \tag{8.60}$$

corresponding to the phase difference between the two signals of δ.

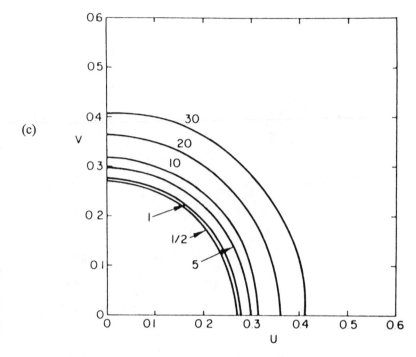

Figure 8.14 (cont.) Metallic grid angular filter geometry: (c) pass-band characteristics of metal grid filter.

This phase shift δ corresponds to the scan angle θ_{max}, where

$$\delta = 2\pi(d/\lambda) \sin \theta_{max} \qquad (8.61)$$

and d is the distance between each subarray of two elements, or between each active control. The amplitude error produces sidelobes at $u = u_0 + \lambda/d$ that have amplitude that grows with increased scan angle. At $\delta = \pi/2$, the value of S is about 0.707 and

$$(d/\lambda) \sin \theta_{max} = 0.25 \qquad (8.62)$$

corresponding to an element use factor of 2 for a one-dimensional array or 4 for a two-dimensional array.

Figure 8.16(b) shows a typical pattern, in this case for a 21-element array at the scan limit and assuming cosine element patterns. With d equal to 0.75 wavelengths, the maximum scan angle is $u_0 = 0.166$ (about 9.6 deg) and the undesirable

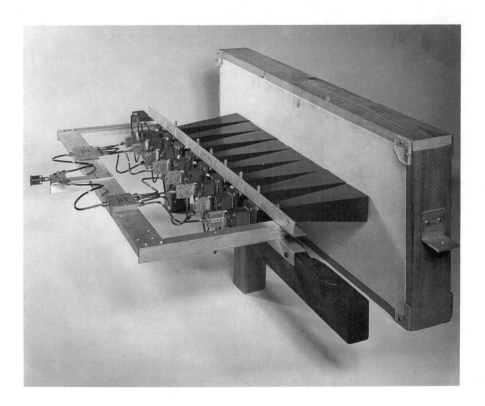

Figure 8.15(a) Limited field-of-view array: test array with dielectric angular filter.

lobes are at about -15 dB. Depending on the element pattern, this can be reduced by a few decibels, but since the illumination is uniform, this level corresponds to a reasonable end-of-scan limit.

In addition to the limit given above, to suppress the grating lobe due to the spacing $d/2$ between each element, one must require that

$$(d/\lambda) < 2.0 \qquad (8.63)$$

Some improvement can be gained by choosing the power division so that the interpolated signal is larger at broadside, but this introduces added complexity that must be weighed against the improved scan.

The use of higher order modes in horn apertures is suggested by Figure 8.10, which shows that the ideal illumination is in phase at the central element (or subarray) and dominantly asymmetrical over the adjacent subarrays. Mailloux [34]

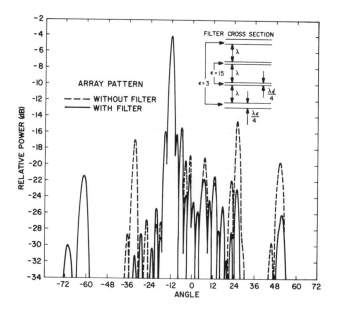

Figure 8.15(b) Limited field-of-view array: scanned array pattern with and without dielectric filter (element spacing 2.9λ).

devised a network to approximate such an illumination using only nearest-neighbor overlap, and showed that a flat-topped subarray pattern could be synthesized in this manner to allow scanning to about

$$(d/\lambda) \sin \theta_{\max} = 0.33 \tag{8.64}$$

The basic circuit, shown in Figure 8.17(a,b), uses sum/difference hybrids to couple into dual-mode horns. An extra 90-deg phase shift, though not shown, is applied to the difference signal. The difference signal amplitude is zero when the array is at broadside, but grows as the array is scanned. This, in effect, automatically tilts the element pattern of each horn to provide improved scan coverage for small angles.

An alternative perspective is gained by considering only one input signal divider into sum/difference components of three elements, with the difference term not present at the central element. The adjacent elements get both the sum and difference contributions, and this nearest-neighbor coupling is used to approximate the required overlapped sin x/x illumination. Figure 8.17(c) shows the theoretical subarray patterns for a typical subarray for several values of the coupling coefficient

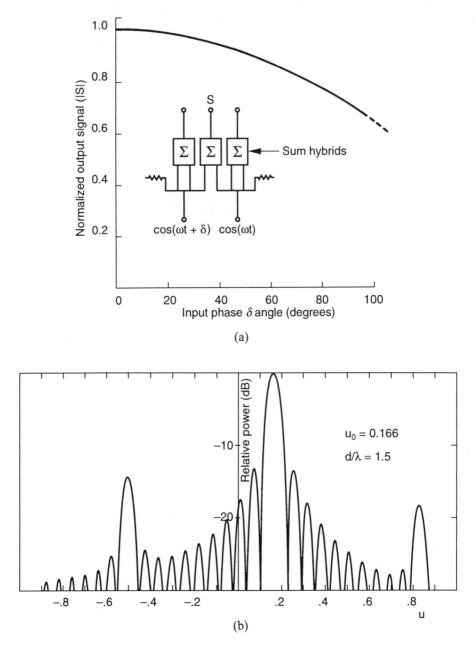

Figure 8.16 Phase interpolation networks for limited field of view: (a) Basic network and signal amplitude at interpolated port (after [41]); (b) end-of-scan pattern for 21-element array ($u_0 = 0.166$ and 1.5λ between subarrays).

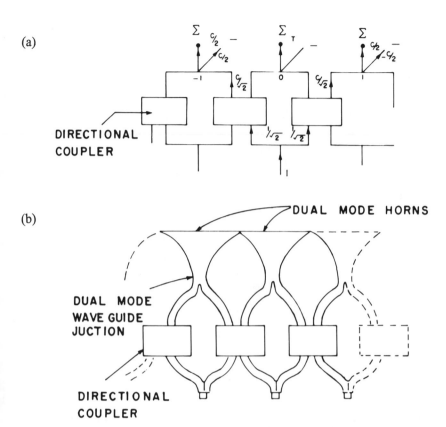

Figure 8.17 Overlapped subarray with higher order mode overlap (after Mailloux [34]): (a) circuit for overlapped array excitation; (b) waveguide network for overlapped array of horns.

C. With the main beam scanned throughout the shaded region ($n = 0$), grating lobes scan through the other shaded regions ($n = \pm 1, \pm 2, \ldots$). The grating lobes can be approximated by the subarray pattern height. The "element" of subarray pattern has the proper flat top form and can be optimized to give good suppression of the nearest grating lobes and some suppression of the second, which ultimately reaches the -15-dB level at the scan limit.

The most generalized constrained networks for forming overlapped subarrays were devised by Dufort [42], who synthesized lossless constrained modular coupled networks that achieve varying degrees of overlap and allow grating lobe suppression with limited angular scanning.

(c)

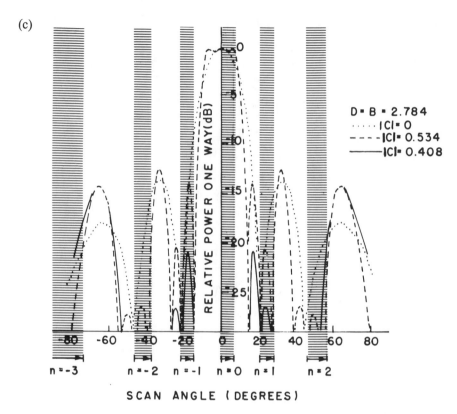

Figure 8.17 (cont.) Overlapped subarray with higher order mode overlap (after Mailloux [34]): (c) subarray pattern of overlapped elements.

Aperiodic Arrays

All of the above schemes are used with fully periodic arrays and so ultimately require the suppression of well-defined grating lobes. A number of investigators have developed arrays for limited field of view using aperiodic array lattices. Circular array lattices have been particularly convenient in breaking up the periodicity and thus reducing the peak grating lobes. In principle, if one could achieve complete randomization, as with the random arrays of Chapter 2, one could reduce the average sidelobe level to $1/N$ for an array of N elements. In most cases, the degree of randomization available with the tightly packed arrays that have been used for limited field-of-view systems is such that sidelobes remain above this level. An example is the structure shown schematically in Figure 8.18(a), which consists of a number of elements with roughly equal area, arranged in a circular grid. This

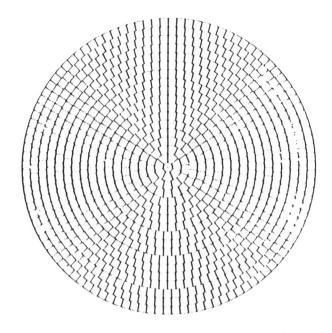

Figure 8.18(a) Aperiodic arrays for limited field of view: aperiodic array (after [30]).

geometry was investigated by Patton [30], who built a 10-foot diameter array at C-band and conducted a theoretical study of a 30-foot array. The circular array consists of dipole subarrays arranged in an aperiodic fashion and excited by an optical power divider feeding a spherical array back face. The subarrays have equal areas, and their size ultimately determines the maximum scan angle of the antenna at the subarray half point, or approximately

$$(d/\lambda) \sin \theta_{\max} = 0.44 \qquad (8.65)$$

corresponding to an element use factor of $0.25/(0.44)^2$ or 1.3. The 30-foot array consisted of 1,000 subarrays, and scanned a 0.36-deg beamwidth throughout approximately a 5-deg cone to obtain the element use factor of 1.3. Peak sidelobes were -15 dB for the 10-foot antenna and predicted to be -20.9 for the 30-foot array. Average sidelobes were high, consistent with a 3-dB loss in gain at the scan limit.

A similar antenna, but using unequal size elements, was developed by Manwarren and Minuti [43]. This antenna was designed to scan a 1-deg pencil beam

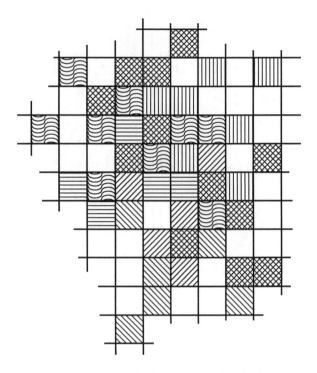

Figure 8.18(b) Aperiodic arrays for limited field of view: interlaced subarray antenna: square grid represents element lattice, different shading represent common subarray elements (after [44]).

over an 8-deg half-angle cone, with −20-dB grating lobes. The antenna consisted of 412 elements and used elements of three different sizes that were arranged in concentric rings to produce the pseudorandom grid. The element use factor was approximately 1.6.

Stangel and Ponturieri [44] studied randomized interlaced subarray configurations that produce low sidelobes because of the aperiodic grid, yet have good aperture efficiency because of the complete filling of the aperture. Figure 8.18(b) illustrates the meaning of the term *interlaced* in this context. The square grid is filled with elements, and elements of common shadings are connected together and fed in phase as a subarray. The intersubarray distance is chosen to give a regular lattice of subarray centers, but the actual subarray configuration is chosen by a random number generating technique. Figure 8.18(c) shows the maximum phase shifter reduction using this technique, as compared with elements on a regular grid with half-wave separation. Although appealing, this technique has not been implemented in a practical array.

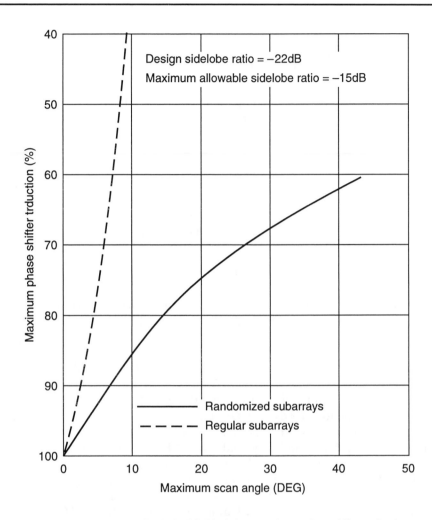

Figure 8.18(c) Aperiodic arrays for limited field of view: maximum phase shifter reduction to achieve a given scan coverage (after [44]).

8.2.3 Constrained Network for Completely Overlapped Subarrays

The network shown in Figure 8.19(a) consists of cascaded multiple-beam matrices and is the most fundamental form of a completely overlapped subarray beam forming network. It often is called a *dual transform network*, since each of the multiple-beam matrices performs a discrete Fourier transform on its set of input signals. The network is shown as a constrained circuit, a combination of Butler or

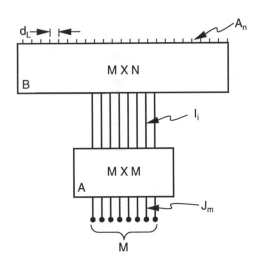

Figure 8.19(a) Cascaded multiple-beam networks of completely overlapped subarray formation: constrained network of cascaded multiple-beam networks.

other multiple-beam networks, but the Fourier transform operations could be achieved with multiple-beam lenses or reflectors, since they are in several of the limited field-of-view systems presented later. The cascaded matrix implementation was proposed by Shelton [45] as a feed for multiple-beam optics, while the matrix-to-lens implementation is the basis of an early development of the transform-fed high-performance subarraying array feed (HIPSAF) [46] lens, referenced in later sections. The overlap is "complete" in that each subarray port excites all of the array elements, and so all of the subarrays overlap.

The sketch of Figure 8.19(a) shows two cascaded multiple-beam systems. At the output, N array elements are fed from the $M \times N$ matrix ($M \le N$). This network is a conventional Butler or other phase shift multiple-beam matrix, with only the circuitry for the (central) M beams included. Excited alone, a signal at an input port of this matrix would provide the phasing for one of the N multiple beams. A key to understanding the operation of this system is to realize that *only* these central beams are accessible at the input ports. In the ideal system, there is no way to put any energy into the remaining $(N - M)$ beams, which represent more rapid phase progression across the array aperture.

The input (network A) beam former is an $M \times M$ Butler matrix, and its M input ports are called *subarray ports*. The subarray input ports are then excited by a set of signals with amplitude-weighted progressive phases.

Before presenting an analysis of the operation of this network, there are several conclusions one can draw from a knowledge of the multiple-beam circuits.

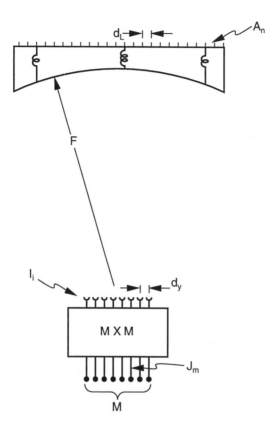

Figure 8.19(b) Cascaded multiple-beam networks of completely overlapped subarray formation: space-fed network.

First, any one input port (say, the *m*th) of the final $M \times N$ matrix (network *B*) excites one of the output beams (the *m*th) of the cluster of *M* beams. Therefore, if the input $M \times M$ matrix (network *A*) were excited in such a way as to provide a nonzero signal at only this *m*th port, then that excitation of network *A* would radiate only the one beam. The required input to produce this output is a uniformly illuminated progressive phase because network *A* is a multiple-beam network now used in reverse as a focusing beam former. Similarly, one can find progressive phases to excite any other of the set of *M* beams. The progressive phase input to the subarray ports can access any of the beams within the limited field-of-view cluster of *M* beams. This description does not explain what happens when a weighted or errored illumination is applied at the subarray input ports, or any further details that result from the detailed analysis to follow.

A signal I_i applied to the ith input port of the upper matrix (the $M \times N$ network) produces a progressive set of phases

$$A_n = I_i \frac{\exp[-ji(n/N)2\pi]}{(N)^{1/2}} \qquad (8.66)$$

at the N array elements spaced d_L apart at the array, and radiates with the pattern below:

$$g_i(u) = \frac{I_i}{(N)^{1/2}} Nf^e(u) \frac{\sin[(N\pi d_L/\lambda)(u - u_i)]}{N \sin[(\pi d_L/\lambda)(u - u_i)]} \qquad (8.67)$$

where $f^e(u)$ is the array element pattern (assumed equal for all elements), and

$$u_i = i\lambda/(Nd_L) \qquad (8.68)$$

This ith beam $g_i(u)$ is one of the Woodward-Lawson beams, and its location is thus frequency dependent, as required to maintain orthogonality at all frequencies.

When the matrix (A) below is used to provide the signals at the input to the $M \times N$ matrix (B), each input J_m excites a set of signals I_{im} at the output on network A and the input of network B. These can be written as

$$I_{im} = \frac{J_m}{M^{1/2}} e^{j2\pi(m/M)i} \qquad (8.69)$$

for $-(M - 1)/2 \leq i \leq (M - 1)/2$.

This illumination is applied to the output network (B) and results in the aperture illumination corresponding to the mth subarray. At the output of network B, the signal at each nth element of the N-element array is

$$\begin{aligned} A_{nm} &= \frac{1}{N^{1/2}} \sum_{i=-(M-1)/2}^{(M-1)/2} I_{im} e^{-j2\pi(n/N)i} \\ &= \frac{MJ_m}{(MN)^{1/2}} \frac{\sin M\pi[(nM - mN)/MN]}{M \sin \pi[(nM - mN)/MN]} \end{aligned} \qquad (8.70)$$

This mth subarray illumination has its maximum at the element with index $n = m(N/M)$, and overlaps all the elements of the array. An example of one such subarray illumination is the dashed curve of Figure 8.20(a) for the subarray ($m = 1/2$) of the array of 64 elements with $\lambda/2$ separation. The array has eight subarrays

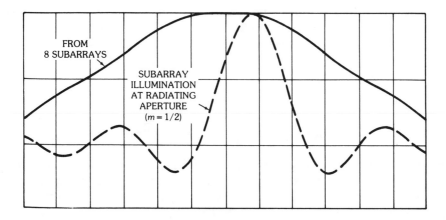

Figure 8.20(a) Characteristics of completely overlapped subarray amplitude illumination for low-side-lobe (-30-dB Chebyshev) pattern, showing one subarray (dashed) for $m = 1/2$, and composite illumination (after [47]).

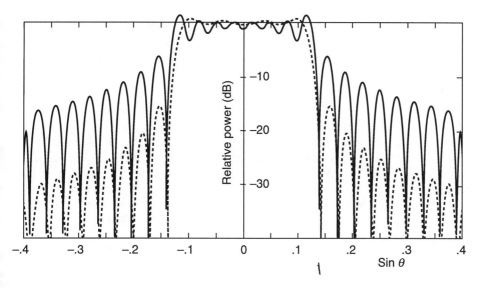

Figure 8.20(b) Overlapped subarray patterns for 64-element array with eight subarrays: central subarray (dashed) $m = 1/2$; edge subarray (solid) $m = -7/2$.

($M = 8$). Each subarray illumination spans the whole N-element array, and so this kind of system is termed a *completely overlapped* subarray system.

Radiated subarray patterns are given in terms of the constituent beams by

$$f_m(u) = \frac{1}{J_m} \sum_i g_i(u)$$

$$= N \frac{f^e(u)}{(MN)^{1/2}} \sum_{i=-(M-1)/2}^{(M-1)/2} e^{j2\pi(m/M)i} \frac{\sin[(N\pi d_L/\lambda)(u - u_i)]}{N \sin[(\pi d_L/\lambda)(u - u_i)]}$$

(8.71)

This expression is the sum of M orthogonal pencil beams arranged to fill the sector to form a flat-topped pattern for the mth subarray. The subarray patterns are similar to each other but not identical. Figure 8.20(b) shows two subarray patterns for the same 64-element array. The selected subarrays are an edge ($m = -7/2$) and one of the two central subarrays ($m = 1/2$). The edge subarray has higher sidelobes and a highly rippled pass region because its illumination is truncated.

Since a total of M constituent beams are used to form the subarray pattern, the peak of the two beams furthest from broadside are at $i = \pm(M - 1)/2$, so $|u_{max}| = (\lambda/2d_x)[(M - 1)/N]$ and the 4-dB points at $(M/N)(\lambda/2d_L)$. The width of the subarray pattern is given by

$$\epsilon = (M/N)(\lambda/d_L)$$

(8.72)

Note that the subarray pattern and its constituent beams move as a function of frequency, and that the subarray width is also a function of frequency.

The array excitation with all subarrays excited is

$$A_n = \sum_{-(M-1)/2}^{(M-1)/2} A_{nm}$$

(8.73)

where the A_{nm} are given in (8.70). If a low-sidelobe illumination is applied at the subarray input ports, that illumination is approximately replicated at the array face. An example of such a composite excitation is shown in the solid curve of Figure 8.20(a).

The radiated array pattern is given in terms of the constituent beams by

$$F(u) = \sum_{-(N-1)/2}^{(N-1)/2} A_{nm} e^{jkud_L n}$$

(8.74)

or equivalently in terms of the output array aperture illumination as

$$F(u) = \sum_{-(M-1)/2}^{(M-1)/2} J_m f_m(u)$$

(8.75)

Application of a steering signal J_m at the subarray input ports selects a combination of the constituent beams both within the subarray pattern and in the array space factor. For example, applying the steering signal vector

$$J_m = |J_m| \exp(-j2\pi m\Delta) \qquad (8.76)$$

can lead to only one constituent beam selected if the $|J_m|$ are all equal and the proper value of Δ is chosen. This can be shown to occur by writing the expression for the total current I_i at the output of the first matrix, with all input signals present and chosen as above. In general,

$$I_i = \frac{1}{(M)^{1/2}} \sum_{-(M-1)/2}^{(M-1)/2} J_m e^{j2\pi(m/M)i} \qquad (8.77)$$

For J_m given above, this reduces to

$$I_i = (M)^{1/2} \frac{\sin[M\pi((i/M) - \Delta)]}{M \sin[\pi((i/M) - \Delta)]} \qquad (8.78)$$

When Δ is chosen equal to i/M, this expression is unity, and all signals at other terminals (I_j) are zero. Only the ith output port has a nonzero signal. This signal becomes the input to matrix B and produces the output phase progression at the nth output port of matrix B:

$$2\pi n(d_L/\lambda)u_i = 2\pi in/N \qquad (8.79)$$

and thus the differential phase between elements of $(2\pi i/N)$.

The array amplitude illumination $[A_{nm}]$, assuming any single input I_{im}, peaks at

$$[nM - mN] = 0 \quad\text{or}\quad n = \frac{mN}{M} \qquad (8.80)$$

and so the number of elements between any two adjacent peaks (for any ith input port) is

$$n_n - n_{n-1} = N/M \qquad (8.81)$$

and the effective subarray size is

$$D = (N/M)d_L \tag{8.82}$$

independent of m. Note that this spacing is independent of frequency for the case of an orthogonal beam (Butler) matrix.

In summary, the input phase progression of $\Delta = 2\pi i/M$ has thus produced an output phase progression of $2\pi i/N$ or the output incremental phase has been reduced by the factor (M/N) compared to the input incremental phase, and the intersubarray distance in the main aperture is d_L increased by the (N/M) ratio. The maximum input incremental phase (to excite the outermost beam) is

$$|\Delta_{max}| = 2\pi \frac{(M-1)}{2M} = \pi(M-1)/M \tag{8.83}$$

and so the maximum output phase is M/N times this, or

$$\delta_{max} = \frac{\pi(M-1)}{N} \tag{8.84}$$

and this corresponds to a maximum scan angle $u_{max} = \sin \theta_{max}$, where

$$\frac{d_L \sin \theta_{max}}{\lambda} = 0.5(M-1)/N \tag{8.85a}$$

or

$$(D/\lambda) \sin \theta_{max} = 0.5 \frac{(M-1)}{M} \tag{8.85b}$$

Since this is the beam peak, the 4-dB point is at $(D/\lambda) \sin \theta_{max} = 0.5$ and the element use factor is unity.

Using the above definitions, one can write the subarray pattern in terms of the subarray aperture distribution A_{nm} from (8.70) as

$$f_m(u) = \frac{1}{J_m} \sum_n A_{nm} \exp(j2\pi n d_L u/\lambda) \tag{8.86}$$

$$f_m(u) = \frac{M}{(NM)^{1/2}} e^{j2\pi u(d_L/\lambda)(N/M)m} \sum_{n=-(N-1)/2}^{(N-1)/2} e^{j2\pi u(d_L/\lambda)[n-mN/M]} \frac{\sin\{(M/N)\pi[n-(mN/M)]\}}{M\sin\{(\pi/N)[n-(mN/M)]\}}$$

This expression gives the pattern of any mth subarray, peaked at the location $n = m(N/M)$ and with phase center at the same point.

Since $nd_L(N/M) = nD$, it is clear that applying the steering vector

$$J_m = |J_m|e^{-j2\pi n(D/\lambda_0)u_0} \tag{8.87}$$

in the array pattern expression (8.75) will scan the peak of the beam to the angle u_0 at the frequency f_0 provided that u_0 is within the subarray angular pass region ($|u_0| < \sin \theta_{\max}$). With this excitation, the array beam will squint with frequency, and applying a time-delayed J_M is necessary for wider band operation.

The flat subarray patterns have the shape required for suppressing grating lobes. If the array pattern were scanned to the peak $u = u_{\max}$, then the first grating lobe of the array is at the angle

$$u = u_{\max} - \lambda/D = \frac{\lambda}{2d_L}[(M-1)/N] - \frac{\lambda M}{Nd_L} = -\frac{\lambda}{2Nd_L}(M+1) \tag{8.88}$$

which is beyond the edge of the subarray pattern (in fact, it is at a zero of the subarray pattern) and so is substantially suppressed.

Figure 8.21 shows an example of limited field-of-view scanning with such a generic overlapped subarray system. The data were computed using the subarray patterns $f_m(u)$ from (8.71) in the pattern expression (8.75), with steering vector in

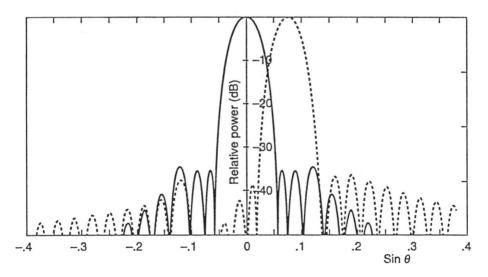

Figure 8.21 Limited field-of-view scanning of constrained overlapped subarray feed (64-element array and 8 subarrays with −40-dB Chebyshev illumination).

(8.87). The results shown in Figure 8.21 demonstrate performance of an array of 64 elements, eight subarrays, and $\lambda/2$ element spacing. The amplitude excitation at the subarray input ports was that of a -40-dB Chebyshev illumination and supported about -35-dB sidelobes at broadside. At $u_0 = 0.75$ (Figure 8.21), the pattern is improved because the near sidelobes on the right are suppressed by the subarray pattern falloff, while those at left are near the center of the subarray pattern and approach the design sidelobe level of -40 dB. The peak at left is the grating lobe onset, and grows to an unacceptable level for $u_0 = 0.1$ (not shown). The data indicate that although the network provides scan for uniformly illuminated subarray ports over the entire set of M output beams with unity element use factor, with a low-sidelobe pattern the beam is so broad that proper grating lobe suppression requires compromise in the element use factor.

8.2.4 Reflectors and Lenses With Array Feeds

Fixed-beam reflector and lens antennas can provide high gain (large aperture) at very low cost relative to array systems. These systems can be scanned over limited angular regions by several different techniques and with varying degrees of success. This section discusses several kinds of array feeds for reflector and lens systems and compares these alternatives to more conventional arrays.

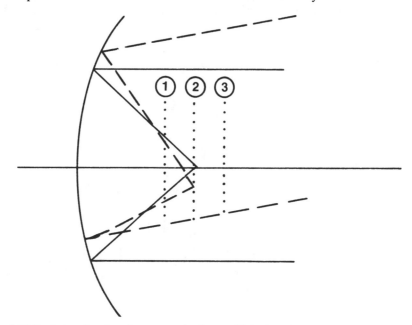

Figure 8.22 Feed plane locations for a scanned reflector: (1) feed at plane closer than focal plane (plane 1); (2) feed at plane passing through focal point; (3) feed at plane beyond focal plane.

The sketch in Figure 8.22 shows three possible feed locations for focus correction of a reflector (or lens). The reflector is shown receiving a plane wave from a direction along the parabola axis (solid lines) and one at an angle θ from that axis (dashed lines). Several alternative feed locations are indicated by straight lines (or planes) but could also be curved three-dimensional surfaces. As the angle of the incident wave changes, the focal point becomes displaced and distorted. The antenna design is selected to best match the off-axis focal spot.

The kind of optical feed required is quite different for each of these three feed locations. In the case of the first type (dashed plane 1), the feed is moved substantially in from the focal plane. The feed is required to match a wavefront that is converging but has nearly constant amplitude and is certainly free of the ripples and sign changes present at the focal plane. At this location, the feed is a relatively large array with fixed amplitude taper and electronic phase shifter control to produce the required phase distribution. Since there is usually no feed amplitude correction, these systems have modest sidelobe levels (usually -20 dB or higher). Clearly, the use of digital beam forming systems will improve this sidelobe level in future systems.

Alternative feed locations are shown at planes 2 and 3. A feed at the focal plane (location 2) is essentially a multiple-beam system, since the received energy is focused to relatively small spot locations in that plane. Electronic control is exercised using a switching matrix to excite the multiple beams separately or in weighted clusters (Figure 8.1).

If a constrained multiple-beam feed is used at location 2, this system is a practical implementation of the completely overlapped subarraying system described in Section 8.2.3 and shown in its ideal form in Figure 8.19(a). Often this type of feed is referred to as a *transform feed*, because the output signals are a discrete Fourier transform of the input signals.

Feeds at location 3 are focusing feeds also and play the same role as the transform feed that might have its front face at location 2. Such feeds will not be distinguished from the transform feeds throughout this section.

Limited Field-of-View Reflector Systems

The earliest forms of limited scan systems consisted of an array used as a transverse feed for a reflector (or lens) antenna [48,49]. The array is located a distance less than the focal length from the reflector (location 1), and so the objective serves to project the incoming received wavefront onto the array face, but does not focus it at the array. The system converts the incident wavefront to another nearly planar wavefront at the array, and a tapered array distribution provides for sidelobe control. A key feature of such systems for the limited field-of-view application is that the objective must be large, because the scanned array illuminates a spot that moves across the main aperture as a function of scan. Design is usually based on the criterion that the array aperture illumination be the complex conjugate of the received field distribution for an incident plane wave [49]. This places a minimum limit on the size of the array, because the usual requirement to scan with phase only requires that the array must be outside of the region of nonuniform fields near the focus. Beginning in the early 1960s, the use of such a transverse feed was investigated by a number of authors and has proven to be an economical means of providing limited sector scanning of reflector antennas.

Winter [49], Tang [50], and Howell [51] have studied the geometrical aspects of scanning and feed blockage with arrays mounted as shown in Figure 8.23(a). Unless extremely large reflectors are used, blockage alone limits the achievable sidelobe level to about −20 dB for on-axis-fed reflector systems. Tang [50] and Howell [51] obtained equations for the size of the reflector based on the geometry of Figure 8.23(a). The figure shows a symmetrical parabolic reflector with an array feed of size $2y_a$. Two rays cross at the top of the array. The lower ray at an incident angle 0 deg is reflected at y_d and passes through the focus. The second ray is incident at an angle θ and hits the parabola at y_e. After reflection, this ray crosses the first ray path at the array edge. The condition for determining array size is to choose the array so that it intercepts all the reflected rays that come from the active region on the reflector for all angles up to the maximum scan angle. Rays at the bottom of the reflector are not considered in this development because they remain on the reflector and add no new constraint.

Reflection from the aperture follows Snell's law:

$$\hat{n}_{REF} = \hat{n}_{INC} - 2(\hat{n}_S \cdot \hat{n}_{INC})\hat{n}_S \qquad (8.89)$$

where $\hat{n}_{REF}$ and $\hat{n}_{INC}$ are unit vectors in the direction of reflected and incident rays, and $\hat{n}_S$ is a unit vector representing the surface outward normal.

Measured from a coordinate system at the parabola focus with z the distance from the y-axis to the reflector surface, the equation of the surface and the outward normal unit vector are

$$z = f[1 - (y/2f)^2]$$

$$\hat{n}_S = \frac{-\hat{z} - \hat{y}(y/2f)}{[1 + (y/2f)^2]^{1/2}} \qquad (8.90)$$

The resulting reflected ray unit vector is

$$\hat{n}_{REF} = \frac{1}{1 + (y/2f)^2} [\hat{y} g_1(y, \theta) + \hat{z} g_2(y, \theta)] \qquad (8.91)$$

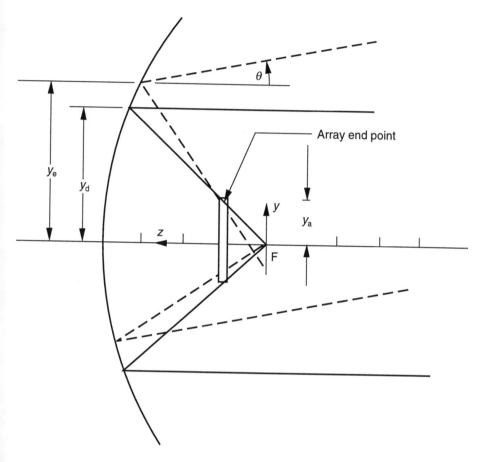

Figure 8.23(a) Reflector scanned by array (after [52]): reflector and array geometry showing ray locations for full array utilization (solid line for on-axis beam, dashed for scanned beam).

where

$$g_1(y, \theta) = [(y/2f)^2 - 1] \sin \theta - (y/f) \cos \theta$$

$$g_2(y, \theta) = [(y/2f)^2 - 1] \cos \theta + (y/f) \sin \theta$$

The resulting size $2y_a$ is given in terms of the other reflector parameters and the given scan angles as [52]

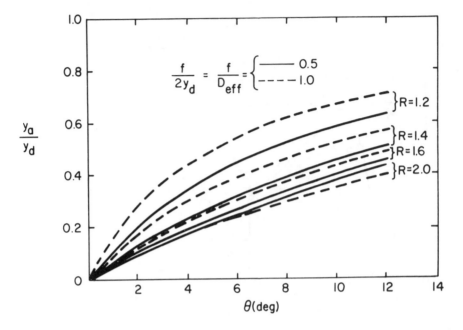

Figure 8.23(b) Reflector scanned by array (after [52]): normalized array size y_a/y_d versus scan angle for oversize ratio $R = y_e/y_d$.

$$\frac{Y_a}{Y_d} = 1 + K_1 \qquad (8.92)$$

where

$$K_1 = \frac{g_2(y_e, \theta)(y_e - y_d)/f + g_1(y_e, \theta)[(y_e)^2 - (y_d)^2]/(2f)^2}{g_2(y_e, \theta)y_d/f + g_1(y_e, \theta)[(y_d/2f)^2 - 1]}$$

Figure 8.23(b) shows the normalized array size (y_a/y_d) as a function of the maximum scan angle θ and several values of allowed oversize ratio or spot motion $R = (y_e/y_d)$ for effective focal length ratios $(f/2y_d)$ of 0.5 and 1.0. The figure shows that the illuminated region must be allowed to move in order to fully utilize the array for all scan angles. The choice of a large y_e tends to make the array smaller.

The result is used to estimate reflector and array size and location for a given coverage sector. It can thus be used to estimate gain reduction, aperture efficiency, and sidelobes due to blockage. One can also obtain a formula for the element use factor, assuming the reflector scans a rectangular angular sector θ_1 by θ_2 radians using effective aperture sizes y_{d1} and y_{d2} and array element spacings a_1 and a_2

$$\frac{N}{N_{\min}} = \frac{0.25}{\theta_1 \theta_2} \frac{y_{a1} y_{a2} \lambda^2}{y_{d1} y_{d2} a_1 a_2} \tag{8.93}$$

The principal advantages of such array-fed reflectors are that they are relatively simple to design, inexpensive relative to array systems, and have low loss feeds. Their disadvantages are that they use an oversize reflector and that the array itself needs to be quite large, with an element use factor for two-dimensional scanning of about 2.5 or 3. Although this number is not much different than that obtained for periodic arrays, there are no grating lobes to be suppressed for the reflector system, so the wide-angle sidelobes (but not the near sidelobes) are improved in comparison with the array case.

Offset-fed reflectors have been used successfully to avoid blockage and improve sidelobe levels, and a number of such array-fed reflectors have found use in airport precision approach radars and other limited field-of-view systems [29].

Rudge and Whithers [53] studied reflector systems fed by Butler matrix feeds, which are completely overlapped subarray systems. They investigated the use of such feeds when the array is moved away from the focal point along a circular arc, as shown in Figure 8.24. The arc chosen is along that circle passing through the focal point and the two extreme edges of the reflector. The key feature of this geometry is that for any point along the arc, the angle $2\theta^*$ subtended by the parabola edges as seen at the feed point is a constant. Rudge and Whithers showed that the transform characteristic of the feed could be used to provide a first-order correction to the off-axis pattern. Figure 8.24 shows that the reflector focus field is a sin x/x-type illumination in response to a received on-axis beam; but when the incident beam comes from some other angle, the focal plane distribution moves off axis, and is, in addition, substantially distorted. The authors showed that the Fourier transform of this distribution is a progressive phase illumination with relatively slow amplitude variation. Since the Butler matrix feed accomplishes this transformation, it needs only to be excited with a corporate power divider and phase shifters and moved so that it intercepts almost the entire focal region in order to correct for off-axis scanning of the reflector. The authors used an eight-element Butler matrix feed and demonstrated good scanning performance over ± 15 beamwidths.

Dual-reflector systems and lens-fed reflector systems with array feeds have been found to have superior scanning characteristics and use smaller, more efficient primary apertures than single-reflector systems. Fitzgerald showed that both near-field Cassegrainian [54] (Figure 8.25(a)) and offset-fed Gregorian [55] (Figure 8.25(b)) confocal paraboloid configurations could scan many beamwidths with good efficiency. The off-axis configuration exhibited better sidelobe performance because of reduced blockage. The element use factor for this geometry was about 2.5. Optimizing the main and subreflector contours can improve scan characteristics

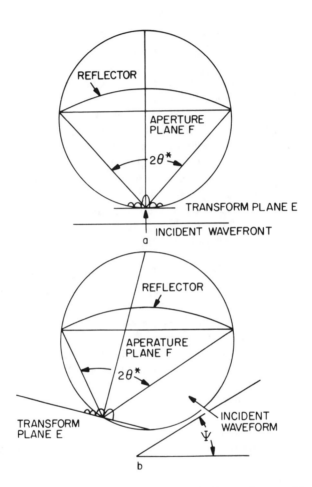

Figure 8.24 The focal region fields of a parabola along a circular arc for several incident wave angles (after [53]).

and reduce the element use factor to about 2. McNee et al. [56] studied a limited scan system consisting of an offset main reflector, a multiple-beam lens feed, and a phased array (Figure 8.26). The system was designed according to the principles of overlapped subarray systems and resulted in an element use factor smaller than that of Fitzgerald because the feed array was made to scan over wider scan angles. The additional scan is possible because the feed lens is made very large (approximately 0.65 the size of the main reflector) compared to the subreflectors of Fitzgerald (0.25 to 0.3 times the main reflector diameter). This allows the array to scan almost to its scan limit ($d/\lambda \sin \theta = 0.5$), but the final structure is bulky and may

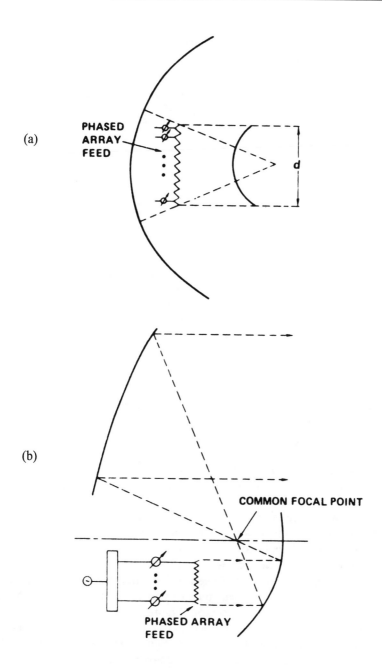

Figure 8.25 Dual-reflector systems for limited field of view: (a) near-field Cassegrain geometry; (b) offset-fed Gregorian geometry (after [54,55]).

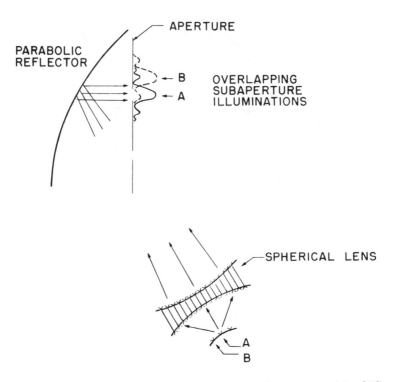

Figure 8.26 Lens/reflector geometry for completely overlapped subarray system (after [56]).

not be suitable for all applications. The analytical results indicate that a 1-deg beam can be scanned over a ±10-deg sector with sidelobes at −20 dB and an element use factor of approximately 1.4.

One important feature of all of the dual-transform systems is that, unlike single-reflector or lens systems in which the scan limit is restricted to some fixed number of beamwidths, the dual-transform systems are basically angle limited, not limited to some given number of beamwidths of scan.

Other dual-reflector systems have been investigated by Bird et al. [57] and Dragone and Gans [58]. The theoretical study of Chang and Lang [59] emphasizes control of maximum gain over a scan sector. Figure 8.27(a) shows the geometry for this study which used point source feeds. The Cassegrainian geometry is a "folded" version of the multiple beam feed configuration with feed elements in the focal plane. The study compared data for a conventional offset Cassegrainian with data for an offset-shaped reflector system with both main and subreflector generated using a bifocal condition. The discrete angles at which the bifocal con-

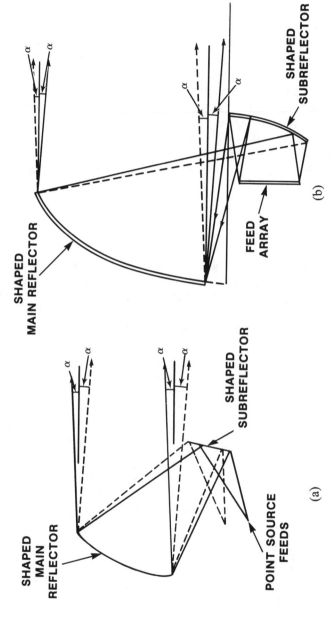

Figure 8.27 Dual-reflector bicollimated systems: (a) multiple-beam Cassegrainial system (after [59]); (b) limited field-of-view scanning system (after [60]).

dition was maintained were only ±2 and ±3 deg off boresight for rotationally symmetrical reflector systems of 0.5-deg beamwidth. The bicollimated versions were found to give improved performance if a scan sector of ±8 beamwidths was required, but at up to about ±5 beamwidths, the Cassegrainian geometry was near optimum.

A recent paper by Rao [60] investigated near-field Gregorian reflector antennas (Figure 8.27(b)) for wide-angle scanning. Data on aperture phase errors ensure that the bicollimated reflector system provides up to 45% more scanning range than an equivalent confocal reflector system. An offset bifocal (two off-axis foci) reflector system designed and studied by Rappaport [61] exhibited over ±12-beamwidth scan with peak gain variation of 3 dB and sidelobes less than −16 dB below the main beam.

Limited Field-of-View Lens Systems

Several dual-lens configurations have been shown to provide high-quality electronic scanning over limited angular sectors. In the configuration of Figure 8.28 investigated by Tang and Winter [62], the array focuses on a small spot on the elliptical rear face of a lens, which transfers the spot to a region on the focal arc of a final lens with a spherical back face. The element use factor is 1.7, and the sidelobes

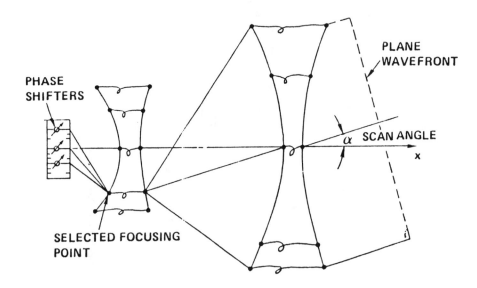

Figure 8.28 Dual-lens limited field-of-view system (after [62]).

are at approximately the −16-dB level for a ±10-deg scan. The beamwidth was kept at 1 deg for this theoretical study. The intermediate lens is about 0.7 times the size of the main lens because of the large scan requirement.

The HIPSAF antenna system was the first implementation of a space-fed overlapped subarray system. This system has application as a feed for a limited field-of-view antenna and for a wide-band array feed. The wide-band properties will be discussed in a later section. The basic HIPSAF geometry, shown schematically in Figure 8.29, consists of a large objective aperture that is a space-fed lens and a Butler matrix feed. The front face of the lens includes phase shifters that are necessary for wide-band wide-angle scanning, but for limited scan application, the phase shifters are fixed to provide the spherical correction so that an incident on-axis wave is focused at the feed. The HIPSAF feed had a spherical front surface, but is shown as a plane in the figure. The figure shows two subarray distributions excited by individual ports of the Butler matrix. The sin x/x form of this illumination

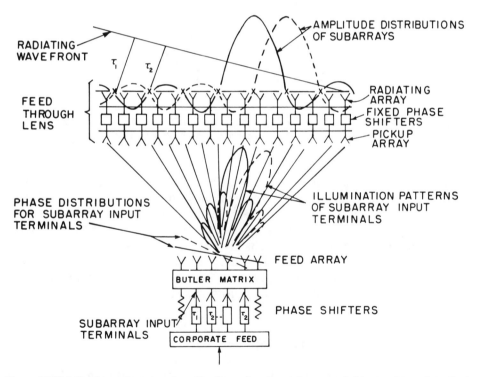

Figure 8.29(a) Dual-transform (overlapped subarray) system using space-fed lens and transform feed (after [28]): geometry showing feed array near-field patterns of several subarray input terminals.

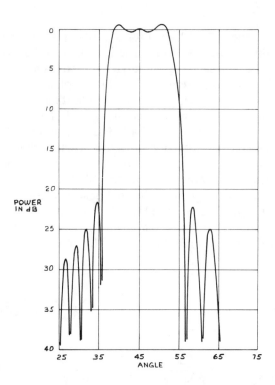

Figure 8.29(b) Dual-transform (overlapped subarray) system using space-fed lens and transform feed (after [28]): subarray pattern (scanned off axis by phase shifters in lens).

results in the flat-topped subarray pattern. The subarray pattern shown at right is scanned so that the center is at 45 deg by means of the phase shifters in the front face. The reason for scanning the subarray pattern has to do with the system's broadband characteristics, and is discussed in the next section. Although the HIP-SAF system was developed as a broadband array feed, the system is a fundamental dual-transform limited scan system. In this mode, the phase shifters are set to zero and the scan controlled by the array feed. Tang [28] showed the feasibility of limited field-of-view control with such transform systems.

Optically Fed Overlapped-Subarray Systems

In recent years, a number of other lens and reflector combinations have been used to produce completely overlapped-subarray patterns. Many of these are discussed in the chapter by Ajioka and McFarlane in the text edited by Lo and Lee [21].

For the purposes of this presentation, consider the generic system of Figures 8.30 and 8.31(a), in which the lens has a circular back face. This geometry was first described in a paper by Borgiotti [63], although the principle of operation for generalized optical systems of this class had been given earlier by Tang [28]. Borgiotti evaluated the scanning performance of this network in detail for the limited field-of-view application. The network has no phase shifters in the array front face, and the number of control elements used is approximately equal to the theoretical minimum. The system uses a hybrid (Butler) matrix and a bootlace lens with a linear outer profile and circular inner profile to perform the two spatial transforms. Borgiotti presents formulas to estimate subarray spacing and other array param-

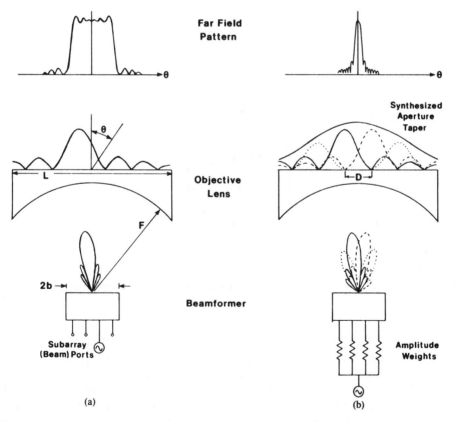

Figure 8.30 Completely overlapped subarray (dual transform) lens perspective (after [64]): (a) feed illumination and radiated subarray pattern; (b) synthesized aperture taper and array radiation pattern.

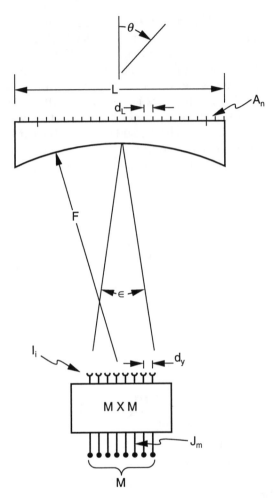

Figure 8.31(a) Geometry of transform-fed lens antenna: two-dimensional cylindrical lens.

eters. The lens is excited by a planar array, which is in turn fed by a Butler matrix or other multiple-beam network. Assume that the network is orthogonal. If a single input port is excited, the multiple-beam network places a progressive phase shift across the array. The array radiates with a sin x/x-type pattern and illuminates the lens back face with this inphase illumination. Exciting any other input port results in a displaced sin x/x-type illumination that is orthogonal to the first, and again completely overlaps the distribution formed by any other input. Each of these overlapped-subarray illuminations is transferred through the main lens and radiates

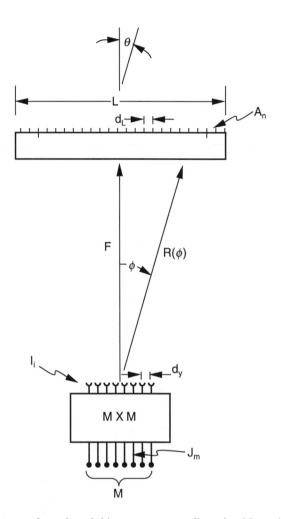

Figure 8.31(b) Geometry of transform-fed lens antenna: two-dimensional lens with flat back face.

to form a flat-topped subarray pattern (since the Fourier transform of a sin x/x function is a pulse).

Each of the subarrays radiate broadside but have their phase centers displaced across the large objective aperture. When all of the subarray inputs are excited with a tapered progressive phase illumination, the main lens radiates a low-sidelobe beam that can be scanned over the width of the subarray pattern.

Figure 8.30 illustrates the operation of the completely overlapped subarray antenna (in this case, for a lens with cylindrical back face). At left, the network is shown excited at a single-beam port. In this sequence, the array is shown radiating a scanned beam that results in a displaced sin x/x illumination at the objective lens and a radiated flat-topped subarray pattern. At right, all subarray ports are excited inphase. The aperture taper is synthesized by selecting the proper weights for the various subarray ports, and the combined illumination radiates with a low-sidelobe broadside pattern. This array pattern can be scanned throughout the width of the subarray pattern using phase weights in addition to the amplitude weights at the beam former input. If the dimensions are chosen for maximum scan, the element factor can be unity. Further details that describe the geometric properties of overlapped subarray lens systems are given in the following sections.

The operation of the system of Figures 8.30 and 8.31(a) is best understood by following the response to a signal applied at the mth port of the Butler matrix (orthogonal beam) feed array. An incident signal at this point results in the progressive phase at the output of the feed:

$$\exp[-j2\pi i(d_y/\lambda)u_m] \tag{8.94}$$

where

$$u_m = \frac{m\lambda}{Md_y} = \sin\phi_m \tag{8.95}$$

Since the lens is cylindrical, with focal length F, the vertical coordinate at the back face of the lens is given by $y = F\sin\phi$.

The feed array radiates to the back face of the main lens with the illumination below, assuming that each feed element has the same spatial element pattern $G(\phi)$. The expression below is shown normalized to unity amplitude and is the feed array pattern (in ϕ-space) or the lens back face illumination (with $y = F\sin\phi$).

$$
\begin{aligned}
A_m(y) &= \frac{G(\phi)}{M}\sum e^{j2\pi i(d_y/\lambda)(\sin\phi - \sin\phi_m)} \\
&= G(\phi)\frac{\sin[M\pi(d_y/\lambda)(\sin\phi - \sin\phi_m)]}{M\sin[\pi(d_y/\lambda)(\sin\phi - \sin\phi_m)]}
\end{aligned} \tag{8.96}
$$

Here the dimension y is measured vertically on the lens back face. $y = F\sin\phi$ and ϕ_m define the center of the mth subarray ($m = +1/2, \ldots$). Since the coordinate $y = nd_L$, this $A_m(y)$ is later denoted by A_{nm}.

The expression (8.96) is the illumination for the mth subarray. In this expression, the beam former has been assumed to be orthogonal, and so ϕ_m is a function of frequency. Each subarray has a similar inphase illumination and radiates to the main lens with the peak at

$$y_m = F \sin \phi_m \tag{8.97}$$

The distance between two adjacent peaks is the intersubarray distance D (a function of frequency) given by

$$D = y_m - y_{m-1} = F \frac{\lambda}{M d_y} \tag{8.98}$$

so $\sin \phi_m$ is also given by

$$\sin \phi_m = \frac{mD}{F} \tag{8.99}$$

In writing the above expression $A_m(y)$, it is assumed that the main lens back face is in the far field of the feed array. This assumption is not severe, and it has been shown by Borgiotti [63] and by Fante [65] that near-field effects can be corrected by adding fixed time-delay units to correct for the curvature of the focused field at the feed array.

The main lens has elements at $y_n = nd_L$, for n, the element index on the main lens, and d_L, the element spacing on the main lens (d_L is nominally about $\lambda/2$ at center frequency).

Each individual subarray pattern $f_m(\theta)$ results from an illumination $A_m(y) = A_m(nd_L)$ at the back face of the lens and having its phase center at $y_m = mD$. The subarray pattern, written with its phase normalized to zero at the subarray phase center y_m and its amplitude normalized to unity is

$$f_m(u) = \frac{K(\theta)}{N} \sum_{-(N-1)/2}^{(N-1)/2} A_{nm} e^{j2\pi n(d_L/\lambda) \sin \theta} \tag{8.100}$$

where $K(\theta)$ is the main lens element pattern.

The above is the most general form for writing the subarray radiation pattern, since it is directly related to the aperture illumination. For the case of isotropic feed array element patterns ($G(\phi) = 1$), one can obtain the convenient form below:

$$f_m(u) = K(\theta) \sum_n A_{nm} e^{j2\pi n(d_L/\lambda) \sin \theta}$$

$$= K(\theta) \sum_{i=-(M-1)/2}^{(M-1)/2} e^{-j2\pi i(d_y/\lambda) \sin \phi_m} \sum_{n=-(N-1)/2}^{(N-1)/2} e^{j2\pi (d_L/\lambda)[(id_y/F)+\sin \theta]n} \qquad (8.101)$$

$$= K(\theta) \sum_{i=-(M-1)/2}^{(M-1)/2} e^{-j2\pi i(d_y/\lambda) \sin \phi_m} \frac{\sin[(N\pi d_L/\lambda)(\sin \theta - id_y/F)]}{N \sin[\pi d_L/\lambda(\sin \theta - id_y/F)]}$$

This expression compares directly with that given in (8.71) for the constrained dual-transform system, except that here the cluster of "constituent" sin x/x-type beams have their peak locations at

$$\sin \theta = id_y/F \qquad (8.102)$$

for $-(M - 1)/2 \le i \le ((M - 1)/2$. Since these beams are formed using an equal path length main lens, their beam peaks are fixed in angle and do not squint. The subarray pattern width is constant, independent of frequency. The above expression, given in terms of the set of constituent beams, allows an estimate of the subarray pattern width. Measured to the 4-dB point of the outermost beam at

$$\theta_{max} = \sin^{-1}[Md_y/(2F)] \qquad (8.103)$$

this width is approximately

$$\epsilon = 2 \text{ arc } \sin[Md_y/(2F)] \sim Md_y/F \qquad (8.104)$$

Figure 8.32(a) shows that the angle subtended by the feed is the same ϵ, so one can readily determine the required feed size and the number of feed elements by equating the subtended feed angle to the desired subarray width.

To obtain an array pattern, it is again convenient to write the subarray pattern directly in terms of the array amplitude distribution $A_m(nd_L)$. The subarray pattern is written

$$f_m(u) = K(\theta) \frac{e^{j2\pi m(D/\lambda)u}}{N}$$

$$\times \left\{ \sum_{n=-(N-1)/2}^{(N-1)/2} G(\phi) e^{j2\pi (u/\lambda)(nd_L - mD)} \frac{\sin[M\pi(d_y/(F\lambda))(nd_L - mD)]}{M \sin[\pi(d_y/(F\lambda))(nd_L - mD)]} \right\} \qquad (8.105)$$

In this form, it is again clear that the array illumination consists of M subarrays evenly spaced across the aperture with frequency-dependent spacing D and having

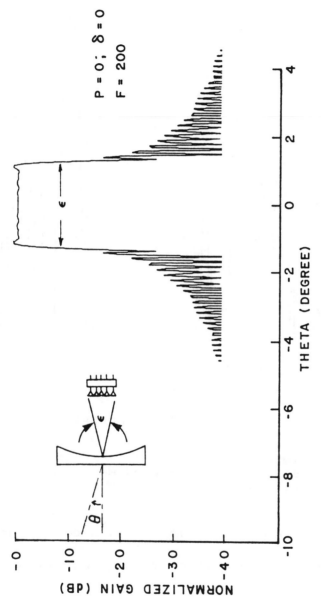

Figure 8.32(a) Limited field-of-view characteristics of cylindrical lens with dimensions (normalized to wavelength) $L = 363.2$: $F = 200$: array size 12 elements (6λ) (after [66]): subarray pattern.

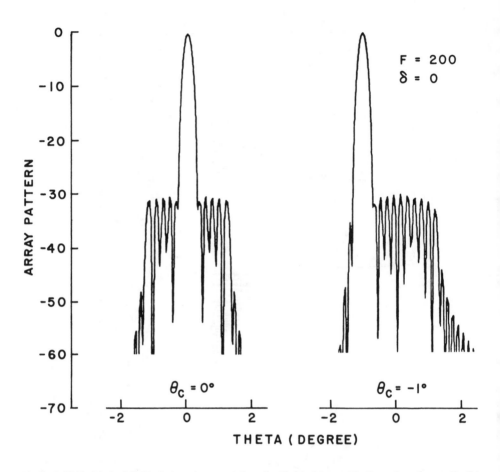

Figure 8.32(b) Limited field-of-view characteristics of cylindrical lens with dimensions (normalized to wavelength) $L = 363.2$: $F = 200$: array size 12 elements (6λ) (after [66]): broadside and scanned patterns.

the peaked sin x/x-type illumination with phase center located at the subarray aperture peaks.

The radiation of the lens system of Figure 8.30(a) with all subarrays excited is given by

$$F(u) = \sum_{m=-(M-1)/2}^{(M-1)/2} J_m f_m(\theta) \qquad (8.106)$$

In this case, the array consists of M subarrays, where M can be even or odd. The J_m are the excitation coefficients for the transform feed. The subarray excitations

J_m for phase scanning the array pattern to the angle θ_0 are given (using (8.105), (8.106) as

$$
\begin{aligned}
J_m &= |J_m| \exp[-j2\pi m(D/\lambda_0)u_0] \\
    &= |J_m| \exp[-j2\pi m(F/Md_y)u_0]
\end{aligned}
\tag{8.107}
$$

If wide-band scanning is required, λ should replace λ_0 in (8.107), and time-delay units must replace phase shifters in the feed input ports.

In place of the orthogonal feed for this lens or reflector, if one were to use a Rotman lens or other true time-delay beamformer to perform the first transformation, then the centers of the subarray aperture illuminations would be fixed in location for all frequencies, and the radiated subarray patterns would be stable and would not move with frequency, but would narrow at the high frequencies and broaden at the low frequencies.

A system with a true time-delay lens feed and lens main aperture is designed at center frequency f_0 such that the multiple-beam antenna (feed) output progressive phase is given by (8.95), but with

$$
u_m = m\lambda_0/(Md_y) = \sin \phi_m
\tag{8.108}
$$

The subarray illumination is given by the same equation, with the above substitution for ϕ_m, and the intersubarray spacing D is then

$$
D = F\lambda_0/(Md_y)
\tag{8.109}
$$

independent of frequency.

The resulting array pattern is given by (8.106) and the beam is phase scanned by currents of the form

$$
\begin{aligned}
J_m &= |J_m| \exp(-j2\pi m(D/\lambda_0) \sin \theta_0 \\
    &= |J_m| \exp\left(-j2\pi m \frac{(F)}{Nd_y} \sin \theta_0\right)
\end{aligned}
\tag{8.110}
$$

If wide-band scanning is required, then again λ_0 should be replaced by λ in (8.110), and time-delay units must replace phase shifters in the network.

Figure 8.32 shows data for a cylindrical lens of diameter 363λ and focal length 200λ. The feed array consists of 12 elements spaced $\lambda/2$ apart. The figure shows that when a transform (focusing) feed is used with a lens primary aperture, exciting any one of the array input ports produces a flat-topped subarray radiation like that shown in the figure. If all the subarrays are excited by an equiphase distribution, then the main lens forms a beam (shown at left in the figure) with near sidelobes at the level determined by the feed input taper (-30-dB Chebyshev for the figure) and far sidelobes further suppressed by the envelope of the subarray pattern. The beam is scanned by applying phased or time-delayed signals at the feed input ports.

Since the lens beamwidth is approximately $\lambda/(Nd)$, or

$$\theta_3 \approx \lambda/(Nd_L) \qquad Nd_L = MD \qquad (8.111)$$

then the beamwidth is also given by

$$\theta_3 = \frac{\lambda}{MD} = \frac{d_y}{F} \qquad (8.112)$$

The subarray pattern width ϵ is thus M times this beamwidth ($\epsilon = Md_y/F$), and so the subarray pattern width is that of the cluster of M constituent beams. As the array is scanned over the extent of the subarray pattern, then the array scans a sector equal to M beamwidths.

In summary, a subarray pattern formed by M elements, and having a width ϵ given by the above, is also seen as formed by a cluster of M orthogonal beams. The maximum scan sector covered by the array feed is equal to the number of elements in the feed (in beamwidths). This corresponds to an element use factor of unity. Figure 8.32 shows an array pattern scanned over the range of the subarray pattern. In this case, the design sidelobe level was a -30-dB Chebyshev pattern and, indeed, the resulting pattern maintains this desired level.

If the lens back face is not cylindrical but is plane, as in Figure 8.31(b), then the subarrays formed by the feed are unequally spaced, will not in general be symmetrical in the angle ϕ, and will need phase corrections within the main lens. Added phase or time-delay corrections are needed at each subarray beam port to account for the path length difference from the array feed to the various subarray phase centers on the main lens. In this case, it is important that the ratio of F/L be large (in excess of 1.5) to minimize distortions. Mathematically, this case is handled with $y_m = F \sin \phi_m$ in (8.97) and subarray centers at y_m replaced by

$$y_m = F \tan \phi_m \qquad (8.113)$$

with F the distance from the planar lens back face to the feed. In addition, the distance between the feed center and points on the lens back face is not F, but

$$R(\phi) = F/\cos \phi \qquad (8.114)$$

and in (8.96), the subarray illumination $A_m(y)$ or A_{nm} must be multiplied by the factor

$$\frac{e^{-j2\pi R(\phi)/\lambda}}{R(\phi)} \tag{8.115}$$

to account for the change in path length across the main lens back face. This effect also needs to be partially compensated for by adding an extra line length to each mth subarray input port to compensate at the phase center of the subarray. The added phase (time-delay) factor is

$$e^{+j2\pi R(\phi)/\lambda} \tag{8.116}$$

which signifies a time advance of the outer subarrays relative to the central subarray. With these changes, the summations indicated in the revised form of (8.96) through (8.101) cannot be written in closed form, but need to be evaluated term by term.

One of the limitations to the use of transform feeds is the need for large array feeds if the system is to be scanned over many beamwidths (see (8.104)). Mailloux [66] extended the reflector scanning work of Rudge and Whithers [53] to investigate wide-angle subarray formation and lens scanning by means of small transform array feeds moved away from the primary focus.

Figure 8.33 shows the transform feed moved off-axis along the circular arc of Figure 8.24 chosen by Rudge and Whithers for the reflector case. This arc has the feature that from any point on the circle, the angle 2Φ subtending the two edges of the cylindrical lens is a constant. This means that the feed array scan sector is fixed. The angle Φ is given by

$$\Phi = \sin^{-1}(L/2F) \tag{8.117}$$

When the feed is moved off axis, the subarray pattern moves off boresight to an angle Δ, measured from the lens back face and given approximately by

$$\sin \Delta = (d/F) \sin \delta \tag{8.118}$$

where δ is the angular displacement of the feed measured from the point P, and d is the diameter of the enclosing circle.

Since the array feed width is Md_y, then if a second array were placed so as to just touch a center mounted feed, then for the array at this position

$$\sin \delta = Md_y/F \tag{8.119}$$

This width is approximately equal to the subarray width ϵ (8.104), and so the subarray patterns of adjacent (touching) feeds would cross at approximately the -4-dB point. Unfortunately, the actual subarray pattern of finite lenses is nar-

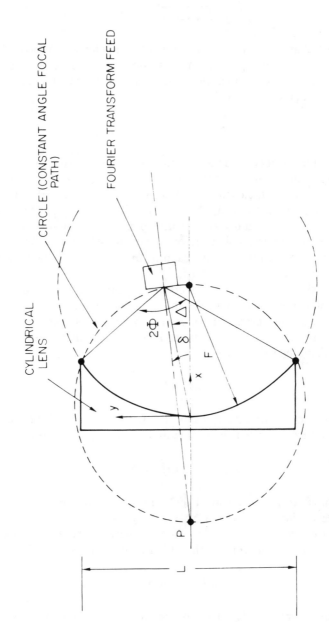

Figure 8.33 Geometry of off-axis feed and two-dimensional cylindrical lens.

rowed, so contiguous feeds produce subarray patterns with a deep trough between them.

As the arrays are moved off axis along the circular arc, the subarray pattern locations also move off axis by an amount approximately equal to the angular feed displacement. Figure 8.34(a) shows subarray patterns for the transform-fed lens as a function of the feed displacement. The feed has 12 elements, and the figure shows subarray patterns corresponding to those at either side of the lens ($p = 6, -5$) and the centrally located subarray ($p = 0$). For the on-axis feed, these two outer subarrays are identical, but as the feed is moved off axis, the subarray patterns become distorted and do not even occupy the same region of space. Low-sidelobe beams can nevertheless be formed out to very-wide-scan angles except at these regions between adjacent subarrays. Figure 8.34(b) shows what level sidelobes can be maintained for a lens with $F = 200\lambda$ and subarray centers as indicated in the figure. Here it is clear that -30 dB can be achieved for subarray centers near broadside, but not at the subarray crossover angles. One can improve the pattern quality for beams at these angles by inserting other feeds at positions between the adjacent ones, as shown for the dashed sidelobe level that corresponds to a feed at $\delta = 3.67$ deg. In brief, by applying no correction other than this simple switching procedure, one can get good pattern control with less than -25-dB sidelobes over the region ± 10.5 deg, or ± 35 beamwidths for the lens data of Figure 8.34. Beyond that angle, one can only get low sidelobes at the subarray center unless more sophisticated control is implemented at the feed.

Another important feature of such transform feeds is that the scan sector for a given F/L ratio is not limited by some maximum number of beamwidths, but is primarily angle limited. This means that two systems with the same F/L can be scanned over the same scan range by merely increasing the number of feeds according to the above condition, independent of the total number of beamwidths scanned.

The main design parameters are chosen as follows:

- The diameter L by the minimum beamwidth;
- The angle 2Φ by the maximum array scan angle and the ratio F/L.

The allowable lower bound on F/L is about 0.58 for a ± 60-deg array scan. In general, increasing F/L makes the array design simpler by decreasing the array scan angle, and reduces phase front curvature in the lens system. This decreases quadratic phase error.

8.2.5 Practical Design of a Dual-Transform System

Recent transform-fed lens studies have gone beyond the conceptual and theoretical level to building and testing actual devices. Southall and McGrath [64] studied and

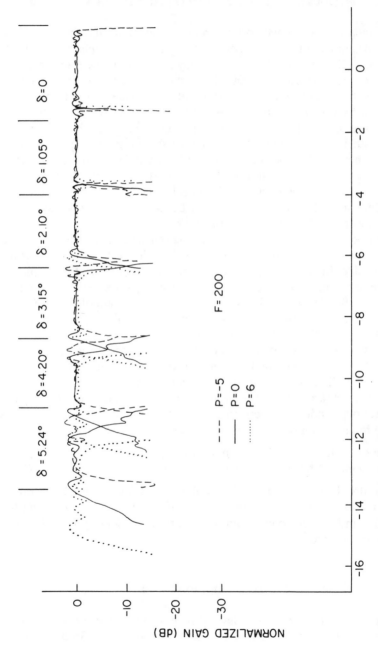

Figure 8.34(a) Radiation characteristics of off-axis transform-fed lens: subarray patterns for off-axis feeds.

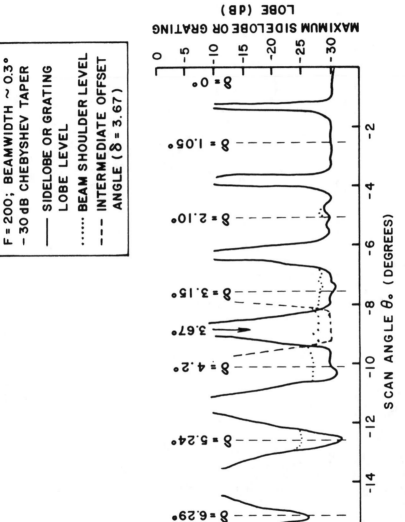

Figure 8.34(b) Radiation characteristics of off-axis transform-fed lens: maximum sidelobe radiation level for scanned feeds at fixed offset angles.

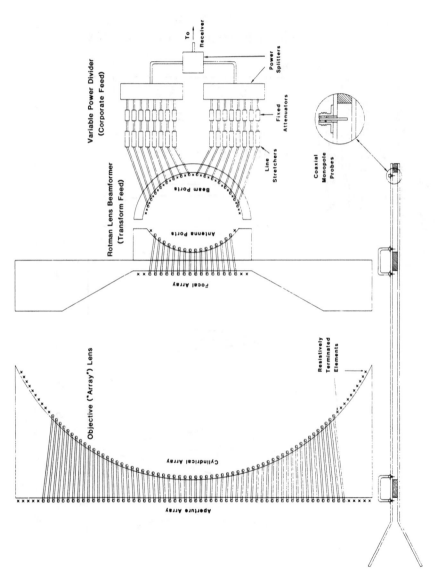

Figure 8.35 Antenna layout drawing of cylindrical lens fed by Rotman lens beam former (after [64]).

built a lens combination like that of Figure 8.35 composed of a probe-fed parallel plane lens with circular back face and fed by a Rotman lens feed.

Although this study was part of the development of a wide-band array feed, the major practical conclusions pertain to both limited field-of-view and wide-band systems. These studies indicated the necessity of careful control of mutual coupling effects in both the main lens and Rotman feed. Among the several innovations introduced in their study was that the main lens circular back face was designed with equally spaced elements. In an early stage of this study, a lens had been constructed with back face elements directly behind their corresponding aperture elements. This forced the cylindrical array element spacing to vary from $0.5\lambda_0$ at the center to $0.72\lambda_0$ at the edges. With this large variation in spacing, there was no possibility of properly matching all of the back face elements, and this resulted in an inverse taper that raised sidelobes. By making the elements equally spaced at $0.52\lambda_0$ across the back face as shown in Figure 8.35, it was possible to match both the front and back faces of the lens. This improved sidelobes considerably.

A second innovation was repositioning of the focal array to account for the fact that the true phase center of an array of probe feeds in front of a back plane does not lie at the back plane (as it would for a single probe and its image). Because of mutual coupling, the phase center may be closer to the probe. This fact impacts the feed array location, which should be at the focus of the lens, and it impacts

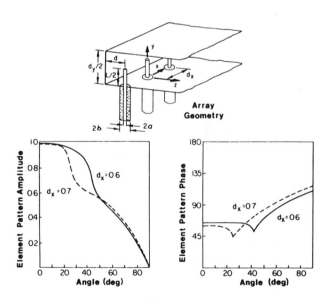

Figure 8.36(a) Element pattern characteristics for microwave lens feeds: monopole characteristics in array (after [67]).

the Rotman lens feed as well. Figure 8.36(a) shows data for an infinite array of probes in a parallel plane region. This figure shows the normalized element pattern for probes spaced 0.4 and 0.6λ apart and clearly indicates significant distortions of the element pattern when the spacing is 0.6λ. The distortion is due to the entrance of a grating lobe at endfire when the scan angle is denoted ϕ_{EGL}. This effect narrows the element pattern and makes the element pattern phase become a complex function of angle. In effect, the element has a nonunique phase center. Reducing the spacing to 0.4λ corrects both of these problems.

Feed design is further complicated because the phase center of an array of probes is not located at the back plane location. Figure 8.36(b) shows the result of phase center location measurements by Southall and McGrath [64] and records the significant change in phase center as a function of frequency. Only at the lower frequencies does the phase center occur close to the back plane. Since the array was to operate in the vicinity of 9 GHz, where the probe phase center was located very near the probe center pin, the array feed was moved so that the center pins were on a line through the main lens focus.

A final innovation was necessary to improve the performance of the Rotman lens feed. The impedance matching problem is more severe for the Rotman lens because of the high degree of curvature within the lens and the need for each lens

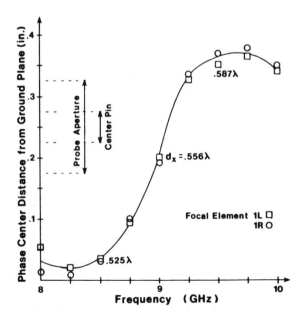

Figure 8.36(b) Element pattern characteristics for microwave lens feeds: phase center location versus frequency (0.556λ spacing) (after [64]).

radiator to have a wide element pattern to properly illuminate the adjacent face. The amplitude curve of Figure 8.36 shows significant variation across the path which would result in a substantial error in illumination and poor sidelobe control. This effect was corrected using additional probe elements with matched loads to broaden the element patterns. Figure 8.35 shows these "dummy" beam ports at the input of the Rotman lens.

8.3 WIDE-BAND SCANNING SYSTEMS

Conventional phased arrays operate over bandwidths that are inversely proportional to the array size. The use of true time delays instead of phase shifts would eliminate the bandwidth restriction due to beam squint, but unfortunately the only viable time-delay technology at the time of this writing consists of switched sections of transmission lines. For example, a large array of 50 wavelengths on a side and scanning to ± 60 deg would need a total time delay from zero to 50 sin 60 deg, or 43λ. To obtain precision equivalent to an N-bit phase shifter, about $N + 6$ bits is necessary. If these units are made with discrete time-delay bits, as is the common practice for phase shifters, the units become too bulky and heavy, and so lossy as to be impractical for most applications, except perhaps for stationary ground-based arrays at relatively low frequencies. The solution to this problem is to devise suboptimum means of providing time delay. Two architectures have been used. The first method consists of matching the time delay at only a fixed number of angles and using phase shifters to scan the array over the small scan ranges between the selected angles of perfect delay. The second method is to divide the array into subarrays and produce true time delay behind each subarray while using phase shift within the subarrays.

8.3.1 Broadband Arrays With True Time-Delayed Offset Beams

The bandwidth of limited field-of-view arrays can be relatively large (1.79), in Chapter 1 because θ_{max} is small. Similarly, if an array is excited by a feed system that produces true time delays at a number of points in space and phase shifters at the array elements to scan between the fixed beam positions, then the bandwidth is given by the same equation, but with the maximum scan angle sin θ_{max} divided by the number of preset time-delayed positions M and the fractional bandwidth multiplied by M.

$$\frac{\Delta f}{f_0} = \frac{0.886 \, B_b M}{(L/\lambda_0) \sin \theta_{max}} \tag{8.120}$$

where B_b is the beam broadening factor, and L is the array length.

There are a number of ways of implementing these offset beams. One method is to use a quasi-optical or constrained multiple-beam array feed and phase shifters at the array face, as in Figure 8.37(a) [68]. Another method that has proven practical is to build discrete time-delay units with M increments of time delay and use one time-delay unit and one phase shifter per element (Figure 8.37(b)). It is thus necessary to construct different time-delay units for each element of the array. However, this array organization has perfect time delay at the chosen offset positions and can multiply the bandwidth by M relative to a conventional array. Most significant is that the array phase progression between adjacent elements is constant, and so the array phase front is continuous and the array sidelobes can be as low as other tolerances will allow. This array time-delay architecture may be costly, but is the standard for low-sidelobe arrays.

8.3.2 Continuous Time-Delayed Subarrays for Wide-Band Systems

The most obvious way of adding time delay to an array is to group elements of the array into subarrays and insert time delay behind each subarray and phase

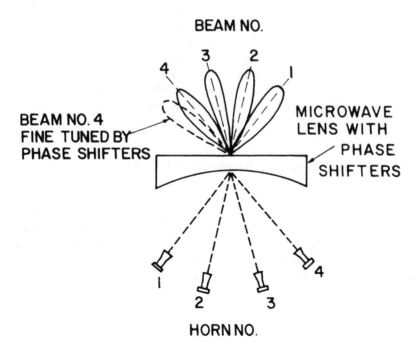

Figure 8.37(a) Wide-band scanning array feeds: array-fed by time-delayed beam former (after [68]).

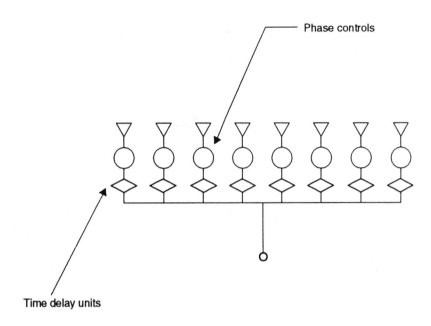

Phase controls

Time delay units

Figure 8.37(b) Wide-band scanning array feeds: array fed by time-delay units and phase shifters.

shifters at the array face to maintain a perfect continuous wavefront at center frequency. This architecture is depicted in Figure 8.37(c). The time-delay units ensure that the center of each subarray is delayed correctly at all frequencies; but as the frequency is changed from the center frequency, the phase progression across each subarray develops the wrong slope and the array incurs a periodic phase error, as indicated in the figure.

The one-dimensional array shown in the figure consists of Q equally spaced subarrays of M elements each, and with time-delay units at the center of each subarray. The resulting array pattern is the product of an array factor that has no frequency dependence times a subarray pattern that represents the phase-shifted subarrays.

$$E(u) = e(u) \frac{\sin[M\pi d_x(u/\lambda - u_0/\lambda_0)]}{M \sin[\pi d_x(u/\lambda - u_0/\lambda_0)]} \frac{\sin[Q\pi D_x/\lambda(u - u_0)]}{Q \sin[\pi D_x/\lambda(u - u_0)]} \tag{8.121}$$

Here $e(u)$ is the element pattern and $D_x = Md_x$ is the subarray size.

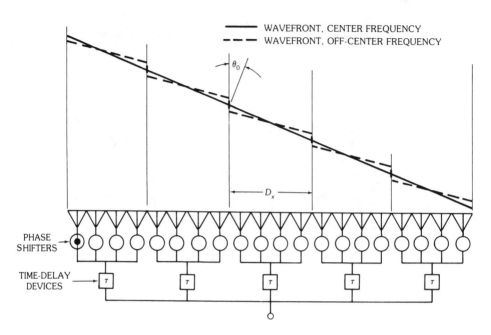

Figure 8.37(c) Wide-band scanning array feeds: array of contiguous subarrays with time delay at the subarray level.

Since the first term of the above equation represents a time-delayed contribution and does not squint with frequency, the only bandwidth limitation is from the second term. Comparing this expression with that of a full phase-steered array of $Q \times M$ elements, it is evident that in terms of power loss alone the subarraying increases bandwidth by the factor M, the number of subarrays.

$$\frac{\Delta f}{f_0} = \frac{0.886 \, B_b M}{(L/\lambda_0) \sin \theta_{max}} \tag{8.122}$$

where $L = QMd_x$.

Although the array of contiguous subarrays provides wide-band operation, the phase discontinuity depicted in the figure results in an increased sidelobe level that may be intolerable for certain applications. The sidelobes that result from this periodic phase error are grating lobes, and the analysis to evaluate their levels follows the development in Chapter 7 for discrete phase shifters and quantization levels [69].

The normalized power in the pth grating lobe is

$$P_S = \frac{(\pi X)^2}{\sin^2[\pi(X + p/M)]} \tag{8.123}$$

where

$$X = \frac{u_0 d}{\lambda_0} \frac{\Delta f}{f_0}$$

and the array has M elements per subarray.

To evaluate the benefits of contiguous subarrays, Figure 8.38(a) shows the broadband characteristics of a 64-element linear array with phase shift steering. The beam is phase scanned to 45 deg ($u = 0.707$) at center frequency. The dashed curve shows squint of the main beam peak to a smaller angle at $f/f_0 = 1.1$. At only

(a)

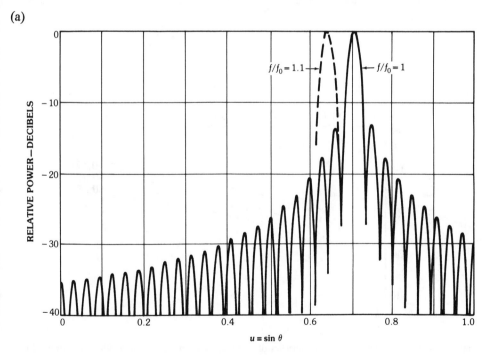

Figure 8.38 Broadband characteristics of an array of contiguous subarrays: (a) pattern of uniformly illuminated array organized with phase shift steering.

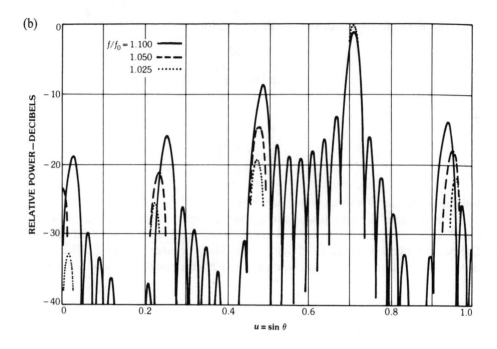

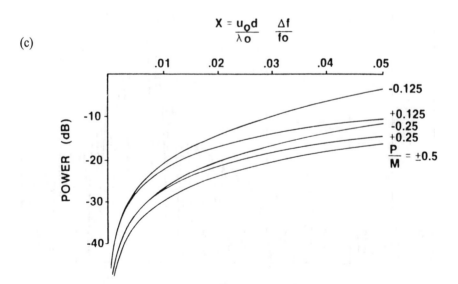

Figure 8.38 (cont.) Broadband characteristics of an array of continguous subarrays: (b) pattern of 64-element array with eight contiguous subarrays (after [69]); (c) grating lobe levels for phased arrays with time-delayed contiguous subarrays (after [69]).

10% off center frequency, only a sidelobe of the pattern radiates in the chosen main beam direction.

Figure 8.38(b) shows the grating lobe power (in decibels) of the 64-element array with time delay at eight contiguous subarrays. The figure plots the power normalized to the main beam versus the generalized variable X for various rations of p/M (grating lobe index divided by the number of elements in the subarray). This curve is general and allows calculation of a wide number of cases.

Figure 8.38(c) shows an example of a uniformly illuminated array with time-delay steering at the subarray level. Although the pattern distortion is substantial, the pattern squint has been eliminated by the time-delay steering. The results of (8.123) are plotted as horizontal lines and are clearly good representations of the computed grating lobe levels for various f/f_0 ratios.

8.3.3 Overlapped Time-Delayed Subarrays for Wide-Band Systems

Constrained Dual-Transform System

The technology of overlapped subarrays, with the resulting flat-topped subarray patterns, presents an excellent means for providing time delay at the subarray level. This application is discussed in the paper by Tang [28], and much of the early work in this area was pioneered by the Hughes Corporation. Consider the ideal subarraying system of Section 8.2.3, but using phase shifters in the array face and time-delay units at the subarray input ports, as shown in Figure 8.38(c). The phase shifters added at the array face produce a progressive phase distribution that scans the center of all subarray patterns from their broadside location in Figure 8.39 to the desired scan angle θ_0 at the center frequency. The subarrays are excited by a true time-delay network that collimates the array to put an array factor peak at the angle θ_0 for all frequencies. The advantage of the overlapped subarray in this case is that, since the subarray pattern moves with frequency, its broad flat-topped shape allows for it to scan substantially without suppressing the main beam radiation at θ_0. Moreover, the steep slope and low sidelobes of the subarray pattern suppress the grating lobes of the periodic phase error.

The equations of Section 8.2.3 are modified to account for the addition of phase shifters at the array face and time-delay units at the subarray input ports. The phase shifters introduce a progressive phase shift across the array face. At the nth element, the phase introduced is

$$\exp[-j2\pi n(d_L/\lambda_0)u_0] \tag{8.124}$$

where $u_0 = \sin\theta_0$ is fixed in frequency. In the absence of any other control signal, this phase scans the center of all subarrays to θ_0 at center frequency f_0.

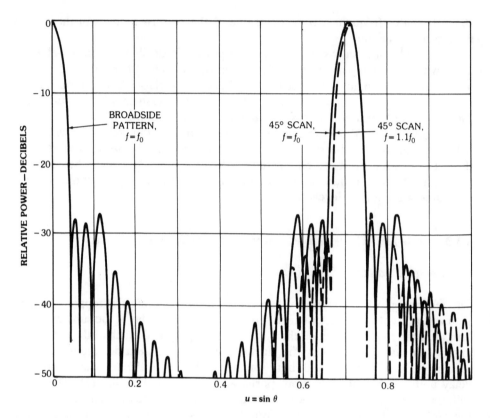

Figure 8.39 Pattern characteristics of 64-element array fed by overlapped subarray beam former (after [48]).

A signal applied to the ith input port of the matrix at right (the $M \times N$ matrix) produces a progressive set of phases at the N array elements and radiates with the scanned pattern below:

$$g_i(u) = \frac{I_i}{N^{1/2}} f^e(u) \frac{\sin[(N\pi d_L/\lambda_0)((u - u_i)f/f_0 - u_0)]}{N \sin[(\pi d_L/\lambda_0)((u - u_i)f/f_0 - u_0)]} \qquad (8.125)$$

where $f^e(u)$ is the array element pattern (assumed equal for all elements), and the constituent beams of the matrix are now displaced by the amount of scan (u_0). Here the u_i are defined as before.

The aperture illumination corresponding to the mth subarray is similar to that given in the previous section, except for an added progressive phase: for each nth element of the N element array,

$$
\begin{aligned}
A_{nm} &= \frac{e^{-j(2\pi/\lambda_0)u_0 n d_L}}{N^{1/2}} \sum_{i=-(M-1)/2}^{(M-1)/2} I_{im} e^{-j2\pi(n/N)i} \\
&= \frac{MJ_m e^{-j(2\pi/\lambda_0)u_0 n d_L}}{(MN)^{1/2}} \frac{\sin M\pi[(nM - mN)/MN]}{M \sin \pi[(nM - mN)/MN]}
\end{aligned}
\tag{8.126}
$$

The scanned subarray patterns are written in terms of the intersubarray distance D using, as before, $md_L = m(M/N)D$ to obtain

$$
f_m(u) = \frac{M e^{j2\pi mD[u/\lambda - u_0/\lambda_0]}}{(MN)^{1/2}} \sum_n \frac{\sin(\pi\Phi_{mn})}{M \sin(\pi\Phi_{mn}/M)} e^{j2\pi D\Phi_{mn}[u/\lambda - u_0/\lambda_0]}
\tag{8.127}
$$

where

$$
\Phi_{mn} = [(M/N)(n - mN/M)]
$$

This expression reveals that the subarray pattern is unchanged from the previous case, except for being phase steered so that it is centered about the direction cosine $u = (\lambda/\lambda_0)u_0$, and therefore at u_0 at center frequency. The subarray pattern is thus phase steered and squints with frequency. The required steering vector to time-delay scan the array to the angle u_{MB}, which may or may not be scanned to the subarray center at u_0, is given by

$$
J_m = |J_m| \exp[-j2\pi mD(u_{MB}/\lambda - u_0/\lambda_0)]
\tag{8.128}
$$

In this expression, the added term $\exp[j2\pi m(D/\lambda_0)u_0]$ is required to remove the excess phase shift at the center of the mth subarray. With this excitation, the subarray pattern is centered at u_0 and squints with frequency, while the array factor is scanned to u_0 by time-delay devices. The array bandwidth is approximately given by (1.79), repeated below, where Δu is interpreted as the subarray pattern width to the 4-dB point at center frequency. This expression does not allow for changes in the subarray width as a function of frequency and does not allow for narrowed bandwidth due to sidelobe growth.

$$
\begin{aligned}
\frac{\Delta f}{f_0} &= \frac{\Delta u}{u_0} \\
&= \frac{M\lambda}{D \sin \theta_0} = \frac{M\lambda}{Nd_x \sin \theta_0}
\end{aligned}
\tag{8.129}
$$

Figure 8.39 shows the pattern characteristics of an overlapped subarray feed for a 64-element array that is fed by a constrained subarray beam former and eight subarrays. In this case, there is no beam squint because of the time-delay steering, and the pattern quality is much improved relative to the contiguous subarray time-delay system. The bandwidth is determined by the growth of grating lobes.

Transform-Fed Lens System

The dual-transform lens system of Figure 8.35 is also used as a wide-band subarraying feed. It was first used in the HIPSAF antenna, and was more recently studied by Fante [65] and Southall and McGrath [64]. The application of the dual transform to reflector scanning was proposed and studied by Chen and Tsandoulas [70]. Parameters for the transform-fed lens system are given below.

The subarray pattern of a lens fed by an orthogonal beam matrix is given by the expression below, which is similar to (8.105):

$$f_m(u) = e^{j2\pi mD(u/\lambda - u_0/\lambda_0)} \sum_{n=-(N-1)/2}^{(N-1)/2} e^{j2\pi F\beta_m(u/\lambda - u_0/\lambda_0)} \frac{\sin[M\pi(d_y/\lambda)\beta_m]}{M \sin[\pi(d_y/\lambda)\beta_m]} \quad (8.130)$$

where $\beta_m = \sin \phi - \sin \phi_m = 1/F(nd_L - mD)$. Like the constrained case, this pattern squints with frequency, but again its broad shape allows for substantial bandwidth relative to the conventional array if time-delay units are used as the feed. The scanning currents at the feed input again need to have the form

$$J_m = |J_m| e^{-j2\pi mD[u_0/\lambda - u_0/\lambda_0]} \quad (8.131)$$

This expression, like (8.128), has a time-delayed exponential component and a phase-shifted component. The array is time-delay steered, but the added phase shift $(\exp(j2\pi mu_0D/\lambda_0)$ term is required to remove the phase shifts introduced at the subarray centers by the phase shifters in the main aperture.

Note that the array aperture is controlled by phase shifters, not time-delay elements, and so although the subarray patterns are centered on the angle θ_0 at center frequency, the subarray pattern squints to move closer to broadside at higher frequencies, and away from broadside (to the right in the figure) at lower frequencies.

The network subarray input ports are then excited with true time delays, and the various subarray patterns are collimated to form a main beam that is fixed in space. As the frequency is changed, the main beam location remains unchanged, but the subarray pattern moves one way or the other until at the frequency end point the beam radiation is cut off by the subarray pattern.

The system bandwidth (for a large array) is given approximately by

$$\frac{\Delta f}{f_0} = \frac{\Delta u}{u_0} = \frac{\epsilon}{u_0} = \frac{Md_y}{F \sin \theta_0}$$

$$= \frac{M}{\sin \theta_0} (\lambda_0/MD) \tag{8.132}$$

$$= M \frac{(\lambda_0)}{(Nd_L \sin \theta_0})$$

or M times the bandwidth of the array without time delay.

A more precise expression for bandwidth of the space-fed lens system, given by Southall and McGrath, accounts for the fact that the subarray width ϵ is independent of frequency, and so the subarray width and array squint only limit bandwidth at the lower frequency band edge. At the upper band edge, the bandwidth is limited by the grating lobe entering into the subarray passband. The resulting fractional bandwidth is thus given by the relationship

$$\frac{\Delta f}{f_0} = \frac{\lambda_0}{D\left(\sin \theta_0 + \dfrac{Md_y}{F}\right)}$$

$$= \frac{M\lambda_0}{(Nd_y)\left(\sin \theta_0 + \dfrac{Md_y}{F}\right)} \tag{8.133}$$

REFERENCES

[1] Woodward, P. M., and J. D. Lawson, "The Theoretical Precision With Which an Arbitrary Radiation Pattern May be Obtained From a Source of Finite Size," *J. AIEE*, Vol. 95, Pt. 3, Sept. 1948, pp. 362–370.
[2] Ricardi, L. J., "Adaptive Antennas," Ch. 22 in *Antenna Engineering Handbook*, R. C. Johnson and H. Jasik, eds., New York: McGraw-Hill, 1984, 1961.
[3] Klein, C. E., "Design of Shaped Beam Antennas Through Minimax Gain Optimization," *IEEE Trans.*, Vol. AP-32, No. 9, Sept. 1984, pp. 963–968.
[4] Galindo-Israel, V., S. W. Lee, and R. Mittra, "Synthesis of Laterally Displaced Cluster Feed for a Reflector Antenna With Application to Multiple Beams and Contoured Patterns," *IEEE Trans.*, Vol. AP-26, No. 2, March 1978, pp. 220–228.
[5] Thomas, D. T., "Multiple Beam Synthesis of Low Sidelobe Patterns in Lens Fed Arrays," *IEEE Trans.*, Vol. AP-26, No. 6, Nov. 1978, pp. 883–886.
[6] Butler, J., and R. Lowe, "Beam Forming Matrix Simplifies Design of Electronically Scanned Antennas," *Elect. Design*, Vol. 9, 12 April 1961, pp. 170–173.

[7] Shelton, J. P., and K. S. Kelleher, "Multiple Beams From Linear Arrays," *IEEE Trans.*, Vol. AP-9, March 1961, pp. 154–161.

[8] Allen, J. L., "A Theoretical Limitation on the Formation of Lossless Multiple Beams in Linear Arrays," *IRE Trans.*, Vol. AP-9, July 1961, pp. 350–352.

[9] Kahn, W. H., and H. Kurss, "The Uniqueness of the Lossless Feed Network for a Multibeam Array," *IEEE Trans.*, Vol. AP-10, Jan. 1962, pp. 100–101.

[10] White, W. D., "Pattern Limitations in Multiple Beam Antennas," *IRE Trans.*, Vol. AP-10, July 1962, pp. 430–436.

[11] Stein, S., "Cross Couplings Between Feed Lines of Multibeam Antennas Due to Beam Overlap," *IEEE Trans.*, Vol. AP-10, Sept. 1962, pp. 548–557.

[12] Johansson, J. F., "Theoretical Limits for Aperture Efficiency in Multi-Beam Antenna Systems," Research Report #161, Dept. of Radio and Space Systems, Chalmers University of Technology, Gothenburg, Sweden, August 1978.

[13] Dufort, E. C., "Optimum Low Sidelobe High Crossover Multiple Beam Antennas," *IEEE Trans.*, Vol. AP-33, No. 9, Sept. 1985, pp. 946–954.

[14] Butler, J. L., "Digital, Matrix, and Intermediate Frequency Scanning," Ch. 3 in *Microwave Scanning Antennas*, R. C. Hansen, ed., Peninsula Publishing, Los Altos, CA, 1985.

[15] Levy, R., "A High Power x-Band Butler Matrix," *Military Microwaves: MM-82*, England: Microwave Exhibitors and Publishers, Ltd., 1992.

[16] Blass, J., "The Multidirectional Antenna: A New Approach to Stacked Beams," *1960 IRE International Convention Record*, Pt. 1, pp. 48–50.

[17] Sletten, C. J., "Multibeam and Scanning Reflector Antennas," Ch. 7 in *Reflector and Lens Antennas: Analysis and Design Using Personal Computers*, C. J. Sletten, ed., Dedham, MA: Artech House, 1988.

[18] McGrath, D. T., "Constrained Lenses," Ch. 6 in *Reflector and Lens Antennas: Analysis and Design Using Personal Computers*, C. J. Sletten, ed., Dedham, MA: Artech House, 1988.

[19] Lee, J. J., "Lens Antennas," Ch. 16 in *Antenna Handbook, Theory, Applications and Design*, Y. T. Lo and S. W. Lee, eds., New York: Van Nostrand Reinhold, 1988.

[20] Rotman, W., and R. F. Turner, "Wide Angle Microwave Lens for Line Source Applications," *IEEE Trans.*, Vol. AP-11, 1963, pp. 623–632.

[21] Ajioka, J. S., and J. L. McFarland, "Beamforming Feeds," Ch. 19 in *Antenna Handbook, Theory, Applications and Design*, Y. T. Lo and S. W. Lee, eds., New York: Van Nostrand Reinhold, 1988.

[22] Rusch, W. V. T., T. S. Chu, A. R. Dion, P. A. Jensen, A. W. Rudge, and W. C. Wong, "Quasi Optical Antenna Design and Applications," Ch. 3 in *The Handbook of Antenna Design*, A. W. Rudge et al., eds., Vol. 1, Peter Peregrinus, Exeter, 1982.

[23] Lo, Y. T., "On the Beam Deviation Factor of a Parabolic Reflector," *IRE Trans.*, Vol. AP-8, 1960, pp. 347–349.

[24] Ruze, J., "Lateral Feed Displacement in a Paraboloid," *IEEE Trans.*, Vol. AP-13, Sept. 1965, pp. 660–665.

[25] Imbriale et al., "Large Lateral Feed Displacement in a Parabolic Reflector," *IEEE Trans.*, Vol. AP-22, No. 6, Nov. 1974, pp. 742–745.

[26] Rusch, W. V. T., and A. C. Ludwig, "Determination of the Maximum Scan-Gain Contours of a Beam-Scanning Paraboloid and Their Relation to the Petzval Surface," *IEEE Trans.*, Vol. AP-21, March 1973, pp. 141–147.

[27] Sletten, C. J., "Reflector Antennas," Ch. 16 in *Antenna Theory, Part 2*, R. E. Collin and F. J. Zucker, eds., McGraw-Hill, 1969.

[28] Tang, R., "Survey of Time-Delay Steering Techniques," *Phased Array Antennas: Proc. 1970 Phased Array Antenna Symp.*, Dedham, MA: Artech House, 1972, pp. 254–260.

[29] Mailloux, R. J., and P. Blacksmith, "Array and Reflector Techniques for Airport Precision Approach Radars," *Microwave J.*, Oct. 1974, pp. 35–64.

[30] Patton, W., "Limited Scan Arrays," *Phased Array Antennas: Proc. 1970 Phased Array Symp.*, A. A. Oliner and G. A. Knittel, eds., Dedham, MA: Artech House, 1972, pp. 254–270.

[31] Stangel, J., "A Basic Theorem Concerning the Electronic Scanning Capabilities of Antennas," URSI Commission VI, Spring Meeting, 11 June 1974.

[32] Borgiotti, G. V., "Degrees of Freedom of an Antenna Scanned in a Limited Sector," *IEEE G-AP Int. Symp.*, 1975, pp. 319–320.

[33] Steyskal, H., private communication, October 1975.

[34] Mailloux, R. J., "An Overlapped Subarray for Limited Scan Application," *IEEE Trans.*, Vol. AP-22, No. 3, May 1974, pp. 487–489.

[35] Tsandoulas, G. N., and W. D. Fitzgerald, "Aperture Efficiency Enhancement in Dielectrically Loaded Horns," *IEEE Trans.*, Vol. AP-20, No. 1, Jan. 1972, pp. 69–74.

[36] Amitay, N., and M. J. Gans, "Design of Rectangular Horn Arrays With Oversize Aperture Elements," *IEEE Trans.*, Vol. AP-29, No. 6, Nov. 1981, pp. 871–884.

[37] Mailloux, R. J., and G. R. Forbes, "An Array Technique With Grating-Lobe Suppression for Limited Scan Application," *IEEE Trans.*, Vol. AP-21, No. 5, Sept. 1973, pp. 597–602.

[38] Mailloux, R. J., L. Zahn, A. Martinez, and G. Forbes, "Grating Lobe Control in Limited Scan Arrays," *IEEE Trans.*, Vol. AP-27, No. 1, Jan. 1979, pp. 79–85.

[39] Mailloux, R. J., "Synthesis of Spatial Filters With Chebyshev Characteristics," *IEEE Trans.*, Vol. AP-24, No. 2, March 1976, pp. 174–181.

[40] Franchi, P. R., and R. J. Mailloux, "Theoretical and Experimental Study of Metal Grid Angular Filters for Sidelobe Suppression," *IEEE Trans.*, Vol. AP-31, No. 3, May 1983, pp. 445–450.

[41] Mailloux, R. J., and P. R. Caron, "A Class of Phase Interpolation Circuits for Scanning Phased Arrays," *IEEE Trans.*, Vol. AP-18, No. 1, Jan. 1970, pp. 114–116.

[42] Dufort, E. C., "Constrained Feeds for Limited Scan Arrays," *IEEE Trans.*, Vol. AP-26, May 1978, pp. 407–413.

[43] Manwarren, T. A., and A. R. Minuti, "Zoom Feed Technique Study," RADC-TR-74-56, Final Technical Report, 1974.

[44] Stangel, J., and Ponturieri, "Random Subarray Techniques," *IEEE G-AP Int. Symp.*, Dec. 1972.

[45] Shelton, J. P., "Multiple Feed Systems for Objectives," *IEEE Trans.*, Vol. AP-13, Nov. 1965, pp. 992–994.

[46] Hill, R. T., "Phased Array Systems, A Survey," *Phased Array Antennas: Proc. 1970 Phased Array Symp.*, A. A. Oliner and G. A. Knittel, eds., Dedham, MA: Artech House, 1972.

[47] Mailloux, R. J., "Periodic Arrays," Ch. 13 in *Antenna Handbook*, op. cit.

[48] Assali, R. N., and L. J. Ricardi, "A Theoretical Study of a Multi-element Canning Feed System for a Parabolic Cylinder," *IRE Trans. PGAP*, 1966, pp. 601–605.

[49] Winter C. E., "Phase Scanning Experiments With Two Reflector Systems," *Proc. IEEE*, Vol. 56, 1968, pp. 1984–1999.

[50] Tang, C. H., "Application of Limited Scan Design for the AGILTRAC-16 Antenna," *20th Annual USAF Antenna Research and Development Symp.*, Univ. of Illinois, 1970.

[51] Howell, J. M., Limited Scan Antennas," *IEEE AP-S Symp. Dig.*, 1974.

[52] Mailloux, R. J., "Hybrid Antennas," Ch. 5 in *The Handbook of Antenna Design, Vol. 1*, A. W. Rudge et al., eds., op. cit.

[53] Rudge, A. W., and M. J. Whithers, "Beam Scanning Primary Feed for Parabolic Reflectors," *Elect. Letters*, Vol. 5, 1969, pp. 39–41.

[54] Fitzgerald, W. D., "Limited Electronic Scanning With a Near Field Cassegrainian System," ESD-TR-71-271, Tech. Rept. #484, Lincoln Laboratory.

[55] Fitzgerald, W. D., "Limited Electronic Scanning With a Near Field Gregorial System," ESD-TR-71-272, Tech. Rept. #486, Lincoln Laboratory.

[56] McNee, F., H. S. Wong, and R. Tang, "An Offset Lens-fed Parabolic Reflector for Limited Scan Applications, *IEEE AP-S Int. Symp. Record*, 1975, pp. 121–123.

[57] Bird, T. S., J. L. Boomers, and P. J. B. Clarricoats, "Multiple Beam Dual Offset Reflector Antenna with an Array Feed," *Elect. Letters*, Vol. 14, 1978, pp. 439–441.

[58] Dragone, C., and M. J. Gans, "Imaging Reflector Arrangements to Form a Scanning Beam Using a Small Array," *Bell System Technical J.*, Vol. 58, No. 2, 1979, pp. 501–515.

[59] Chang, D. C. D., and K. C. Lang, "Preliminary Study of Offset Scan-Corrected Reflector Antenna System," *IEEE Trans.*, Vol. AP-32, No. 1, Jan. 1984, pp. 30–35.

[60] Rao, J. B. L., "Bicollimated Gregorian Reflector Antenna," *IEEE Trans.*, Vol. AP-32, No. 2, Feb. 1984, pp. 147–154.

[61] Rappaport, C. M., "An Offset Biconical Reflector Antenna Design for Wide Angle Beam Scanning," *IEEE Trans.*, Vol. AP-32, No. 11, Nov. 1984, pp. 1196–1204.

[62] Tang, C. H., and C. F. Winter, "Study of the Use of a Phased Array to Achieve Pencil Beam Over Limited Sector Scan," AFCRL-TR-73-0482, Final Report, Contract F19628-72-C-0213, July 1973.

[63] Borgiotti, G. V., "An Antenna for Limited Scan in One Plane: Design Criteria and Numerical Simulation," *IEEE Trans.*, Vol. AP-25, Jan. 1977, pp. 232–243.

[64] Southall, H. L., and D. T. McGrath, "An Experimental Completely Overlapped Subarray Antenna," *IEEE Trans.*, Vol. AP-34, No. 4, April 1986, pp. 465–474.

[65] Fante, R. L., "Systems Study of Overlapped Subarrayed Scanning Antennas," *IEEE Trans.*, Vol. AP-28, No. 5, Sept. 1980, pp. 668–679.

[66] Mailloux, R. J., "Off-Axis Scanning of Cylindrical Lenses," *IEEE Trans.*, Vol. AP-31, No. 4, July 1983, pp. 597–602.

[67] Tomasic, B., and A. Hessel, "Linear Phased Array of Coaxially Fed Monopole Elements in a Parallel Plate Guide," RL-TR-91-124. Rome Laboratory in-house report, April 1991.

[68] Rotman, W., and P. Franchi, "Cylindrical Microwave Lens Antenna for Wideband Scanning Application," *IEEE AP-S Int. Symp. Dig.*, 1990, pp. 564–567.

[69] Mailloux, R. J., "Array Grating Lobes Due to Periodic Phase, Amplitude and Time Delay Quantization," *IEEE Trans.*, Vol. AP-32, No. 12, Dec. 1984, pp. 1364–1368.

[70] Chen, M. H., and G. N. Tsandoulas, "A Dual-Reflector Optical Feed for Wideband Phased Arrays," *IEEE Trans.*, Vol. AP-22, 1974, pp. 541–545.

List of Symbols

α	angle (deg or rad), attentuation constant (nep/m)
β	phase constant (equals 2p/l)
$\bar{\delta}^2$	amplitude error variance, normalized to unity
ε	permittivity (dielectric constant, F/m)
ε_A	aperture efficiency
ε_P	polarization efficiency
ε_L	loss efficiency
ε_T	taper efficiency
Γ	reflection coefficient
η	characteristic impedance
η_B	number of scanned beamwidths
θ	angle (deg or rad)
$\hat{\theta}$	unit vector in the q direction
θ_3	3-dB bandwidth
λ	wavelength
μ	permeability (H/m)
p	3.1415927...
ρ	unit vector in the r direction, polarization unit vector
σ	radar cross section
$\bar{\sigma}^2$	sidelobe level variance
$\hat{\phi}$	angle (deg or rad)
Φ^2	phase error variance (rad^2)
ϕ	unit vector in the ϕ direction
Ω	ohm
ω	angular frequency (equals 2pf, rad/s)
A	ampere, magnetic vector potential (Wb/m), area (m^2)
A	magnetic vector potential (Wb/m)

a	area (m^2)
a_n	element excitation
B	magnetic flux density (Wb/m^2), susceptance (mhos)
B	magnetic flux density (Wb/m^2), susceptance (mhos)B_b
B_b	beam broadening factor
C	a constant, velocity of light (m/s), capacitance (F),
°C	degree Celsius
c	a constant, velocity of light (m/s)
D	electric flux density (F/m^2)
D	electric flux density (F/m^2)
$D(\theta,\phi)$	directive gain
D_O	directivity
d	distance (m)
dB	decibel equals 10 log (P_2/P_1)
dB_i	decibel over isotropic
E	electric field intensity (V/m)
E	electric field intensity (V/m)
F	farad, noise figure, electric vector potential (cuolombs/meter)
F	electric vector potential (coulombs/meter)
$F(\theta,\phi)$	array factor
$f(\theta,\phi)$	element pattern
G	conductance (mhos), gain
g	circuit gain (g>1) or loss (g<1)
H	henry
H	magnetic field (A/m)
I	current (A)
I	current (A)
i	current (A)
J	joule
J	current density (A/m^2)
$\mathbf{J_s}$	surface current density (A/m)

K	Kelvin, Boltzman's constant
k	wave number (equals $2\pi/\lambda$, m^{-1})
L	inductance (H)
M	magnetic current (V/m^2), covariance matrix
M$_s$	magnetic surface current (V/m)
N	number (integer), circuit noise (W)
N$_A$	noise power (W)
n	number (integer)
n	unit vector normal to surface
P	power (W)
Q	charge (C)
R	resistance (ohms)
S	signal power (W), Poynting vector (W/m^2)
S	Poynting vector (W/m^2)
s	second (of time)
T	temperature (°K)
T$_A$	antenna temperature (°K)
T$_B$	brightness temperature (°K)
u	direction cosine
V	volt
v	direction cosine
W	watt
W$_b$	Webers
X	reactance (ohms)
$\hat{x}$	unit vector in x direction
Y	admittance (mhos)
$\hat{y}$	unit vector in y direction

Z_L	load impedance
Z_O	characteristic impedance (of free space)
$\hat{z}$	unit vector in z direction

Index

Radiowave Propagation and Antennas for Personal Communications, Second Edition, Kazimierz Siwiak

Solid Dielectric Horn Antennas, Carlos Salema, Carlos Fernandes, and Rama Kant Jha

Understanding Electromagnetic Scattering Using the Moment Method: A Practical Approach, Randy Bancroft

For further information on these and other Artech House titles, including previously considered out-of-print books now available through our In-Print-Forever® (IPF®) program, contact:

Artech House
685 Canton Street
Norwood, MA 02062
Phone: 781-769-9750
Fax: 781-769-6334
e-mail: artech@artechhouse.com

Artech House
46 Gillingham Street
London SW1V 1AH UK
Phone: +44 (0)20 7596-8750
Fax: +44 (0)20 7630 0166
e-mail: artech-uk@artechhouse.com

Find us on the World Wide Web at:
www.artechhouse.com